AF664180

CONCEPTS OF TROPICAL AGRICULTURE

CONCEPTS OF TROPICAL AGRICULTURE

Arvind Kumar

2024

Daya Publishing House®

A Division of

Astral International Pvt. Ltd.

New Delhi – 110 002

First Published, 2006
Reprinted, 2024
ISBN: 978-81-7035-447-5 (HB)

Published by : **Daya Publishing House®**
A Division of
Astral International Pvt. Ltd.
– ISO 9001:2015 Certified Company –
4736/23, Ansari Road, Darya Ganj,
New Delhi-110 002
Ph. 011-43549197, 23278134
E-mail: info@astralint.com
Website: www.astralint.com

Printed at : **Replika Press Pvt. Ltd.**

Preface

Agriculture has the responsibility of meeting the basic food and fibre requirements of the society. As the society enlarge and becomes more sophisticated, the requirements also increase and becomes more varied, therefore, agricultural practices in a society must keep on improving if the society has to continue its upward growth. The agro ecological diversity in the country makes it necessary to study ambient environment of the place and analyse it for food grain storage requirement. Agriculture in India has made excellent progress after independence because it is basically an agriculture country and about 70 per cent of its people live in villages. Indian agriculture has made spectacular advances since independence.

Agricultural production in India has made miraculous progress through Green Revolution during last two or three decades in our country, which was once importer of food grains has now become an exporter. The technological advancement in agriculture has been brought about through increased production of new high-yielding varieties of crops by means of the application of chemical fertilizers and pesticides. Technology induced farming for feeding the surging population needs to harvest potential crop yields with excessive use of chemical fertilizers and pesticides. However, it has also polluted the environment and caused slow deterioration of soil health. To combat these problems and in the light of sustainable agriculture, green technology is now being greatly used.

Thus, keeping in view of highlighting some of the lesser-known, I have ventured to compile current research article from eminent agricultural scientists.

My special thanks and appreciation go to the scientists whose contributions have enriched this volume. I wish to express my sincere gratitude to Dr. P.C. Hembram Hon'ble Vice Chancellor, S.K.M. University, Dumka who has been a source of constant inspiration. I owe my special thanks to Professor M. C. Dash, Hon'ble Vice Chancellor of Sambalpur University, Professor H.R. Singh, Hon'ble Vice Chancellor of Allahabad University, Professor K.C. Pandey, Former Vice Chancellor of Meerut University, Professor N.C. Datta of Calcutta University, Dr. S.N. Kaul of NEERI, Professor U.S. Bagde of Mumbai University, Professor S.K. Konar of Kalyani University, Professor D.K. Belsare of Bhopal University,

Professor U.C. Goswami of Gauhati University, Professor P.C. Mishra of Sambalpur University, Professor P. Natarajan of Kerala University, Professor P.S. Murthy of Bangalore University, Professor Ajit Varma of J.N.U., New Delhi, Professor P.P. Sood of Saurashtra University, Professor A.L. Bhatia of Jaipur University, Professor A.K. Mittal of B.H.U., Professor S.P. Hosmani of Mysore University, Professor K. Kapoor of Udaipur University, Professor K.B. Reddy of Nagarjuna University, Professor M. Vikram Reddy of Pondicherry University, Professor B.K. Tiwari of NEHU, Professor G. Tripathi of Jodhpur University, Professor R. Ramalingam of Annamalai University, Professor Sharif U. Ahmad of Nagaland University, Professor G.K. Kulkarni of Aurangabad University, Professor S.U. Mehram of Nagpur University, Professor S. K. Battish of PAU, Professor B. Manihar Sharma of Manipur University, Professor B.D. Joshi of Hardwar University, Professor G.C. Pandey of Faizabad University, Professor K.C. Sharma of Ajmer University, Professor A.K. Dobryal of Garhwal University, Professor Kaushal Kumar of Gorakhpur University, Professor M. Raziuddin of Hazaribag University, Professor Gurdeep Singh of I.S.M., Dhanbad, Dr. P.K. Goel of Karad, Dr. Ram Kumar of Delhi University, Professor S.P. Roy and Shri Tribhuwan Poddar of Bhagalpur University, Bhagalpur as well as Dr. Chandan Bohra of B.S.K. College, Barharwa (Jharkhand).

I also express my deep sense of gratitude to my parents whose blessings have always prompted me to pursue academic activities deeply. I am also thankful to my sweet wife, *Kumari Bimla* and my two lovely sons, *Kumar Pallav Shivshankaran* and *Kumar Prasun Ramakrishnan* whose natural smiles extended to me relief all through this tiresome endeavour.

Last but not the least, I am also thankful to Mr. Anil Mittal, Proprietor, Daya Publishing House, Delhi for taking keen interest in bringing out of this book. Finally, I will always remain a debtor to all my well-wishers for their blessings, without which this book would not have come into existence.

Dr. Arvind Kumar

Contents

Chapter 1

Alternative Agricultural Practices: A Sustainable Approach for Conserving Biodiversity as well as Developing an Eco-friendly Environment

Arvind Kumar

Environmental Science Research Unit, Post Graduate Department of Zoology, S.K.M. University, Dumka – 814 101, Jharkhand, India

Introduction

While the world has been changing during the last few years both politically and economically in unexpected and remarkable way food security remains an unfulfilled dream of billions of people. With the world population crossing six billion mark, the demand for food grains is increasing at an alarming rate. Increasing population, on the other hand, has led to decrease in cropping area. In order to meet the global food demand we have to increase the productivity.

Because of intensive cash grain production practices, agricultural lands worldwide have become vulnerable to degradative processes such as soil erosion from wind and water, nutrient depletion and loss of organic matter, and have suffered a consequent decline in soil productivity. Insects have always been a major cause of yield loss in agriculture. They account for 20-30 per cent of crop-loss. Several approaches *e.g.* physical, chemical and biological have been used with varying degree of success to reduce such losses. To increase the net productivity by preventing such losses in an eco-friendly manner is the need of the hour. The restoration and rehabilitation of the degraded soils to an acceptable level of productivity can be enhanced by using various off-farm sources of organic wastes, including sewage sludge, municipal solid wastes and agricultural and industrial processing wastes (USDA, 1978; Hornick and Parr, 1987; Parr and Hornick, 1992a and b). However, the quality and acceptability of these materials as soil amendments can be greatly improved through composting, or even by co-composting them with available on farm wastes.

Natural resistance that provides reasonable protection against insect pests is available in some crops. It means that crop or variety has a specific gene, which confers the plant resistance through production of certain protein. Through breeding the resistance can be transferred to different varieties of the same crop. But such type of transfer of resistance from one species to another is not possible through breeding programme. With the advancement of technologies in the recent past, it is now possible to identify and isolate a particular gene of one organism, and then to introduce it into another. Such techniques are referred to as genetic engineering. Such gene transfer can be done across the species boundaries. The organism, which receives the gene, is called transgenic. So, it is now possible to introduce the insect resistance gene of one species into a desired plant, which never had such type of resistance gene in its gene pool. Once introduced, the new trait will be inherited generation after generation and can also be transferred to other varieties of that crop by hybridization. This new technology is seen as an additional tool for the control of insect pests and could offer certain advantages over conventional use of insecticides, such as more effective targeting on specific insects, greater resilience to weather condition, no question of seasonal application, operator exposure and most importantly eco-friendly system.

So, in the light and need of sustainable agriculture we will discuss the dynamics of soil productivity; concepts of soil quality and sustainable agriculture as they relate to organic recycling; some guideline on the composting of agricultural and municipal wastes for safe and beneficial use as soil conditioners and biofertilizers; some perspectives on the use of microbial inoculants; and developing transgenic insect resistant plants.

Proper and regular additions of on-farm organic wastes such as animal manures and crop residues are of utmost importance in maintaining the tilth, fertility and productivity of agricultural soils; protecting them from wind and water erosion; and preventing nutrient losses through runoff and leaching. These materials usually have predictable beneficial effects on a broad spectrum of soil properties (Allison, 1973). We should also note that there is a growing shortage of good quality on-farm organic wastes for use as soil conditioners and Biofertilizers because of competitive uses as fuel, fodder and fiber (USDA, 1978).

Soil Erosion and the Loss of Productivity

Soil erosion by wind and water, and the associated decline in soil productivity, and the adverse effects on water quality, continue to be most serious agricultural and environmental problems (Larson *et al.*, 1990). Much of this has been the result of improper and exploitive farming practices related to intensive cash grain production. Brown and Wolf (1984) concluded that the soil erosion crisis must be considered in a global context because the production, distribution and consumption of food is part of the global economy. They estimated that the mean annual loss of topsoil worldwide is about 0.7 per cent . This is of great concern, because the loss of productivity may not easily be restored, even with the application of chemical fertilizers. It has been studied that when the topsoil is removed, or where it has been severely eroded, crop yields are from 20 to 65 per cent lower compared with non-eroded soils (Langdale *et al.*, 1979; Massee, 1990).

For most agricultural soils, degradative processes such as soil erosion, nutrient runoff losses, and organic matter depletion are going on simultaneously with beneficial effects of conservation practices such as crop rotations, conservation tillage, and recycling of animal manures and crop residues (Hornick and Parr, 1987). As soil degradative processes proceed and intensity, soil productivity decreases concomitantly. Conversely, soil conservation practices tend to slow degradative processes and increase soil productivity. Thus, the potential productivity of a particular soil at any point in time is the result of ongoing degradative processes and applied conservation practices. Generally, the most serious degradative processes are soil erosion and associated depletion of plant nutrients and organic matter.

On the best agricultural soil that is gently sloping, medium textured, well structured soils with a deep, well drained profile, a high level of productivity can be maintained by relatively few, but essential, conservation practices that readily offset most degradative processes. However, on marginal soils of limited capability, such as steeply sloping, coarse textured, poorly structured soils depleted of nutrients, and with a shallow, poorly drained profile, soil conservation practices must be maximized to counteract further degradation.

Thus, a truly sustainable farming system is one in which the beneficial effects of various conservation practices are equal to or exceed the adverse effects of degradative processes. Organic wastes and residues offer the best possible means of restoring the productivity of severely eroded agricultural soils or of reclaiming marginal soils (Hornick and Parr, 1987; Parr and Hornick, 1992a and b).

Concept of Sustainable and Alternative Agriculture

Sustainable agriculture is increasingly viewed as a long-term goal that seeks to overcome problems and constraints that confront the economic viability, environmental soundness, and social acceptance of agricultural production system globally. There are many definitions of sustainable agriculture but most of them encompass the same elements of productivity, profitability, conservation, health, safery, and the environment, differing only in the degree of emphasis. Furthermore, "sustainable" implies a time dimension and the capacity of a farming system to evolve and endure indefinitely (Lockeretz, 1988). Sustainable agriculture can be defined as: "Agriculture that for the foreseeable future will be productive, competitive profitable, conserve natural resources, protect the environment, and enhance public health, food quality, and safety". The U.S. Congress (1990) in drafting the 1990 Farm Bill defined sustainable agriculture as: "An integrated system of plant and animal production practices having site-specific application that will, over the long-term:

1. Satisfy human food and fiber needs
2. Enhance environmental quality and the natural resource base.
3. Make efficient use of non-renewable resources
4. Utilize natural biological cycles and controls
5. Improve the economic viability of farming systems.
6. Enhance the quality of life for farmers and society as a whole".

The National Research Council (NRC, 1989) concluded that the ultimate goal of sustainable agriculture is to develop farming systems that are productive and profitable, conserve the natural resource base, protect the environment, and enhance health and safety over the long-term. Alternative agricultural practices provide the best means of achieving this goal. The National Research Council defined alternative agriculture as:

"A system of food and fiber production that applies management skills and information to reduce input costs, improve efficiency, and maintain production levels through such practices and principles as:

1. Crop rotation in lieu of monoculture.
2. Integrated crop/livestock systems
3. Nitrogen-fixing legumes.
4. Integrated pest management.
5. Conservation tillage.
6. Integrated nutrient management.
7. Recycling of on-farm wastes as soil conditioners and Biofertilizers".

It is also important to mention here that the single most important component of a sustainable farming system is skilled management.

Alternative agriculture seeks to optimize the use of internal production inputs (*i.e.* on-farm resources) and skilled management in ways that provide acceptable levels of sustainable crop yields and livestock production, and result in economically profitable net returns (Parr *et al.*, 1990; Reganold *et al.*, 1990 and Handa and Shylla, 2004). This approach emphasizes such cultural and management practices as crop rotations, use of animal and green manures, and conservation tillage to control soil erosion and nutrient losses. Currently, achieving a more sustainable agriculture has become the ultimate goal. This goal will depend on creative and innovative conservation and production practices that provide farmers with economically viable and environmentally sound alternatives or options in their farming systems. While low-input/sustainable farming systems may be feasible in developed countries, it is likely that inputs in many developing countries will have to increase substantially to raise the production potential above a subsistence level before agricultural sustainability can be achieved. From this discussion, it is readily apparent that alternative agricultural practices and the ultimate goal of a long-term sustainable agriculture depend largely, and vitally, upon regular additions of various organic amendments to soils.

Concept of Soil Quality

Various physical, chemical and biological properties of soils interact in complex ways to determine their potential fitness or capability for sustained production of healthy, nutritious crops. The integration of growth- enhancing factors that make a soil productive has often been referred to as "soil quality". The Soil Science Society of America (SSSA, 1984) defines soil quality as an inherent attribute of a soil, which is, inferred soil from characteristics or indirect observations (e. g. compatibility, erodibility, and fertility). Thus, soil quality has traditionally focused on, and has been equated with, soil productivity. More recently, the concept of soil quality has been broadened to include attributes of food safety and quality, human and animal health, and environmental quality (Parr *et al.*, 1992). Thus, soil quality, in view of this, might be defined as:

"The capacity or capability of a soil to produce safe and nutritious crops in a sustained manner over the long-term, and to enhance human and animal health, without impairing the natural resource base or adversely affecting the environment".

Soil quality also plays a major role in plant health and in the nutritional quality of the food that is produced. Thus, if properly characterized, soil quality should serve as a measure or indicator of changes in (a) the soil's capacity to produce optimum levels of safe and nutritious food, and (b) its structural and biological integrity, which can relate to the status of certain degradative processes, as well as environmental and biological plant stresses.

Soil quality is directly related to soil degradation, which can also be defined as the time/rate of change in soil quality (Parr *et al.*, 1992). The maintenance or restoration of soil quality is highly dependent on organic matter and the diversity of beneficial macro-organisms and microorganisms that it supports. The proper and regular addition of organic amendments such as animal manures and crop residues can effectively offset many of these degradative processes. It is also the best and most expedient means of developing a biologically active soil that requires less energy for producing crops; increases the resistance of plants to pests and diseases; and enhances the decomposition of toxic substances such as residual pesticides (Sampson, 1981; Hornick and Parr, 1987 and Parr *et al.*, 1992).

Soil Quality: The Linkage Between Alternative Agriculture and Sustainable Agriculture

The concept of soil quality includes the attributes of food safety and quality, human and animal health, and environmental quality. The best means of improving and maintaining soil quality are alternative agricultural practices such as crop rotations, recycling of crop residues and animal manures, reduced input of chemical fertilizers and pesticides, and increased use of cover crops and green manure crops, including N_2-fixing legumes. All of these help to maintain a high level of soil organic matter that enhances soil tilth, fertility and productivity and protect the soil from erosion and nutrient runoff. Effective implementation of these alternative agricultural practices using a holistic or systems approach requires skilled management and innovativeness by the farmer (Papendick and Parr, 1989; Parr *et al.*, 1983 and Parr *et al.*, 1989).

According to Parr *et al.* (1992), the attributes of soil quality provide a vital link between the strategy of alternative agriculture and the ultimate goal of sustainable agriculture. Soil quality occupies a pivotal position in this concept and is the "Key" to agricultural sustainability.

Composting of Organic Wastes

Composting is a viable means of transforming various organic wastes into products that can be used safely and beneficially as biofertilizers and soil conditioners. During the composting process, organic wastes are decomposed, plant nutrients are mineralized into plant-available forms, pathogens are destroyed and malodors are abated (Parr and Honick, 1992a).

Composting is a practice that farmers have used for centuries to convert organic wastes into useful biofertilizers and soil amendments. More specifically, it is a microbiological process that depends on the growth and activity of mixed populations of bacteria, actinomycetes and fungi that are indigenous to the wastes being composed. Composting can be conducted by either aerobic or anaerobic methods. However, the aerobic method is generally preferred, since it proceeds more rapidly and provides greater pathogen reduction because higher temperatures are attained. According to Parr *et al.* (1978), composting "The aerobic, thermophilic decomposition of organic wastes by mixed populations of indigenous micro-organisms under controlled conditions which yields a partially stabilized residual organic material that decomposes only slowly when conditions again become favourable for microbiological activity".

Some organic wastes may have chemical, physical and/or microbiological properties that would greatly limit the extent to which they could be composted alone. Some wastes may have an extremely acidic or alkaline pH, others may have an unusually high or low C:N ratio, and still others may vary widely in their solid contents. In such cases, selective co-composting of these wastes with sewage sludge, pit latrine waste or night soil, municipal solid waste, crop residues, animal manures, food processing wastes and certain industrial wastes, may alleviate these deficiencies and provide a readily compostable mixture and higher quality product. For example, rice straw has a very high C:N ratio and is slow to compost on its own. If it is combined with a low C:N material such as poultry manure, a more favourable ratio is achieved for rapid composting, and a compost is produced that has a higher nutrient content (Parr and Colacicco, 1987; Wilson, 1989 and Rynk, 1992).

Organic Wastes as Biofertilizers and Soil Conditioners

The simplest and most common method of estimating the value of organic wastes is to consider them as substitutes for chemical fertilizers. Assessing the current market value of their plant nutrient content, usually limited to the macronutrients N, P and K, while excluding all others, does this. Crop residues and municipal solid waste both have a comparatively lower nutrient content, and thus a lower market value as fertilizer substitutes. Unfortunately, the benefits of the organic component are often not considered. In addition to the fertilizer value of organic wastes, it is often important to distinguish between their agronomic and economic values.

The important consideration here is that good quality organic wastes and residues, composted or otherwise that are used as biofertilizers and soil conditioners have a far greater value than just their macronutrient content. These materials have a much greater residual effect on soil tilth and fertility than most chemical fertilizers, because of the slow release character of their nitrogen and phosphorus components. Thus, a significant portion of the economic value of organic wastes is their capacity to elicit crop yield responses over time. This response must be accounted for to assess the true value of the material. Barbarika *et al.* (1980) and Calacicco (1982) estimated that the cumulative agronomic and economic value of some organic materials applied to agricultural soils could be more than five times greater in the post-application period than the value realized during the year of application.

Municipal Wastes as Soil Conditioners and Biofertilizers

During the course of municipal development, two principal types of wastes originated sewage sludge and effluents and solid waste or garbage. For centuries, municipal wastes have been used as soil conditioners and biofertilizers to enhance the productivity of agricultural land. During the last two decades, considerable research has been conducted to ensure that sewage sludge could be used safely and beneficially on cropland (USEPA, 1989 and Chaney, 1990a and b). In USA, the current annual production of sewage sludge is about 8.0 million mt, and more than 30 per cent of it is being utilized beneficially on agricultural and non-agricultural land as compost, or by land spreading and liquid sludge infection (Parr and Hornick, 1992a).

In many countries, the production and disposal of municipal solid waste (MSW) has become an increasingly difficult problem, far greater than the disposal of sewage sludge. USA alone produces about 180 million mt of MSW each year. Only 10 per cent of this MSW is being recycled, while 80 per cent goes into landfills and remaining 10 per cent incinerated (USEPA, 1989). Paper garden wastes, food waste and wood waste account for more than 60 per cent of our MSW. Thus, under proper conditions this biodegradable fraction could be composted or co-composted for beneficial use as soil conditioners and biofertilizers (Parr and Hornick, 1992a).

The following research should be given high priority to optimize the safe and beneficial use of municipal wastes in agriculture:

1. Develop technology for Co-composting sewage sludge, municipal solid wastes, and agricultural wastes that will ensure product quality and acceptance by farmers.

2. Improve the agronomic value of municipal, industrial, and processing wastes as composted organic fertilizers by enriching or "spiking" them with chemical fertilizer.
3. Inoculate finished compost or co-compost with cultures of beneficial microorganisms, to promote plant growth and suppress soil borne pathogens and diseases.
4. Develop reliable methods of determining the maturity and stability indices for composts and co-composts from municipal wastes. These include simple chemical and physical measurements and plant bioassay procedures.

Microbial Inoculants

The idea of controlling and manipulating the soil microflora through the use of inoculants, organic amendments, and cultural and management practices, to create a more favourable soil microbiological environment for optimum crop production and protection, is not new. For at least five decades, microbiologists have known that organic wastes and residues, including animal manures, crop residues green manures, and municipal wastes (both raw and composted) contain their own indigenous populations of micro-organisms, and that these often have broad physiological capabilities. It is also known that when such organic wastes and residues are applied to soil, many of these introduced micro-organisms which can function as biocontrol agents by controlling or suppressing soil borne plant pathogens through their competitive and antagonistic activities (Baker, 1968; Papavizas and Lumsden, 1980 and Kloepper *et al.*, 1989).

Microbiologists have tried to culture beneficial micro-organisms for use as soil inoculants for many years, to overcome the harmful effects of phytopathogenic organisms, including bacteria, fungi and nematodes. However, such attempts have often involved single application of pure culture of micro-organisms and have been largely unsuccessful for several reasons. When various organic amendments containing mixed cultures of indigenous micro-organisms are applied as soil amendments and Biofertilizers, some species tend to survive longer than others. However, most of them tend to "die away". Following the peak of growth and activity resulting from utilization and depletion of available substrate carbon by the soil microflora. Effectively the inoculated micro-organisms are overwhelmed by the native soil micro-flora through competition and antagonisms.

Higa and Wididana (1991) and Higa (1994) reported a unique approach in maximizing the beneficial effects of microbial inoculants. Through repeated applications of the organic amendments, or mixed culture inoculants, a high inoculums density of the inoculated micro-organisms for an extended period can be maintained. This approach can help to ensure that the members of beneficial microbial cultures that promote plant growth and protection will remain high during the first three to four weeks after planting a crop.

Research Needs to Enhance the Effectiveness of Microbial Inoculants in Developing an Eco-Friendly Cropping Environment

The following research should be given high priority in the attempt to optimize the effectiveness of microbial inoculants in agriculture:

1. Ensure a high level of consistency in performance and benefits.
2. Determine the effect of pure culture and mixed culture inocula.
3. Determine modes of action of inocula on the indigenous soil micro-flora and on plant growth, yield, quality and protection.

4. Ensure quality control for producing inocula with respect to inoculums density.
5. Determine the practicability and feasibility of using microbial inoculants to enhance and improve the species and genetic diversity of marginal soil.
6. Determine and monitor the survival and disposal of inocula in treated soils.

Transgenic Insect Resistant Plants

Transgenic insect resistant plants can be developed through following steps:

1. Submission and approval of bio-safety protocols for transgenic crop development from the appropriate government and other environmental agencies along with its disposal.
2. Identification of insect resistance gene.
3. Isolation of the desired gene.
4. Construction of recombinant DNA with a suitable vector and the desired gene or preparation of a construct.
5. Transfer of the vector construct to the callus or protoplast of the desired plant.
6. Regeneration of transform ants.
7. Study regarding expression of insect resistant character in transgenic plant.
8. Examination of the stability of the trans gene.
9. Bio-safety protocol confirmation and proper disposal of products.
9. Large scale field trial at multiple locations for several generations.
10. Proper notification to the bio-safety agencies about the transgenic product along with failure, if any.

Genes for insecticides proteins like δ-endotoxin, protease inhibitors, λ-amylase inhibitors, lectins, cholesterol oxidase, isopentenyl transferage and vegetative insecticides proteins have been identification and transgenic plants with these genes have been developed (Table 1.1).

Several other proteins with insecticidal property are being identified and characterized and will provide valuable options for insect management using transgenic crops.

δ-Endotoxin

Bacillus thuringiensis is a gram positive soil borne facultative bacterium. It undergoes sporulation along with the formation of a bipyramidal crystal protein in adverse climatic condition. This crystal protein is toxic to insects and known as Cry protein, Bt toxin or δ endotoxin (Koziel *et al.*, 1993). The spores can survive for many years but the protein (δ endotoxin) remains functional for a few days to a few months depending on the environmental condition. Different sub-species of the bacterium produce different crystal proteins and they are very specific in their action. This protein results in pore formation of columnar cells of mid gut by osmotic lysis and subsequently leads to paralysis and death of the larvae. There are different types of Cry proteins *i.e.* Cry I, Cry II, Cry III etc which have their own specific receptor in the columnar cells of the midgut.

The Bt is being used as insecticidal spray since 1957 under different trade names like Thuricide, Agree, Dipel, Javelin, Condor, Foil and Biobit. But high cost of production and low stability in the field due to UV and sunlight have resulted in its limited use. With the advancement in the field of genetic engineering, the δ endotoxin gene of *B. thuringiensis* became an attractive candidate to be among the first genes transferred into plants. In 1987, transgenic tobacco and tomato plants with δ endotoxin gene were developed (Barton *et al.*, 1987). Transgenic cotton with modified CryIA (C) gene have been found resistant to cotton bolloworm, leaf perforator, pink of bolloworm and beet armyworm. Monsanto, an American company, has developed cotton bolloworm and pink bolloworm resistant transgenic cotton varieties called "Bollgard" and Ingard". Transgenic potato plants with CryIIIA gene show resistance to Colorado potato beetle.

Table 1.1: Some Transgenic Plants Developed with Insect Resistance Genes

(A) Gene Product	Target Insect	Transformed plants
Transgenic Plants with crystal protein gene of *Bacillus thuringiensis*		
Cry I A b	Lepidoptera	Apple, Cotton, Maize, Rice, Potato, Tomato, Tobacco, White clover
Cry I H	Lepidoptera	Maize
Cry 3 A	Coleoptera	Tobacco, Potato
Cry 9 C	Lepidoptera	Maize
(B) Transgenic plants with protease inhibitor gene		
C II (Soyabean serine Protease inhibitor)	Coleoptera Lepidoptera	Potato, Tobacco, Oil seed, Rape
Cp TI (Cowpea trypsin inhibitor)	Coleoptera Lepidoptera	Apple, Potato, Rice, Sunflower, Sweet Potato, Tobacco, Tomato
OC-I (Rice cysteine protease inhibitor)	Coleoptera	Tobacco, Oilseed
Pot PI-II (Potato protease Inhibitor II)	Lepidoptera	Tobacco, Rice
(C) Transgenic plants with Lectin gene		
Snowdrop lectin (GNA)	Homoptera Lepidoptera	Tobacco, Potato, Rice, Sweet Potato, Sugarcane, Sunflower, Tomato
Pea Lectin (Plee)	Homoptera	Potato, Tobacco
Wheat germ agglutinin (WGA)	Lepidoptera	Maize
Rice Lectin	Lepidoptera	Maize
(D) Transgenic plants with α-Amylase inhibitors		
α-amylase inhibitor of common bean (α AI-Pv)	Coleoptera	Pea, Tobacco
I amylase inhibitor from cereals (WMAI-I)	Lepidoptera	Tobacco

Protease Inhibitor

Many plants have natural defense mechanisms against insects based on the production of certain protease inhibitors. These protease inhibitors generally act on the enzymes which are responsible for protein catabolism in insect midget. In contrast to δ endotoxin these inhibitors have antimetabolic activity against a wide range of insects. Lepidopterans have serine proteases and coleopterans have cysteine proteases.

Transgenic tobacco plants with cowpea trypsin inhibitor (Cp TI) genes are resistant to larvae of tobacco budworm. Transgenic tobacco with tomato or potato protease inhibitor II is resistant to tobacco homoworm. However the major disadvantage of protease inhibitor is that it needs very high level of the inhibitor for insect mortality. Transgenic plants with serine protease inhibitor genes from tobacco hornworm are resistant to a number of lepidoptera and homoptera.

α Amylase Inhibitor

Insecticidal activity associated with the common bean seed is due to an inhibitor of α amylase (a digestive enzyme) activity. Studies have shown that λ amylase inhibitors present in seeds are active against insects and mammals but not against plant. Thus, they are highly specific Transgenic pea plants with α amylase inhibitor gene have been found to produce 1.2 per cent of α amylase inhibitor protein in seeds and confer resistance to larvae of two seed feeding beetles- cowpea weevil and azukibean weevil (Shade *et al.*, 1994).

Lectins

Lectins are carbohydrate binding proteins available in plenty in plants, animals and bacteria. Lectins found in the seeds of plants are involved in defense against the attach of bacteria, fungi, insect and other animals. Lectin binds to midgut epithelial cells of insects and leads to dismption of critical cellular of critical cellular functions. Transgenic maize with lectin genes like wheat germ agglutinin or jacalin or rice lectin gene is found to be resistant to European corn borer and southern corn rootworm. Transgenic tobacco, sweet potato and mustard with lectin gene are also resistant to various insects (Schuler *et al.*, 1998).

Cholesterol Oxidase

Cholesterol is required for integrity and normal function of virtually all cellular membranes. Thus, any interference in its availability leads to toxicity. Cholesterol oxidase (CO) catalyses the oxidation of cholesterol, already in corporated into the membrane to produce Ketosteroids and hydrogen peroxide. Midgut epithelium is the primary target for CO and thus it interferes with the process of digestion and absorption (Purcell *et al.*, 1993). Enzymatic activity of CO has been detected in protoplast of transgenic tobacco with native CO gene (Estruch *et al.*, 1997).

Vegetative Insecticidal Proteins (VIPs)

It is discovered that culture supernatant fluid collected during vegetative growth of *Bacillus* sp. is a rich source of insecticidal substances. Supernatant fluid of *Bacillus cereus* isolates possesses acute bioactivity against Northern and Western corn rootworm Similarly, supernatant fluid of certain *B. thuringiensis* cultures has potent insecticidal properties against black cut worm and fall armyworm.

Agriculture in Asia has changed considerably in the last century. Forming practices now focus more on intensive food and cash grain production and less on crop rotations. This requires heavy and frequent applications of chemical fertilizer and pesticides. In order to survive, farmers of Asia must give a higher priority to short term economic gain rather than long-term conservation and sustainability. Alternative agriculture practices are vitally dependent upon the proper and regular additions of organic wastes and residues to soil. The rapidly increasing amounts of municipal wastes can provide an alternative, off-farm source of organic amendments that can be used safely and beneficially to increase the productivity, tilth, fertility and sustainability of our agricultural soils. Research is needed to ensure a high level of consistency in performance and benefits and to ensure adequate quality control with respect to inoculum density and activity.

Insect pests are the major cause of loss to the world's commercially important crops. Current strategies to reduce crop losses due to pests rely primarily on chemical pesticides. Alternatively, transgenic crops with intrinsic pest resistance proteins offer a promising alternative and need to be developed.

References

Allison, F.E. 1973. *Soil Organic Matter and its Role in Crop Production*. Elsevier Science Publ., New York, USA.

Baker, R. 1968. Mechanisms of biological control of soil borne pathogens. *Annual Reviews of Phytopathology*, 6: 263–294.

Barbarika, A., Colacicco, A. D. and Bellows, W. J. 1980. The value and use of organic wastes. Maryland Agri-Economics, May 1980. Cooperative Extension Service, Univ. of Maryland, College Park Maryland, USA.

Barton, K. A., Whiteley, H. R. and Yang, N. 1987. *Plant Physiology*. 85: 1103–1115.

Brown, L. R. and Wolf, E. C. 1984. *Soil Erosion: Quiet Crisis in the World Economy*. World Watch Paper 60. World Watch Institute, Washington, D. C., U. S. A.

Chaney, R. L. 1990a. Twenty years of land application research. *Biocycle*, 31: 55–59.

Chaney, R. L. 1990b. Public health and sludge utilization. *Biocycle*, 31: 68–73.

Colacicco, D. 1982. Economic aspects of composting..*Biocycle*, 23: 26–30.

Estruch, J. J., Carozzi, N. B., Desai, N., Duck, N. B., Warren, G. W. and Koziel, M. G. 1997. *Nat. Biotech*, 15: 137–147.

Handa, A. and Shylla, B. 2004. Alternative agricultural practices-Novel concept for conserving biodiversity and developing an eco-friendly environment. In: *Biodiversity and Environment* (Ed.) A. Kumar. APH Publ. Corp., New Delhi, p. 125–137.

Higa, T. 1994. Effective microorganisms: A new dimension for nature farming. In: *Proceedings of the 2nd International Conference on Kyusci Nature Farming*, (Eds.) Parr, J. F., Hornick S. B. and Simpson, M. E. U. S. Dept. of Agril., Washington, USA, p. 20–22.

Higa, T. and Wididana, G. N. 1991. The concept and theories of effective microorganisms. In: *Proceedings of the 1st International Conference on Kyusci Nature Farming*, (Eds.) Parr, J. F., Hornick, S. B. and Whitman, C. E.). USDA, Washington, DC, USA, p. 118–124.

Hornick, S. B. and Parr, J. F. 1987. Restoring the productivity of marginal soils with organic amendments. *American Journal of Alternative Agriculture*, 2: 64–68.

Kloepper, J. W., Lifeshitz, R. and Zoblotowicz, R. M. 1989. Freeliving bacterial inocula for enhancing crop productivity. *Trends in Biotechiology*, 7: 39–44.

Koziel, M. G., Carozzi, N. B., Currier, T. C., Wamen, G. W. and Evola, S. V. 1993. *Biotechnol and Genet. Eng. Rev.*, 11: 171–195.

Langdale, G. W., Box, J. E., Leonard, R. A., Bamelt, A. P. and Fleming, W. G. 1979. Corn yield reduction on eroded Southern Piedmont soils. *Journal of Soil and Water Conservation*, 34: 226–228.

Larson, W. E., Foster, G. R. and Smith, C. M. 1990. *Research Issues in Soil Erosion/Productivity-Executive Summary*. Published by Univ. of Minnesota St. Paul, Minnesota, USA. Pp. 35.

Lockeretz, W. 1988. Open questions in sustainable agriculture. *Amer. Jour. of Alt. Agril.*, 3: 174–181.

Massee, T. W. 1990. Simulated erosion and fertilizer effects on winter wheat cropping inter-mountain dry land area. *Soil Sci. Soc. of Am. Jour.*, 54: 1720–1725.

National Research Council. 1989. Alternative Agriculture Committee on the role of alternative arming methods in modern production, Agriculture Board on Agriculture. National Academic Press, Washington DC, USA. Pp. 448.

Papavizas, G. C. and Lumsden, R. D. 1980. Biological control of soil borne fungal propagules. *Annual Reviews of Phytopathology*, 18: 389–413.

Papendick, R. I. and Parr, J. F. 1989. The value of crop residues for water conservation. In: *Proceedings International Workshop on Soil, Crop and Water Management Systems for Rainfed Agriculture in the Sudano-Sahelian Zone*, (Eds.) Renard, C. Van Den Beldt, R.J. and Parr, J.F. ICRISAT, Patancheru, A.P., India, p. 183–190.

Parr, J. F. and Colacicco, D. 1987. Organic materials as alternative nutrient sources. In: *Energy in Plant Nutrition and Pest Control*, (Ed.) Z.R. Helsel. Elsevier Sci. Publ. Amsterdam, p. 81–99.

Parr, J. F. and Hornick, S. B. 1992a. Utilization of municipal wastes. In: *Soil Microbial Ecology: Applications in Agricultural and Environmental Management*, (Ed.) F.B. Metting. Marcel Dekker Inc., New York. USA, p. 545–559.

Parr, J. F. and Hornick, S. B. 1992b. Agricultural use of organic amendments: A historical perspective. *Am. Jour. of Alt. Agril.*, 7: 181–189.

Parr, J. F., Papendick, R. I. and Youngberg, I. G. 1983. Organic farming in the United States: Principles and Perspectives. *Agro-Ecosystems*, 8: 183–201.

Parr, J. F., Papendick, R. I., Hornick, S. B. and Colacicco, D. 1989. Use of organic amendments for increasing the productivity of arid lands. *Arid Soil Research and Rehabilitation*. 3: 149–170.

Parr, J. F., Papendick, R. I., Hornick, S. B. and Meyer, R. E. 1990. Sustainable agriculture in the United States. In: *Sustainable Agricultural Systems*, (Eds.) Edwards, C.A., Lal, R., Madden, P., Miller, R. H. and House, G. Soil and Water Conservation Society, Ankeny Jowa, USA. Pp. 50–67.

Parr, J. F., Papendick, R. I., Hornick, S. B. and Meyer, R. E. 1992. Soil Quality: Attributes and relationship to alternative and sustainable agriculture. *Am. Jour. of Alt. Agril.*, 7: 5–11.

Parr, J. F., Wilson, G. B., Chaney, R. L., Sikora, L. J. and Tester, C. F. 1978. Effect of certain chemical and physical factors on the composting process and product quality. In: Proceeding of the Nat. Conf. on Design of Municipal Sludge Compost Facilities. Hazardous Materials Control Research Institute. Silver Spring, Maryland, USA. Pp. 137.

Purcell, J. P., Greenplate, J. T., Jennings, M. G., Ryerse, J. S. and Stonard, R. J. 1993. *Biochem. Biophys. Res. Commun.*, 196: 1406–1415.

Reganold, J. P., Papendick, R. I. and Parr, J. F. 1990. Sustainable Agriculture. *Scientific American*, 262: 112–120.

Rynk, R. 1992. On farm Composting Hand Book North-east Regional Agricultural Engineering Service, Cooperative Extension Service, Cornell University, Ithaca, New York, USA.

Sampson, R. N. 1981. Losing Soil Quality. In: *Farmland or Wasteland*. Rodale Press, Emmaus, Pennsylvania, USA. p. 133–152.

Schuler, T. H., Poppy, G. M., Kerry, B. R. and Denholm, I. 1998. *Trends Biotechnol.*, 16: 168–178.

Shade, R. E., Schroeder, H. E. and Chrispeels,, M. J. 1994. *Biotechnology*, 12: 793–805.

Soil Science Society of America. 1984. *Glossary of Soil Science Terms*. Soil Science Society of America, Madison, Wisconsin, USA. Pp. 37.

U. S. Congress. 1990. Food, Agriculture Conservation and Trade Act of 1990. Public Law. Pp. 101–624.

U. S. Department of Agriculture. 1978. Improving Soils with Organic Wastes Report to the Congress in response to section 1461 of the Food and Agril. Act. if 1977. U. S. Govt. printing Office, Washington, D. C., USA.

U. S. Environmental Protection Agency. 1989. EPAs Policy promoting the beneficial, Use of Sewage sludge, and the New Proposed Technical Sludge Regulations Office of Municipal Pollution Control, Washington, D. C., USA.

Willson, G. B. 1989. Combining raw materials for Composting. *Biocycle*, 30: 82–85.

Chapter 2

Biological Nitrogen Fixation in Rice: A Review

M. Senthil Kumar and S.P. Sundaram

Department of Agricultural Microbiology, Tamil Nadu Agricultural University, Coimbatore – 641 003, Tamil Nadu, India

ABSTRACT

Rice is the most important cereal crop. The nutrient input for a growing rice crop can largely be met by promoting the activities of physiologically diverse microorganisms in the aerobic, anaerobic and interfaces zones in the ecologically important flooded soils. Considerable efforts have been made so far in selecting efficient bacterial strains as inoculants. The engineering of plants capable of fixing their own nitrogen is an extremely complex task, requiring the coordinated and regulated expression of *nif* genes. Knowledge gained so far on how the nitrogen-fixing symbiosis and rice interacts by using molecular and genetic approaches is underway, albeit at a slow pace. Here we review the potentials of BNF system in rice agriculture.

Introduction

Rice is the staple food of approximately one half of the world's population. About 75 per cent of the 143 million hectares of rice lands are lowlands where rice grows in flooded fields. In Asia, one of the constraints to high yields is the limited availability and high prices of nitrogen and phosphorus fertilizers. The idea of utilizing Biological Nitrogen Fixation (BNF) as an alternative or supplementary source of nitrogen for rice is not new. Nitrogen fixing green manures have been used availability and high prices of nitrogen and phosphorus fertilizers. The idea of utilizing Biological Nitrogen Fixation (BNF) as an alternative or supplementary source of nitrogen for rice is not new. Nitrogen fixing green manures have been used for centuries in some rice growing areas and research on biofertilizers, including algal and bacterial inoculants, began in early 1930s. However, the productivity of these sustainable systems is low, with yields less than four tonnes per hectare. Modern agriculture has increased rice yields to five to eight tonnes per hectare, but requires the input of 60 to 100 kg ha^{-1} of fertilizer to supply additional nitrogen. Decreasing the amount of industrially produced fertilizer nitrogen needed in agricultural systems is an important goal of agricultural scientists in general (Bohlool *et al.*, 1992).

In case of sustainable rice production, in particular, one important aim is to replace industrially fixed nitrogen with BNF. Experience since the plant-associating diazotrophs were discovered by Doberenier and her co-workers has shown that this objective will be elusive and that achievement of nitrogen fixation with cereals is still some distance away (Baldani *et al.*, 1983). But the strategies of enabling rice to fix its own nitrogen are complex and of a long-term nature. However, if achieved, they could enhance rice productivity, resource conservation and environmental security (Table 2.1).

Table 2.1: Potential Effect of Biological Nitrogen Fixation on Environmental Parameters

BNF Process per Season	*Effect of Reduced Fertilizer Production and Application*	*Effect of BNF Crops on N Cycling of Cropping Systems*	*Non-N Effects of BNF Organisms*
Soil acidification	**Fossil energy use**	**Pre-cropping/ cropping period**	**Soil structure**
CO_2 fix/mol N assimilation, Soil uptake	CO_2 emission, NO_2 emission	N_2O emission, NH_3 volatilization N leaching risk, Risk of N loss from green manure than N fertilizer	Break crop effect, Soil erosion control, Deep rooting, Carbon sequestration, Bio diversity
		Post harvest period	
		N_2O emission, NH_3 volatilization, N leaching risk in intensive system, N benefit to next crop	
		Long term effect of BNF	
		Soil fertility building, Soil N supply cover, Risk of N losses in intensive systems	

The rate of obtaining success would of course benefit tremendously from concerted efforts from a critical mass of committed scientists around the world, as well as a constant and continued funding support from the donor community.

Inorganic Nitrogen Fertilizer: A Key Input for Rice Production

Nitrogen is the most critical nutrient for rice predominantly. Rice takes one kg of nitrogen to produce 15–20 kg of grain (Kannaiyan, 2000). In traditional, wetland rice culture yields of 2–3 t grain ha^{-1} seem to be sustainable indefinitely even where no N fertilizer is applied, for such yields to be sustained between 60 and 80 kg of nitrogen are required for each crop (Bennett and Ladha, 1992). It requires approximately 18.5 m calories of fossil energy to produce one kg of fertilizer nitrogen and even though, unlike other fertilizers, there is an unlimited supply of this element in the air, which is more than 6 times the energy required to produce either phosphate or potassium fertilizer (Da Silva *et al.*, 1978).

With the inevitable price rises of fossil fuels that must occur over the next few decades due to the depletion of petroleum reserves and increased production of costs of other fuel, now is the time that alternative strategies for nitrogen supply should be developed before these increased cost force farmers to cut N inputs which will result in drastic yield reduction in the staple cereal crops which feed the burgeoning human population of the third world (Boddey *et al.*, 1995).

Indigenous BNF Systems

Indigenous free-living and associative BNF systems, which are ubiquitous in paddy soils, offer an alternative to the input-based exogenous BNF technologies (Roger and Watanabe, 1986). Indigenous BNF and soil organic N (native N) together meets more than half of N requirement of rice in both low and high input production systems. Nitrogen supplied by both these processes has often been sufficient for sizeable yields, sometimes equivalent to yields obtained from using moderate amounts of fertilizer N (Bouldin, 1986). The aquatic condition in rice fields provides the optimum environment for BNF by diverse free-living autotrophs and heterotrophs (Roger and Watanabe, 1984).

The major BNF systems are cyanobacteria and photosynthetic bacteria that inhabit in floodwater and soil surfaces, as well as heterotrophic bacteria in the root zone and free-living in the soil (Table 2.2).

BNF occurs at all sites in the soil-plant-flood water ecosystem, throughout rice growth and between crops under optimum moisture, However, due to the free living nature and long duration of indigenous BNF, methods to accurately estimate total BNF contribution by free-living diazotrophs in a rice field are not available (Roger and Ladha, 1992). Nevertheless, semi quantitative data suggests that free-living diazotrophs seem to have a low to moderate potential to supply N to rice. The contributions of cyanobacterial BNF, based on acetylene reduction assays (ARA), are estimated to be 10–80 kg N ha^{-1} $crop^{-1}$, averaging about 30 kg N $crop^{-1}$ (Roger and Ladha, 1992). Heterotrophic bacterial BNF estimated from N-balance studies averaged 7 kg N ha^{-1} (App *et al.*, 1986). Using available N of a stabilized ^{15}N-labeled soil as a control, Zhu (1989) calculated that in the absence of fertilizers and photo dependent BNF, heterotrophic BNF contributed 16–21 per cent of rice N, or 11–16 kg N ha^{-1} $crop^{-1}$. Incorporation of straw into the soil increased BNF by 2–4 kg N kg^{-1} straw applied (Santiago-Ventura *et al.*, 1986). Soil and plant N balance studies (sum of all gains and losses) in the greenhouse revealed significant differences in the ability of rice cultivars to support a positive N balance (App *et al.*, 1986). Most rice lines maintained the soil N but some even increased soil N after three crops.

Although the N balance study of App *et al.* (1986) included N contributions from the soil, it indicated for the first time that rice influences BNF, and that the ability differs widely between lines. Using ^{15}N stabilized soil in pots, Shrestha and Ladha (1996) quantified N in the plant derived from the atmosphere (Ndfa, per cent of plant N content) for 69 rice lines coming from diverse backgrounds (*indica, japonica,* and *javonica;* traditional and improved) and duration (early, medium, and long). Based on ^{15}N enrichment of a reference cultivar grown under representative conditions that included available soil N, Ndfa was estimated. Ndfa differed significantly among the various lines, ranging from 1.3 to 20 per cent. The possible mechanisms by which rice genotypic differences associated with ^{15}N dilution may influence diazotrophs in the rhizosphere remain a mystery. It has been suggested that diazotroplis associate with roots in response to a specific attractant excreted by the roots, and favorable environmental conditions in the rhizosphere. Malic acid in the root exudates has been identified as the most important attractant for dtazotrophs associated with C_4 crops (Neyra and Hagerman, 1978). In rice, a C_3 crop planted under flooded conditions, the main components of organic acids in root exudates are citric acid and oxalic acid (Lin and You, 1989), but the nature of the main chemotactic attractants in the rice have not yet been identified.

Table 2.2: List of Selected Nitrogen-fixing Microorganisms in Flooded Rice Soils (Rao *et al.*, 1998)

Genus	*Selected Physiological and Morphological Characters*
	Nitrogen Fixers Under Aerobic Conditions
Azotobacter	Gram negative, large blunt rods/cocci, unusual high respiratory rate
Beijerinckia	Gram negative, rods peritrichous flagella
Derxia	Gram negative, rods polar flagella, autotrophic with H_2 as energy
	Nitrogen Fixers Under Microaerophilic Conditions
Azospirillum	Gram negative, vibrioid, aerobic if supplied with a source of fixed N_2
Pseudomonas	Gram negative, straight/curved rods, catalase positive
Methylococcus	Gram negative, rods/cocci, utilize CH_4 as carbon and energy source
Methylosinus	Gram negative, vibrios, form exospores, CH_4 as carbon and energy source
	Facultative Anaerobic Nitrogen Fixers
Klebsiella	Gram negative, rods usually encapsulated
Bacillus	Gram positive, rods, sporetormers, fix N_2 under anaerobic conditions.
Escherichia	Gram negative, rods, has enterobacterial common antigen.
Enterobacter	Gram negative, rods grow best at 30°C
	Anaerobic Nitrogen Fixers
Clostridium	Gram positive, rods, spore formers, fermentative metabolism
Desulfovibrio	Gram negative, vibriod/helical, anaerobic respiratory metabolism using inorganic sulfur as electron acceptors
Desulfotomaculum	Gram positive, rods, anaerobic respiration with sulphate as terminal electron acceptors
	Phototrophic Bacteria
Rhodospirillum	Gram negative, helical, purple non sulfur, anoxygenic photoorganotrophic
Rhodopseudomonas	Gram negative, rods, nonprosthecate
Rhodomicrobium	Gram negative, ovoid, form buds at the end of prosthecae
Chromatium	Gram negative, ovoid, purple sulfur, anoxygenic photolithotrophic
Chlorobium	Gram negative, rods, greensulfur, photolithotrophs
	Cyanobacteria
	Heterocystous Filamentous
Anabaena, Aluosira, Nostoc, Calothrix, Tolypothrix, Westicella	Oxygenic phototrophic, contain chlorophyll-a, carotenoids, phycobilins heterocysts lack photo system
	Non-heterocystous Filamentous
Gloeothece, Cyanothece, Synechococcus	Oxygenic phototrophic contain chlorophyll-a, carotenoids, phycobilins

Furthermore, although it has long been reported that most bacteria (80 per cent) found in rice roots are N_2-fixing (Watanabe *et al.*, 1979), the physiological and biochemical mechanisms that operate in these associations have yet to be elucidated (Ladha, 1986). Hence, questions that must be addressed are:

1. What types of diazotrophs proliferate in the rhizosphere and within the roots?
2. What component(s) of rice root exudates are chemotactic attractants for diazotrophs?
3. What is the extent of genotypic variation in the excretion of these attractants?
4. What is the genetic basis of excretion?

The difficulties encountered in investigating these questions reside in the complexity of analyzing root exudates of the whole plant in soil conditions. Molecular approaches may help in identifying the genetic factors underlying these mechanisms. Wu *et al.* (1995) mapped the genes underlying rice genotypic variability in their ability to stimulate N_2 fixation. Seven restriction fragment length polymorphism (RFLP) marker loci on four chromosomes were significantly associated with the segregation of atom per cent ^{15}N excess in an F_2 population derived from Palawan × IR 42 rice lines.

These results indicate that the ability of rice to enhance N_2 fixation in the rhizosphere is controlled by multiple genes. This is consistent with a previous report by Iyama *et al.* (1993). The identification of quantitative-trait loci provides the first real evidence for the presence of genetic factors that interact with diazotrophs in the rice rhizosphere. Associative N_2 fixation occurs mostly during the heading stage in rice (Ladha *et al.*, 1988), when the availability of both soil and fertilizer-N is low. Associative N_2 fixation, unlike in legume-*Rhizobium* symbiosis, is due to non-specific associations with a variety of diazotrophs, and is greatly affected by genotype × environment (G × E) interactions. Considering the narrow range of genetic differences in associative N_2 fixation, particularly in soils with high N and high G × E interactions, selection and breeding is unlikely to dramatically increase BNF (Malarvizhi and Ladha, 1999). A larger contribution of N_2 fixation may be possible if endophytic system, like that in sugarcane, is found in rice.

Rice-Endophyte Interactions

As long as 1929, an Indian research worker suggested that wetland rice plants were able to obtain some contribution of nitrogen from nitrogen-fixing bacteria associated with the plant roots (Sen, 1929). His evidence was based on the isolation of *Azotobacter* spp. from the rhizosphere and roots of rice including species of *Beijerinckia* (Dobereiner and Ruschel, 1961), *Azospirillum* (Baldani and Dobereiner, 1980), *Alcaligenes* (Qui *et al.*, 1981), *Pseudomonas* (Barraquio *et al.*, 1983), *Klebsiella* and *Enterobacter* (Ladha *et al.*, 1983), *Flavobacterium* (Bally *et al.*, 1983), *Agromonas* (Ohta and Hattari, 1983) and *Serratia marcescens* (Gynashwar *et al.*, 2000). However the presence of nitrogen fixing bacteria associated with rice does not necessarily mean that the plants obtain significant contributions from BNF (Boddey *et al.*, 1.995).

Azospirillum spp. have been isolated in considerable numbers from the rhizosphere and histosphere of wetland rice (Baldani *et al.*, 1993) and a new species of *Pseudomonas* (*P. diazotrophicus*) was reported to dominate the rhizosphere bacterial population (Watanabe and Lin, 1984). However, it has been pointed out by several authors, N_2-fixing bacteria are distant from the main sources of carbon substrates in the root (the vascular tissue) and are in competition with other soil microorganisms for these substrates (Kennedy and Techan, 1992). On the other hand N_2-fixing bacteria found within rice roots or aerial tissues are unlikely to suffer from these disadvantages.

A further possible candidate for an endophytic diazotrophs which will infect rice plants are bacteria of the newly denominated genus *Azoarcus*. Originally isolated from Kallar grass (*Leptochloa fusa* Kunth) growing in saline-sodic soils of Pakistan, a strain of gram-negative nitrogen-fixing bacterium *Azoarcus* sp BH72 has been described by (Reinhold-Hurek *et al.*, 1993) that also colonizes grasses such rice in laboratory experiments (Hurek *et al.*, 1994). It is microaerophilic and has strict aerobic metabolism, grew well on salts of organic acids and colonizes root cortex and xylem. Five species have been identified, belonging to the beta sub group of Proteobacteria (Hurek and Reinhold-Hurek, 1995).

Diazotrophs such as *Gluconacetobacter diazotrophicus* and *Herbaspirillum* spp. grow endophytically in the stems and leaves of sugarcane. In addition, *Azospirillum*, *Burkholdaria* and perhaps other diazotrophs also inhabit sugarcane. Even though it is still not clear which of these endophytes is the main contributor to the observed BNF in sugarcane, evidence, nevertheless, suggest that the total endophytic BNF, which according to N-balance studies can be as high as 150 kg N ha^{-1} $year^{-1}$ in sugarcane (Boddey *et al.*, 1995). By inhabiting the interior of the plants; these bacteria are thought to (1) avoid competition with bacteria of the rhizosphere and (2) derive nutrients directly from host plants (James and Olivares, 1998). In return, the plant interior (which is low in O_2 and relatively high in carbon) provides an environment conducive to N_2 fixation allowing the bacteria to efficiently transfer fixed N products to the host (Sprent and James, 1995). Dobereiner *et al.* (1993) speculated that endophytic diazotrophs in certain rice genotypes may, in fact, be responsible for the substantial contributions of BNF to rice as reported earlier (App *et al.*, 1984).

Rice-Rhizobia Interactions

Past and present, this proves that BNF has an assured place in agriculture. Of the various naturally occurring nitrogen-fixing microbes-plant associations, the symbiosis between rhizobia and legume plants is the best understood scientifically. Moreover, this association is also the most successful agriculturally, providing fixed nitrogen for many of the world's major crop species. Therefore, rhizobia-legume symbiosis form natural models for efforts to develop a sustainable nitrogen fixing system in rice. However, it is recognized that this is likely a long-term approach (Ladha *et al.*, 1997). This long term goal could be achieved in a variety of ways, including the transfer to and expression of nitrogen fixation (*nif, fix*) genes in rice itself (Dixon *et al.*, 1997). Alternatively, naturally occurring diazotrophic endophytic bacteria could be isolated and genetically modified to efficiently provide the plant with fixed nitrogen (Kennedy *et al.*, 1997).Yet another possibility is to try to establish a functional symbiotic interaction of rhizobia with rice via genetic manipulation of both plant and microbe. Extensive knowledge of the legume-rhizobia symbiosis would need to be obtained to design strategies for extending the symbiosis to rice or other cereals.

The first essential step in this process would be critically evaluate and understand the responses of rice plants to rhizobia and vice versa, in comparison with the *Rhizobium*-legume symbiosis (Reddy and Ladha, 1995). Moreover, a detailed assessment of root morphogenesis in legumes versus cereals and the distinct response of these systems to microbial infection would need to be completed (Rolfe *et al.*, 1997). In the past several researches have examined rice-rhizobia interactions and reported a variety of responses, such as the ability of rhizobia to attach to rice root (Terouchi and Syono, 1990), elicit the deformation of root hairs (Plazinski *et al.*, 1985) and to form nodule like structures/ hypertrophies (Jing *et al.*, 1990) or thick short lateral roots on rice plants (Cocking *et al.*, 1993). Moreover recently a report has appeared that examined a range of diverse rice and rhizobia genotypes to determine the variability of responses and interactions between the two partners.

There are potentially many approaches to the development of symbiosis in cereals. The legume-*Rhizobium* or *Frankia* root nodule symbiosis may be used directly as models, seeking to identify key features that would be regarded as essential for the development of analogous systems in cereals. The earlier approach used by Cocking and associates with wild *Parasponia* rhizobia and wheat and the recognizations plant of a moderate degree of intercellular and intracellular infection associated with shortened, thickened lateral roots (Cocking *et al.*, 1992) could be recognized as seeking a more primitive stage of legume nodule development. More recently this group has focused their attention on association between azorhizobia and cereals, also studied with some success by Chen *et al.* (1992) seeking evidence of possible chemical signalling between the bacteria and host plants that could influence success in establishing successful symbiosis based on colonization associated with the points of emergence of lateral roots (Webester *et al.*, 1997). Using lacZ marker gene, they showed that the flavanone naringenin at 10^{-5} M stimulated significantly the colonization of wheat roots by azorhizobia.

The choice of the rhizobial strain ORS 571, which forms stem and root nodules on the tropical legume *Sesbania rostrata* was the most extensively, used *Azorhizobium caulinodans* strains for rice. The basic characteristic features for this strain are–ORS 571 has unusual ability among rhizobia to fix nitrogen in free-living conditions without differentiation into bacteroids. Secondly, it can tolerate upto 12 µM dissolved oxygen level while fixing nitrogen in culture. Thirdly ORS 571 forms nodules on its host plant following crack entry infection (*i.e.*) intercellularly. This type of rhizobial infection does not involve the formation of infection threads at the tip of roots hair, which is considered to be an advanced character confined to certain legume tribes (Sprent *et al.*, 1987).

Para-nodule Induction with Synthetic Auxin

During the past few years since the status of BNF in non-leguminous field crops several papers have described the laboratory model of associative nitrogen fixation known as the para-nodule (Katupitya *et al.*, 1995). The approach is based on the initial observations at Shandong university that 2,4-D acting as an auxin could stimulate colonization by rhizobia in "nodular structures" modified from lateral roots of many plant species. These structures derived from the induction of the initials of lateral roots, are quite dissimilar to legume nodules (Kennedy *et al.*, 1997). Different concentrations of 2,4-D for paranodulation have been observed to suit different species of plant.

For example 0.5, 1.5 and 3 ppm (at different growth stages) were used for paranodualtion in rice, maize and sorghum in the presence of *A. caulinodans.* A large number of plant species *viz.*, wheat, rice, maize, sweet potato, oil seed rape, sunflower and vegetables have been shown to form paranodules by 2,4-D treatment along with inoculation with diazotrophs. Colonization and nitrogen fixation by *A. caulinodans* in 2,4-D induced paranodules of rice grown in sterile media has been reported (Zhiguo *et al.*, 1994).

Mechanism of 2,4-D Activity in Para Nodule Induction

Bacteria including rhizobia, *Azospirillum* and free-living diazotrophs are capable of making forced entry into root cells and root hairs by 2,4-D. Thus 2,4-D is endowed with ability to overcome the barrier of root cell wall. The exact effect of 2,4-D on agglutination, recognition or removal and elimination of receptors on cell wall is not known 2,4-D has two functions inhibition of some metabolic process in plants such as nutrient assimilation, respiration, decomposition, photosynthesis, nitrogen metabolism and deep changes in colloid nature of protoplasm in higher concentration. In lower concentrations stimulation of plant growth and development are observed. Sugar utilization is enhanced and consumption of starch and stored carbohydrates is accelerated. To stabilize the structure and activities of paranodules 2,4-D should be added 2–3 times in field, otherwise paranodules will revert back to roots.

2,4-D molecules switch on or activate genes for cell division by interacting with them. 2,4-D may combine with receptors on plant cells causing configurational changes in receptors. These altered receptors then penetrate into nucleus of cell and promote binding to promoters of genes for cell division. This leads to formation of a paranodule primordium, which provided the optimum environment for bacteria to penetrate into cells to form paranodules. A series of *NOD* genes may also be related to production of hormones or physiologically active substances. The possible molecular mechanism of 2,4-D mode of action in forming P-nodules in rice as mentioned by Kalia and Gupta (2002) is given in the Figure 2.1.

2,4-D

↓

Switch cell division genes

↓

Change (Configurational) in receptors

↓

Altered receptors, penetrate nuclei

↓

Binding to cdc promoters

↓

Paranodule primordium formation

Figure 2.1: Molecular Mechanism of 2,4-D Action

nif Gene Transfer to Rice

An alternative approach to achieve N_2 fixation in rice involves transferring the N_2 fixation (*nif*) genes into rice. Genes that would need to be transferred include *nifH*, *nifD*, and *nifK* that encode the three structural polypeptides of nitrogenase (*nifH* codes for iron sulfur protein and *nifD*, and *nifK* determine two subunits of molybdenum and iron containing protein) as well as *nifB*, *nifN*, *nifE*, and *nifV* (involved in the synthesis of the FeMo cofactor of the Fe-Mo protein), *nifM*, *nifS*, *and nifU* (needed for synthesis of active Fe protein). In addition to ensuring that the *nif* genes would be expressed, nitrogenase must also be protected from inactivation by oxygen. In plant cells, potential locations for introducing foreign genes included the nucleus, the mitochondrion, and the chloroplast. Of these three locations, the chloroplast appears to provide the most suitable environment for *nif* gene expression (Dixon *et al.*, 2000). As DNA-sequence studies have demonstrated that signals for transcription and translation in chloroplasts most closely resemble those of prokaryotes (Whitfeld and Bottomley, 1983).

Furthermore, preliminary evidence indicates that *nif* encoded polypeptides are more stable when expressed in the chloroplast compared with the plant cytosol. For effective transcription of the *nif* genes in chloroplasts, however, it may be necessary to substitute each *nif* gene promoter with an appropriate chloroplast promoter, or some other suitable analogous promoter recognized by chloroplast RNA polymerase. The localization of N_2 fixation in chloroplasts may allow some of the energetic costs of N assimilation to be met through the use of photosynthetically produced reductant. To avoid deactivation of nitrogenase by oxygen in chloroplasts it may be necessary to establish temporal separation of N_2 fixation and photosynthesis by regulation that limits *nif* genes expression to dark periods, or spatial separation could be achieved by expressing nitrogenase in non-photosynthetic root amyloplasts, Recently, Dixon *et al.* (2000) in their studies aimed at assessing the feasibility of developing autonomous N_2-fixing eukaryotes demonstrated the expression of *nifH* gene within plastids of the model photosynthetic eukaryote *Chlamydomonas reinhardtii*. With the current technology, it should be possible to introduce *nif* genes directly into rice chloroplasts as well. In fact, a method for stable chloroplast transformation in higher plants has been reported (Staub and Maliga, 1992).

Physiological Requirements for Nitrogenase Activity

Nitrogen fixation is an energy intensive process and requires a suitable reducant to support electron transport to nitrogenase. Fortunately, photosynthetically generated ATP and NADPH are plentiful in the chloroplast, but the energy demand of nitrogenase could compete with the demands for photosynthate provided by the Calvin cycle. This problem can be exacerbated by the requirement for synthesis of high levels of nitrogenase component proteins to support efficient nitrogen fixation, since nitrogenase is a relatively slow enzyme (Thomeley and Lowe, 1983). Since nitrogenase is a metalloenzyme, its synthesis requires the assimilation of transition metal atoms-Fe and Mo. Iron is readily available in the chloroplast, while the molybdenum is translocated from root to leaves as it is relatively a mobile element to act as a cofactor in nitrate reductase.

Both of the nitrogenase components are extremely oxygen sensitive and are irreversibly denatured by oxygen. Free living diazotrophs achieve the latter by a variety of strategies. Cyanobacteria provide the best models for the reconciliation of photosynthesis with diazotrophy, using either physical barrier as in heterocystous species or temporal separation in non-heterocystous forms (Gallon, 1992). It is possible that photorespiration could contribute to protection of nitrogenase from oxygen, but this would result in significant drain in ATP and reductant. Conformational protection of nitrogenose, by introduction of the gene for the *Azotobacter vinelandii* Shethna (FeS II) protein (Moshiri *et al.*, 1994) into plastids, would help to temporarily stabilize the enzyme in the presence of oxygen and prevent degradation. There are conditions in which nitrogenase itself can remove oxygen, resulting in autoprotection. In the presence of excess reductant the Fe protein of *A. vinelandii* nitrogenase can actually reduce dioxygen to H_2O_2 and ultimately to water (Thornely and Ashby, 1989).

The engineering of a specialized compartment for nitrogenase such as the heterocyst would be an extremely complex and difficult task. So that non-heterocystous cyanobacteria provide the best model for the development of diazotrophic organelles. These organisms are able to cope with both photosynthesis and nitrogen fixation, even though nitrogenase is not spatially separated from photosynthetic oxygen evolution.

The Way Forward

The complexity of plant-microbe interactions, particularly in associative relationships, poses further stiff challenges in working out possible exploitation strategies. Table 2.3 illustrates the comparison of conventional and future potentials of BNF systems in rice.

The major limitation is the energy diversion and budgeting for an effective and efficient management of biological nitrogen fixation associated in rice. Having developed the tools to mark and track some of the promising stable endophytes and monitor the physiological environment in their colonization niches, it is now able to test the hypothesis that natural or genetically modified rice endophytes may be useful to achieve the long term goal of developing rice-microbe interactions capable of supplying biologically fixed nitrogen for higher yields.

Table 2.3: Conventional and Future Biological Nitrogen Fixation Systems for Sustainable Rice Production, (Ladha and Reddy, 1995)

BNF System	*N Supply Potential (kg ha^{-1})*	*Rice Grain Yield Potential (t ha^{-1})*	*Essential Trait in Rice Genotype**	*Technology Available*	*Economic Feasibility and Farmers Adoption*
Conventional BNF Systems					
Free living/Associative (Phototrophs and Heterotrophs)	50–100	3–6	ANFS, NAE, NUE	3–5 years	High
Green manure (*Azolla* and *Sesbania*)	100–200	5–8	NAE, NUE	Available	Low
Future BNF Systems					
Endophytic	?	?	Endo$^+$ fix$^+$	3–5 years	High
Induced Symbiosis (Rhizobia, Frankia etc.)	>200	>8	NUE Nod$^+$ fix$^+$ NUE	>5 years	High
Nif gene transfer	>200	>8	Nif$^+$ fix$^+$ NUE	>5 years	High

* ANFS: Associative nitrogen fixation stimulation; NAE: Nitrogen acquisition efficiency; NUE: Nitrogen utilisation efficiency; Endo$^+$: Endophytic; fix$^+$: N_2 fixation ability; nif$^+$: N_2 fixation genes; nod$^+$: Nodulation ability.

References

App, A., Santiago, T., Daez, C., Menguito, C., Ventura, W., Tirol, A., Po, J., Watnabe, I., De Datta, S.K. and Roger, P. 1984. Estimation of the nitrogen balance for integrated rice and the contribution of phototrophic nitrogen fixation. *Field Crops Res.*, 9: 17–27.

App, A., Watanabe, I., Ventura, T.S., Bravo, M. and Jurey, C.D. 1986. The effect of cultivated and wild rice varieties on the nitrogen balance of flooded soil. *Soil Sci.*, 141: 448–452.

Baldani, V.L.D., Baldani J.I. and Dobereiner, J. 1983. Effect of *Azospirillum* inoculation on root infection and nitrogen incorporation in wheat. *Can. J. Microbiol.*, 29: 924–929.

Baldani, V.L.D. and Dobereiner, J. 1980. Host-plant specificity in the infection of cereals with *Azospirillum* spp. *Soil Biol. Biochem.*, 12: 433–439.

Bally, R., Thomas-Bauzon, D., Heulin, T. and Balandreau, J. 1983. Determination of the most frequent N_2-fixing bacteria in a rice rhizosphere. *Can. J. Microbiol.*, 29: 881–887.

Barraquio, W.L., Ladha, J.K. and Watanabe, I. 1983. Isolation and identification of N_2-fixing *Pseudomonas* associated with wetland rice. *Can. J. Microbiol.*, 29: 867–873.

Bennett, J. and Ladha, J.K. 1992. Introduction: feasibility of nodulation and nitrogen fixation in rice. In: *Nodulation and Nitrogen Fixation in Rice: Potential and Prospects*, (Eds.) G.S. Khush and J. Bennett. The International Rice Research Institute, Manila, Philippines, p. 1–44.

Boddey, R.M., de Oliveira, O.C., Urguiaga, S., Reis, V.M., de Olivares, F.L., Baldani, V.L.D. and Dobereiner, J. 1995. Biological nitrogen fixation associated with sugarcane and rice: Contributions and prospects for improvement. *Plant Soil*, 174: 195–209.

Bohlool, B.B., Ladha, J.K., Garrity, D.P. and George, T. 1992. Biological nitrogen fixation for sustainable agriculture: A perspective. *Plant Soil*, 141: 1–11.

Bouldin, D.R. 1986. The chemistry and biology of flooded soils in relation to the nitrogen economy of rice fields. *Fertil. Res.*, 9: 1–14.

Chen, T.W., Scherer, S. and Boger, P. 1992. Nitrogen fixation of *Azorhizobium* in artificially induced root para-nodules in wheat. *Science China (B)* 35: 1463–1470.

Cocking, E.C., Davey, M.R., Kothari, S.L., Srivastava, J.S., Jing Y., Ridge R.W. and Rolfe, B.G. 1992. Altering the specificity control of the interaction between rhizobia and plants. *Symbiosis*, 14: 123–130.

Cocking, E.C., Srivastava, J.S., Cook, J.M., Kothari, S.L. and Davey, M.R. 1993. Studies on nodulation of maize, wheat, rice and oilseed rape. Interaction of rhizobia with emerging lateral roots. *In: Biological Nitrogen Fixation-Novel Associations with Non-legume Crops* (Eds.) N. Yanfu, I.R. Kennedy and C. Tingwei. Quingdao Ocean University Press, p. 53–58.

Da Silva, J.G., Serra, G.E., Moreira, J.R., Goncalves J.C. and Goldenberg, J. 1978. Energy balance for ethyl alcohol production from crops. *Science*, 210: 903–906.

Dixon, R., Cheng, Q. and Day, A. 2000. Prospects for constructing nitrogen-fixing cereals. In: *Quest for Nitrogen Fixation in Rice*, (Eds.) J.K. Ladha and P.M. Reddy. International Rice Research Institute, Makati City, Philippines, p. 327–336.

Dixon, R., Cheng, Q., Shen, G.F., Day, A. and Day, M.D. 1997. *Nif* gene transfer and expression in chloroplasts, prospects and problems. *Plant Soil*, 194: 193–203.

Dobereiner, J., Reis, V.M., Paula, M.A. and Olivares, F. 1993. Endophytic diazotrophs in sugarcane, cereals and tuber plants. In: *New Horizons in Nitrogen Fixation*, (Eds.) R. Palacios, J. Mora and W.E. Newton. Kluwer Academic Publ. Dordrecht, The Netherlands, p. 671–676.

Dobereiner, J. and Ruschell, A.P. 1961. Inculacao do arroz com bacterias fixadoras de nitrogenio do genero Beijerinckia Derx. *R. Bras. Biol.*, 21: 397–407.

Gallon, J. 1992. Tansely reviews no: 44 Reconciling the incompatible nitrogen fixation and oxygen. *New Phytol.*, 122: 571–609.

Gynaeshwar, P., Reddy, P.M. and Ladha, J.K. 2000. Nutrient amendments influence endophytic colonization of rice by *Serratia marcescens* IRBG500 and *Herbaspirillum seropedicae* Z67. *J. Microbial Biotechnol.*, 10:694–699.

Hurek, T., Hurek, B.R., Van Montagu, M. and Kellenberger, E. 1994. Root colonization and systemic spreading of *Azoarcus* sp. strain BH 72 in grasses. *J. Bacteriol.*, 176: 1913–1923.

Hurek, T. and Reinhold-Hurek, B. 1995. Identification of grass associated and toluene-degrading diazotrophs, *Azoarcus* spp., by analyses of partial 16S ribosomal DNA sequences. *Appl. Environ. Microbiol.*, 61: 2257–2261.

Iyama, S., Sano, Y. and Fuji, T. 1993. Diallel analysis of nitrogen fixation in rhizosphere of rice. *Plant Sci. Lett.*, 30: 427–435.

James, E.K. and Olivares, F.L. 1998. Infection and colonization of sugar cane and other graminaceous plants by endophytic *diazotrophs. Crit. Rev. Plant Sci.*, 17: 77–119.

Jing, Y., Li, G., Jin, G., Shan, X., Zhang, B., Guan, C. and Li, J. 1990. Rice root nodules with acetylene reduction activity. In: *Nitrogen Fixation: Achievements and Objectives*, (Eds.) P.M. Gresshoff, L.E. Roth, G. Stacey and W.I. Newton, Chapman and Hall, New York, p. 779–780.

Kalia, A. and Gupta, R.P. 2002. Nodule induction in non-legume by rhizobia. *Indian J. of Microbiol.*, 42: 183–193.

Kannaiyan, S. 2000. Biofertilizers-key factor in organic farming. In: *Survey of Indian Agriculture*, p. 165–174.

Katupitiya, S., New, P.B., Elmerich, C. and Kennedy, I.R. 1995. An improved nitrogen fixation in 2,4-D treated wheat roots associated with *Azospirillum lipoferum:* colonization using reporter genes. *Soil Biol. Biochem.*, 27: 447–452.

Kennedy, I.R. and Techan, Y.T. 1992. Biological nitrogen fixation in non-leguminous field crops: Recent advances. *Plant Soil*, 141: 93–118.

Kennedy, I.R., Pereg-Gerk, L.L., Wood, C., Deaker, R., Gilchrest, K. and Katupitiya, S. 1997. Biological nitrogen fixation in non-leguminous field crops. Facilitating the evolution of an effective association between *Azospirillum* and wheat. *Plant Soil*, 194: 65–79.

Ladha, J.K and Reddy, P.M. 1995.Extension of nitrogen fixation to rice: Necessity and possibilities. *Geo Journal*. 35: 363–372.

Ladha, J.K., Barraquio, W.L. and Watanabe, I. 1983. Isolation and identification of nitrogen-fixing *Enterobacter cloacac* and *Klebsiella planticola* associated with rice plants. *Can. J. Microbial.*, 29: 1301–1308.

Ladha, J.K., de Bruijn, F.J. and Malik, K.A. 1997. Introduction: Assessing opportunities for nitrogen fixation in rice–a frontier project. *Plant Soil*, 194: 1–10.

Ladha, J.K, Watanabe, I. and Saono, S. 1988. Nitrogen fixation by leguminous green manures and practices for its enhancement in tropical lowland rice. In: *Sustainable Agriculture: Green Manure in Rice Farming*. International Rice Research Institute, Los Banos, Manila, Philippines, p. 165–183.

Ladha, J.K. 1986. Studies on nitrogen fixation by free-living and rice-plant associated bacteria on wetland rice field. *Bionature*, 6: 47–58.

Lin, M. and You, C.B. 1989. Root exudates of rice (*Oryza sativa* L.) and its interaction with *Alcaligenes faecalis. Sci. Agric. Sinica*, 22: 6–12 (English abstract).

Malarvizhi, P. and Ladha, J.K. 1999. Influence of available nitrogen and rice genotype on associative dinitrogen fixation. *Soil Sci. Soc. Am. J.* 63: 93–99.

Moshiri, F., Kim, J., Fi, C. and Maier, R. 1994. The FeSII protein of *Azotobacter vinelandii* is not essential for aerobic nitrogen fixation but confers significant protection to oxygen-mediated inactivation of nitrogenase *in vitro* and *in vivo*. *Mol. Microbial.*, 14: 101–114.

Neyra, C.A. and Hagerman, R.H. 1978. Relationship between CO_2, malate and nitrate accumulation and reduction in corn (*Zea mays* L.) seedlings. *Plant Physiol.*, 58: 726–730.

Ohta, H. and Hattori, T. 1983. *Agromonas oligotrophica* gen. nov., anitrogen-fixing oligotrophic bacterium. *Antonie van Leeukwen-hoek*, 49: 429–446.

Plazinski, J., Innes, R.W. and Rolfe, B.G. 1985. Expression of *Rhizobium trifolli* early nodulation genes on maize and rice plants. *J. Bacteriol.*, 163: 812–815.

Qui, Y.S., Zhou, S., Mo, X., Ye Cai. S., Chen, C.Y., He, S. and Deng, R. 1981. Study of nitrogen fixing bacteria associated with the rice root. The characteristics of nitrogen fixation by *Alcaligenes faecalis* strain A-15 and *Enterobacter cloacae* E26. *Acta. Microbiol. Sinica*, 21: 473–476.

Rao, V.R., Ramakrishnan, B., Adhya, T.K., Kanungo, P.K. and Nayak, D.N. 1998. Review: Current status and future prospects of associative nitrogen fixation in rice. *World J. Microbial. and Biotechnol.*, 14: 621–633.

Reddy, P.M. and Ladha, J.K 1995. Can symbiotic nitrogen fixation be extended to rice? In: *Nitrogen Fixation: Fundamentals and Application*, (Eds.) I.A. Tikhonovich, N.A. Provorov, V.I. Romanov, and W.E. Newton. Kluwer-Acad. Publ., The Netherlands, pp. 629–633.

Reinhold-Hurek, B., Hurek, T., Gills, M., Hoste, B., Vancanney, M., Kesters, K. and De Ley, J. 1993. *Azoarcus* gen. nov., a nitrogen fixing Proteobacteria associated with the roots of Kallar grass (*Leptochloa fusca* (L.) Kunth.). and description of two species *Azoarcus indigens* sp. nov. and *Azoarcus communis*. sp. nov. *Int. J. Syst. Bacteriol.* 43: 574–588.

Roger, P.A. and Ladha, J.K. 1992. Biological N_2 fixation in wetland rice fields. Estimation and contribution to nitrogen balance. *Plant Soil*, 141: 41–55.

Roger, P.A. and Watanabe, I. 1984. Nitrogen fixation in wetland rice field. *In: Current Developments in Biological Nitrogen Fixation*, (Ed.) N.N. Subba Rao. Oxford and IBH Pub. Co., New Delhi, India, p. 237–276.

Roger, P.A. and Watanabe, I. 1986. Technologist for utilizing biological nitrogen fixation in wetland rice: potentials, current usage and limiting factors. *Fertil. Res.*, 9: 39–77.

Rolfe, B.G., Djordjevic, M.A., Weinman, J.J., Mathesius, U., Pittock, C., Gartner, E., Dong, Z., McCully, M. and Mclver, J. 1997. Root morphogenesis in legumes and cereals and the effect of bacterial inoculation on root development. *Plant Soil*, 194: 131–144.

Santiago-Ventura, T., Bravo, M., Daez, C., Ventura, W., Watanabe, I. and App, A.A. 1986. Effects of N fertilizers, straw and dry fallow on the nitrogen balance of a flooded soil planted with rice. *Plant Soil*, 93: 405–411.

Sen, J. 1929. Is bacterial association a factor in nitrogen assimilation by rice plants? *Agri. J. India*, 24: 229–231.

Shrestha, R.K and Ladha, J.K. 1996. Rice genotypic variation promotion of rice dinitrogen fixation as determined by ^{15}N dilution. *Soil Sci. Soc. Am. J.*, 60: 1815–1821.

Sprent, J.I. and James, E.K. 1995. N_2-Fixation by endophytic bacteria: Questions of entry and operation. In: *Azospirillum VI and related Microorganisms*, (Eds.) I. Fendrik, M. del Gallo, J. Vanderleyden, M. de Zamaroczy. Springer, Berlin, Heidelberg, Germany, p. 15–30.

Sprent, J.J., Sutherland, J.M. and de Faria, S.M. 1987. Some aspects of the biology of nitrogen fixing organisms. *Phil. Trans. R. Soc., Lond. B.* 317: 111–129.

Staub, J.M. and Malliga, P. 1992. Long regions of homologous DNA are incorporated into the tobacco plastid genome by transformation. *Plant Cell*, 4: 39–45.

Terouchi, N. and Syono, K. 1990. *Rhizobium* attachment and curling in asparagus rice and oat plants. *Plant Cell Physiol.*, 31: 119–127.

Thorneley, R.N.F. and Ashby, G.A. 1989. Oxidation of nitrogenase iron protein by dioxygen without inactivation could contribute to high respiration rates of *Azotobacter* species and facilitate nitrogen fixation in other aerobic environments. *Biochem. J.*, 261: 181–187.

Thorneley, R.N.F. and Lowe, D.J. 1983. Nitrogenase of *Klebsiella pneumoniae*, Kinetics of the dissociation oxidized iron protein from molybdenum-iron protein: identification of the rate-limiting step for substrate reduction. *Biochem. J.*, 215: 393–403.

Watanabe, I. and Lin, C. 1984. Response of wetland rice to inoculation with *Azospirillum lipoferum* and *Pseudomonas* sp. *Soil Sci. Plant Nut.*, 30: 117–124.

Watanabe, I., Barraquio, W.L., de Guzman, M. and Cabrea, D.A. 1979. Nitrogen fixation (acetylene reduction) activity and population of aerobic heterotrophic nitrogen fixing bacteria associated with wetland rice. *Appl. Environ. Microbial.*, 37: 813–819.

Webster, G.C., Gough, C., Vasse, J., Batchelor, C.A., O' Callaghan, K.J., Kothari, S.L., Davey, M.R., Denarie, J. and Cocking, E.C. 1997. Interactions of rhizobia with rice and wheat. *Plant Soil*, 194: 115–122.

Whitfeld, P.R. and Bottomley, W. 1983. Organization and structure of chloroplast genes *Ann. Rev. Plant Physiol.*, 34: 297–310.

Wu, P., Zhang, G., Ladha, J.K., McCouch, S. and Huang, N. 1995. Molecular marker-facilitated investigation on the ability to stimulate nitrogen fixation in the rhizosphere by irrigated rice plants. *Theor. Appl. Genet.*, 91: 117–118.

Zhiguo, M., Zhigang, L. and Qingyuan, X. 1994. On rice's induced nodulation and transfer of the nodule bacterium into the plant. In: *Biological Nitrogen Fixation: Novel Associations with Non-Legume Crops*, (Eds.). N. Yanfu, I.R. Kennedy and C. Tingwei. Qingdao Ocean Univ. Press, Qingdao, China, p. 53–58.

Zhu, Z. 1989. Dynamics of soil nitrogen and its management. In: *Progress in Irrigated Rice Research.* International Rice Research Institute, Makati City, Philippines, p. 151–164.

Chapter 3

Indirect Organogenesis through Pseudostem Callus in Turmeric Variety Suguna of *Curcuma longa* L.

M.C. Gayatri, V. Roopadarshini and R. Kavyashree

Plant Biotechnology Unit, Annexe-Department of Botany, Bangalore University, Bangalore – 560 056

ABSTRACT

Profuse, compact, creamish white callus was induced from *in vitro* pseudostem explant of turmeric variety Suguna when cultured on Linsmaier and Skoog's Basal Medium (LSBM) fortified with 2,4-D (3 mg^{-1}). On transfer of callus to LSBM fortified with BAP (4 mgl^{-1}), shoot and root primordia were seen to differentiate from the callus. Upon subculture, this morphogenetic callus developed 28–30 multiple shoots with well developed roots on the same medium. The complete plantlets were hardened and successfully transferred to the field with 92 per cent survival frequency. Statistical analysis revealed significant difference between and within the treatments with regard to fresh and dry weight of the calli. For the first time, repeatable protocol has been developed for differentiation of complete plantlets from the pseudostem callus in turmeric variety Suguna.

Keywords: Turmeric, Pseudostem, Callus, Differentiation, Multiple shoots.

Abbreviations

LSBM: Linsmaier and Skoog's Basal Medium

BAP: 6-Benzyl Amino Purine

NAA: α-Naphthalene Acetic Acid

2,4-D: 2,4-Dichlorophenoxy Acetic acid

Kn: Kinetin

Introduction

Turmeric (*Curcuma longa* L.) is one of the oldest spices cultivated for its culinary and medicinal properties in the world scenario (Srimal, 1997; Ishita Chattopadhyay *et al.*, 2004). Turmeric is propagated vegetatively, since conventional breeding programmes are ineffective due to rare flowering and seed setting. In addition, turmeric is highly susceptible to various diseases (Sarma and Anandaraj, 2000) and pests (Abdulla Koya *et al.*, 1991) thereby resulting in considerable crop loss. Hence, conservation of genetic resources in the field gene bank is very difficult because none of the existing cultivars of turmeric are resistant to these diseases and pests. Therefore, production and supply of disease free planting material to the farmers for commercial exploitation is very scanty.

During the last three decades, considerable work has been carried out with turmeric varieties for *in vitro* propagation using different non-embryogenic propagules, vegetative bud (Nadgauda *et al.*, 1978; Sato *et al.*, 1987; Yasuda *et al.*, 1988; Keshavachandran and Khader, 1989; Winnaar, 1989, Balachandran *et al.*, 1990, Raghurajan, 1997, Shirgurkar *et al.*, 2001, Sunitibala *et al.*, 2001, Salvi *et al.*, 2002, Prathanturarug *et al.*, 2003), leaf-base (Salvi *et al.*, 2001) and embryogenic propagule immature inflorescence (Salvi *et al.*, 2000). In the present investigation, an attempt has been made to develop a repeatable protocol for large scale multiplication of turmeric variety Suguna through indirect organogenesis of pseudostem explant.

Material and Methods

Turmeric variety Suguna is a monocot belonging to the family Zingiberaceae, subfamily Zingiberoideae and order Zingiberales. It is a perennial herb growing upto a height of 107 cm with a well developed rhizome. The non-embryogenic vegetative propagule vegetative buds measuring 0.5–1 cm were dissociated from the dried scales and washed thoroughly in running tap water, surface sterilized using liquid detergent Labolene–5 per cent (v/v), given a rinse in 70 per cent ethanol for 30 secs, disinfected in 0.1 per cent Mercuric chloride ($HgCl_2$) for 3 mins, treated with 0.1 per cent Streptomycin for 1 min to avoid bacterial contamination and washed several times in sterile water to remove the traces of sterilant. The sterilized vegetative buds were cultured on LSBM fortified with BAP (4 mgl^{-1}) to obtain *in vitro* complete plantlets.

The pseudostem explants measuring 0.5–1 cm were excised from *in vitro* raised plants and inoculated on LSBM fortified with different concentrations of 2,4-D and NAA ranging from 1.0–4.0 mgl^{-1} for callus induction (Table 3.1). The callus thus obtained was subcultured on LSBM supplemented with different concentrations of BAP and Kn ranging from 1.0–5.0 mgl^{-1} to study their morphogenetic response.

The cultures were maintained at a temperature of 25±2°C under white fluorescent light at a photon density of 30–50 $\mu Em^{-2} s^{-1}$ under a photoperiodic regime of 16 hours light and 8 hours dark cycles with Relative Humidity (RH) 55–60 per cent. The multiple shoots with well developed roots were taken out from the culture bottles and washed thoroughly to remove the traces of agar. They were then planted in plastic pots containing potting mixture peat: perlite: vermiculate 50 : 25 : 25 (v/v) and maintained in the Hardening Chamber with controlled temperature, light and Relative Humidity. The acclimatized complete plantlets were then transferred to the field and its survival frequency was recorded.

The data obtained in the present study pertaining to fresh and dry weight of the calli and its differentiation were analyzed statistically by one-way analysis of variance (ANOVA) to determine the variation between the treatments and Critical Difference (CD) to determine variations within the treatments.

Table 3.1: Effect of Different Concentrations or Auxins on Callus Initiation from *in vitro* Pseudostem Explant of Turmeric Variety Suguna (*Curcuma longa* L.)

Basal Medium	*Auxin Concentration (mgl⁻¹)*	*Culture Duration*	*Response (%)*
	2,4-D		
LS	1.0	15 days	42
LS	2.0	15 days	65
LS	3.0	15 days	84
LS	4.0	15 days	70
	NAA		
LS	1.0	15 days	16
LS	2.0	15 days	36
LS	3.0	15 days	45
LS	4.0	15 days	42

Results

Callus initiation was noticed at the cut ends of aseptic pseudostem explant (Figures 3.1 and 3.2) on LSBM fortified with 2,4-D and NAA after 15 days of culture on all the concentrations tried with varying frequency of response (16–84 per cent) (Table 3.1). The highest (84 per cent) and lowest (16 per cent) frequency of callusing was observed on LSBM fortified with 2,4-D (3 mg^{-1}) and NAA (1 mg^{-1}) respectively. Upon two passages of subculture, profuse compact creamish white callus (Figure 3.3) was obtained on the same medium after 45 days of culture.

The statistical analysis of the data pertaining to the amount of calli revealed that there exist highly significant differences between and within the treatments. The highest amount of compact creamish white callus was observed on LSBM fortified with 2,4-D (3 mgl^{-1}) with fresh weight 9.86 g and dry weight 1.90 g when compared to that of other concentrations of 2,4-D and NAA tried (Table 3.2).

The profuse compact creamish white callus differentiated into shoot and root primordia (Figure 3.4) on LSBM fortified with BAP and Kn after 15 days of subculture on all the concentrations tried with varying frequency of response (22–86 per cent). The highest (86 per cent) and the lowest (22 per cent) frequency of shoot and root initiation response was noticed on LSBM supplemented with BAP (4 mgl^{-1}) and Kn (1 mgl^{-1}) respectively (Table 3.3).

The shoot and root primordia developed into multiple shoots (28–30) with well developed roots (Figure 3.5) after 30 days of subculture on LSBM supplemented with BAP (4 mgl^{-1}) after a single passage of subculture, thereby eliminating an *in vitro* step for rooting.

The statistical analysis of the data pertaining to the number of shoots and roots formation revealed that there exist highly significant differences between and within the treatments. The highest number of shoots (30.33) and roots (20.56) from the morphogenetic calli was observed on LSBM fortified with BAP (4 mgl^{-1}) when compared to that of other concentrations of BAP and Kn tried (Table 3.3).

The well developed complete plantlets were hardened (Figure 3.6) using potting mixture peat : perlite : vermiculate in the ratio 50 : 25 : 25 (v/v) and successfully transferred to the field with 92 per cent survival frequency.

Figures 3.1–3.6: Indirect Organogenesis through Pseudostem Callus of *Curcuma longa* L. Variety Suguna

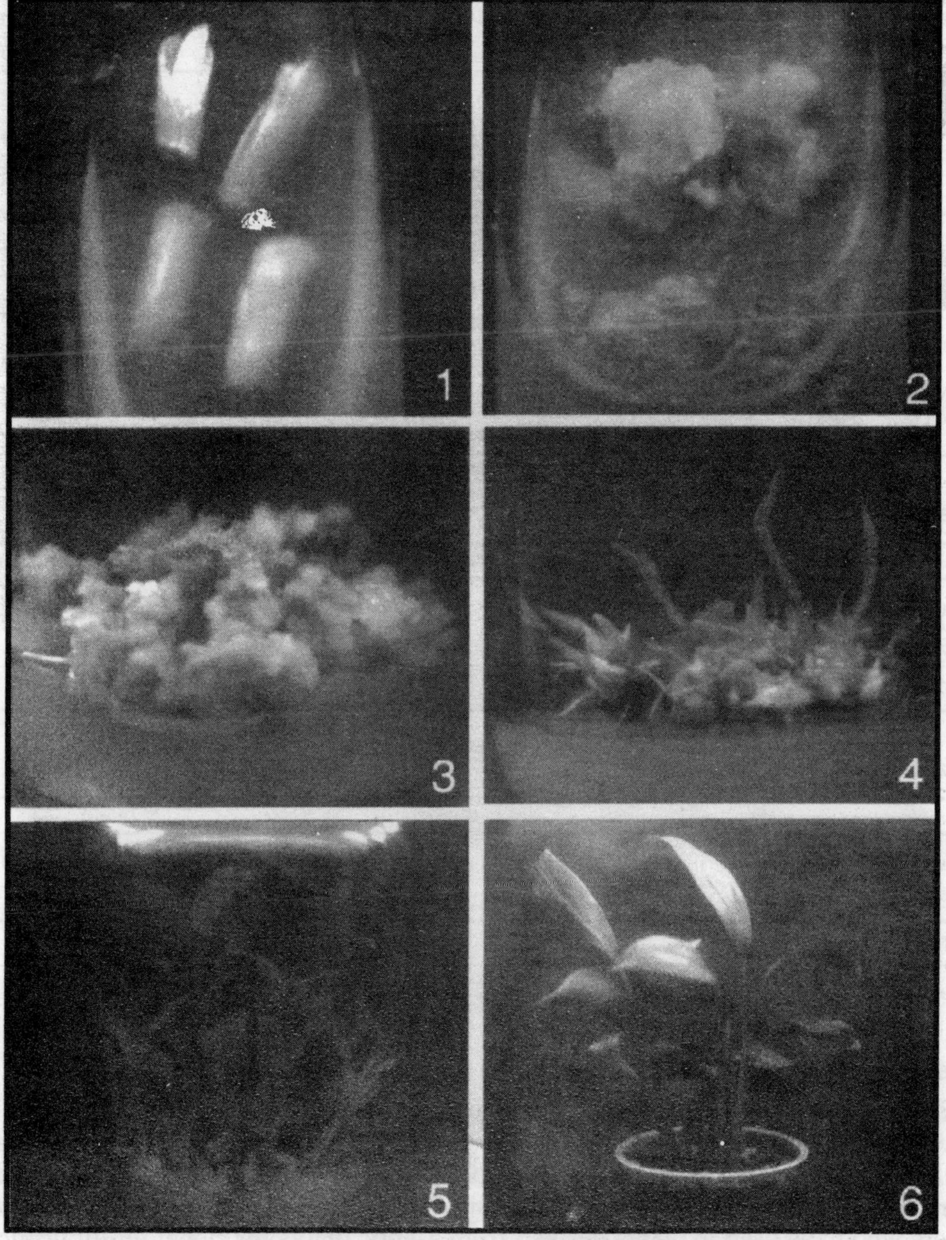

Figure 3.1: Pseudostem Explant at Culture

Figure 3.2: Callusing of Explant at Cut Ends

Figure 3.3: Profuse Compact Creamish White Callus

Figure 3.4: Differentiation of Shoot and Root Primordia

Figure 3.5: Multiple Shoots with Well Developed Roots

Figure 3.6: Hardened *in vitro* Plant

Table 3.2: Effect of Different Concentrations of Auxins on Callus Growth from *in vitro* Pseudostem Explant of Turmeric Variety Suguna of *Curcuma longa* L.

Basal Medium	*Auxin Concentration (mgl⁻¹)*	*Culture Duration*	*Fresh Weight (g) $\bar{X}$*±SD*	*Dry Weight (g) $\bar{X}$*±SD*
	2,4-D			
LS	1.0	45 days	7.16±0.11	1.10±0.09
LS	2.0	45 days	8.56±0.28	1.50±0.05
LS	3.0	45 days	9.86±0.02	1.90±0.10
LS	4.0	45 days	8.68±0.34	1.62±0.24
	NAA			
LS	1.0	45 days	3.65±0.18	0.70±0.02
LS	2.0	45 days	4.20±0.25	0.83±0.02
LS	3.0	45 days	5.80±0.11	1.01±0.07
LS	4.0	45 days	4.45±0.29	0.84±0.02

ANOVA Table (for Fresh Weight)

SV	*DF*	*SS*	*MSS*	*F_{cal} Ratio*	*F_{tab}** Tatio*	*CD*
Treatments	7	388.43	55.49	967.79	2.154	0.363
Error	72	4.13	0.057			
Total	**79**	**392.56**				

ANOVA Table (for Dry Weight)

SV	*DF*	*SS*	*MSS*	*F_{cal} Ratio*	*F_{tab}** Tatio*	*CD*
Treatments	7	13.12	1.87	152.52	2.154	0.168
Error	72	0.88	0.012			
Total	**79**	**14.01**				

Note: *: Mean of 10 replications; **: Significant F value @ 5 per cent level.

Discussion

In vitro non-embryogenic propagule pseudostem derived from direct regeneration of *in vivo* vegetative bud was chosen for retrieval of plants through callus phase. The advantage of using aseptically derived juvenile explants is essential due to the fact that it is free of phenolic compounds, the accumulation of which inhibits organogenesis. The present investigation coincides with many researchers who have used explants derived from aseptic plants for clonal multiplication (Bonga 1987; George and Sherrington, 1984; Torres, 1989; Salvi *et al.*, 2001).

It was noticed that 2,4-D (3 mgl^{-1}) was found to be the best auxin in effective induction of callus when compared to NAA from *in vitro* pseudostem of turmeric variety Suguna. This observation corroborates with the findings of Ilahi and Jabeen (1987), Malamug *et al.* (1991) and Babu *et al.* (1992)

in Ginger. It was found that the statistical analysis of the data pertaining to the fresh weight and dry weight of the calli revealed significant differences between and within the treatments.

Table 3.3: Effect of Different Concentrations of Cytokinins on Initiation and Development of Shoots and Roots from Pseudostem Callus of Turmeric Variety Suguna (*Curcuma longa* L.)

Basal Medium	Cytokinin Concentration (mgl^{-1})	Response (%)	No. of Shoots/ Explant $\bar{X}$*±SD	No. of Roots/ Explant $\bar{X}$*±SD
	Kn			
LS	1.0	22	10.75±0.81	6.23±0.25
LS	2.0	28	12.81±1.28	7.18±0.12
LS	3.0	34	16.28±0.60	10.75±0.56
LS	4.0	39	20.10±0.46	12.75±0.68
LS	5.0	36	16.24±0.69	9.92±0.50
	BAP			
LS	1.0	42	17.96±0.72	13.56±0.57
LS	2.0	56	21.60±1.42	15.98±0.71
LS	3.0	70	26.80±1.16	17.31±0.33
LS	4.0	86	30.33±0.57	20.56±0.72
LS	5.0	68	25.30±0.64	17.10±0.11

ANOVA Table (for No. of shoots/explant)

SV	DF	SS	MSS	F_{cal} Ratio	F_{tab} ** ratio	CD
Treatments	9	3527.33	391.92	543.85	2	1.210
Error	90	64.86	0.72			
Total	**99**	**3592.19**				

ANOVA Table (for No. of roots/explant)

SV	DF	SS	MSS	F_{cal} Ratio	F_{tab} ** Tatio	CD
Treatments	9	1959.27	217.69	754.15	2	0.766
Error	90	25.98	0.29			
Total	**99**	**1985.25**				

Note: *: Mean of 10 replications; **: Significant F value @ 5 per cent level.

In the present study, the best morphogenetic response of 86 per cent with induction of organogenesis and subsequent development of shoots (28-30) with well developed roots after one passage of subculture was observed on HAP (4 mgl^{-1}) when compared to Kn. Similar results have been reported by Balachandran *et al.* (1990). However, the present findings is in contrast with the results of Salvi *et al.* (2000) who have reported that NAA (0.1 mgl^{-1}) is essential for rooting. The significant difference between and within the treatment was observed with regard to formation of shoots and roots from the morphogenetic callus. The retrieved plantlets were hardened and successfully transferred to the field.

The protocol developed for regeneration of plants through callus phase from pseudostem explant of turmeric variety Suguna, as outlined in this paper can be used as an viable and valuable tool to evolve high yielding varieties in turmeric which is not possible through conventional hybridization techniques.

Acknowledgement

Thanks are due to Department of Science and Technology, Ministry of Science and Technology, Government of India, New Delhi for research grant to carry out this work in the project entitled "*Isolation of somaclones and disease resistant varieties in turmeric*" sanctioned to Prof. M.C. Gayatri.

References

Abdulla Koya, K.M., Devasahayam, S. and Prem Kumar, T. 1991. Insect pests of ginger (*Zingiber officinale* Rosc.) and turmeric (*Curcuma longa* Linn.) in India. *J. Plant Crops*, 19(1): 1–13.

Babu, K. N, Samsudeen, K. and Ratnambal, M.J. 1992. *In vitro* plant regeneration from leaf derived callus in ginger (*Zingiber officinale* Rosc.). *Plant Cell Tiss. Org. Cult.*, 29(2): 71–74.

Balachandran, S.M., Bhat, S.R. and Chandel, K.P.S. 1990. *In vitro* multiplication of turmeric (*Curcuma* spp.) and ginger (*Zingiber officinale* Rosc.). *Plant Cell Rep.*, 8(9): 521–524.

Bonga, J.M. 1987. Clonal propagation of mature trees: Problems and possible solutions. In: *Cell and Tissue Culture in Forestry*, (Eds.) Bonga, J.M. and Durzan, D.J. Martinus Nijhoff Publishers, Dordrecht, p. 249–271.

George, E.F. and Sherrington, P.D. 1984. *Plant Propagation by Tissue Culture: Handbook and Dictionary of Commercial Laboratories*. Exegetics Ltd., England.

Ilahi, I. and Jabeen, M. 1987. Micropropagation of *Zingiber officinale* L. *Pak. J. Bot.*, 19(1): 61–65.

Ishita Chattopadhyay, Kaushik Biswas, Uday Bandyopadhyay and Ranjit Banerjee. 2004. Turmeric and curcumin: Biological applications and medicinal properties. *Curr. Sci.*, 87(10): 44 –53.

Keshavachandran, R. and Khader, A. 1989. Tissue culture propagation of turmeric. *South Ind. Hort.*, 37(2): 101–102.

Malamug, J.J.F, Inden, H. and Asahira, T. 1991. Plant regeneration and propagation of ginger callus. *Scientia Horticulture (Netherlands)*, 48(1): 89–97.

Nadgauda Rajini, S., Mascarenhas, A.F., Hendré, R.R. and Jagannathan, V. 1978. Rapid multiplication of turmeric (*Curcuma longa* Linn.) plants by tissue culture. *Indian J. Expt. Biol.*, 16(1): 120–122.

Prathanturarug, S., Soonthornchareonnon, N., Chuakul, W., Phaidee, Y. and Saralamp, P. 2003. High-frequency shoot multiplication of *Curcuma tonga* L. using thidiazuron. *Plant Cell Rep.*, 21: 1054–1059.

Raghurajan, V. 1997. Micropropagation of turmeric (*Curcuma longa* L.) by *in vitro*–microrhizomes. In: *Biotechnology of Spices, Medicinal and Aromatic Plants*, (Eds.) Edison, S., Ramana, K.V., Sasikumar, B., Nirmal Babu, K. and Santosh Eapen, J. *Proceedings of the National Seminar on Biotechnology of Spices and Aromatic Plants*, IISR, Calicut, p. 25–28.

Salvi Neeta, D., Leela George and Susan Eapen. 2000. Direct regeneration of shoots from immature inflorescence cultures of turmeric. *Plant Cell Tiss. Org. Cult.*, 62: 235–238.

Salvi Neeta, D., Leela George and Susan Eapen. 2001. Plant regeneration from leaf base callus of turmeric and random amplified polymorphic DNA analysis of regenerated plants. *Plant Cell Tiss. Org. Cult.*, 66: 113–119.

Salvi Neeta, D., Leela George and Susan Eapen. 2002. Micropropagation and field evaluation of micropropagated plants of turmeric. *Plant Cell Tiss. Org. Cult.*, 68: 143–151.

Sarma, Y.R. and Anandaraj, M. 2000. Diseases of spice crops and their management. *Indian J. Arecanut Spices and Medicinal Plant*, 2(1): 6–20.

Sato, M., Kuroyanagi, M., Veno, A., Shimomura, K. and Satake, M. 1987. Plant tissue culture of Zingiberaceae (II): *In vitro* propagation of turmeric (*Curcuma longa* Linn.). *Plant Tissue Cult. Lett.*, 4: 86–88.

Shirgurkar Mridula, V., John, C.K. and Rajani Nadgauda, S. 2001. Factors affecting *in vitro* micro rhizome production in turmeric. *Plant Cell Tiss. Organ Cult.*, 64(1): 5–11.

Srimal, R.C. 1997. Turmeric: A brief review of medicinal properties. *Fitoterapia.*, 68(6): 483–493.

Sunitibala, H.M., Damayanti and Sharma, G.J. 2001. *In vitro* propagation and rhizome formation in *Curcuma longa* Linn. *Cytobios*, 105(409): 71–82.

Torres, K.C. 1989. *Tissue Culture Techniques of Horticultural Crops*. Van Norstrand Reinhold, New York.

Winnaar, W.D. 1989. Turmeric successfully established in tissue culture. In: *Information Bulletin, Citrus and Subtropical Fruit Research Institute*, South Africa, 193: 1–2.

Yasuda, K., Tsuda, T., Shimizu, H. and Sugaya, A. 1988. Multiplication of *Curcuma* species by tissue culture. *Plant Med.*, p. 75–79.

Chapter 4

Resource Productivity and Allocation Efficiency in the Production of Sunflower and Groundnut in Andhra Pradesh

Y. Sudhakar Reddy* and G.P. Reddy**

**Agricultural Officer (FCO) Lab, Government of Andhra Pradesh, Rajendranagar, Hyderabad – 560 030*

***Senior Scientist (Agricultural-Economics), National Academy of Agricultural Research Management, Rajendranagar, Hyderabad – 560 030*

ABSTRACT

India has the largest area (21 million hectares in 2002–2003) in the world under oil seeds and yet it has emerged as the largest importer of edible oils. It is well known that only about 20 years ago India was a net exporter of vegetable oils. But, the increased internal demand, due to improved standards of living and rapid population growth have led to a ban of exports of edible oil seeds and oils. The tremendous increase in the demand for oil seeds and vegetable oils, coupled with the drain on the exchequer of the country by way of imports and the need to build up the oil economy of the country to a self supporting level necessitated the identification of alternative sources, to bridge the gap between demand and supply of oil seeds and the cultivation of sunflower was one of the means sought for this purpose. In Andhra Pradesh, sunflower is cultivated largely in *kharif* as a rain fed crop and is mainly grown in Kurnool district of Rayalaseema region and Medak district, parts of Mahaboobnagar and Nalgonda districts of Telangana region. In this context it would be relevant to study the resource productivity and allocation efficiency in the production of sunflower and groundnut. Hence, this study was designed to evaluate empirically the relative profitability of sunflower and groundnut crops in Adoni taluk of Kurnool district and to analyse the constraints, if any, in the production of the sunflower. Data on quantity of inputs used and yields obtained with their respective value were recorded for both sunflower and groundnut production from the respondents, in 1999–2000. The functional analysis revealed that seeds, labour and fertilizers significantly influence production in both small and large farms. The ratio of MVP to MFC was more than unity for these factors indicating the possibilities for increased use of these resources to increase the yield of sunflower and groundnut. The results revealed that land was not significant in both small and large farms, which indicates that the allocation of additional area to sunflower and groundnut crops is not economical. The study indicated overall higher value of MVP to MFC ratios in sunflower production. It is inferred that the farmers would be better placed if they shift the resources (particularly seed to and fertilizers)/sunflower production from groundnut to a limited extent. The same holds true in both the size groups. Hence, the concerned agencies should educate the farmers on scientific method of the cultivation of sunflower by providing other infrastructure facilities like supply of quality seeds, fertilizer, credit, etc., at right time. At the same time, we cannot afford to neglect the cultivation of groundnut. Hence, extension agencies should still intensify their educational activities to educate the farmers on scientific method of cultivation of groundnut also, which is an important oil seed crop in the country.

Introduction

Oil seeds occupy an important position in India's national economy, next to food grains, accounting for about 10 per cent of the cultivated area, as well as the value of all agricultural produce. They occupy very prominent position in the human dietary system. The bulk of vegetable oil production in India is derived from nine oil seeds; namely, groundnut, rape seed mustard, sesame, safflower, niger, soybean and sunflower, forming the edible group and linseed and castor, being the non-edible group. In addition, cottonseed, rice bran, seeds of some tree species etc., are also being exploited for vegetable oils. Both the area and production of oil seeds in India have been increasing (Anonymous, a). Both the area and production of oil seeds in India have been increasing (Anonymous, 1987a). The area has increased from an average area of 17.3 million hectares in 1974–1975 to 21.1 million hectares in 2002–2003 *i.e.*, by 22 per cent. The production has increased by about 63 per cent from 9.2 million tonnes in 1974–1975 to 15 million tonnes in 2002–2003.

The commercial cultivation of sunflower crop was started in India during 1972–1973, in the southern states *viz.*, Andhra Pradesh, Karnataka and Tamil Nadu. Among all the sunflower varieties tried, BC-68414, EC-68415 and modern were found adaptable to various agro-climatic conditions and different cropping systems. Varietal enovation programme, initiated in 1977, helped in stabilization of yield and oil content in these varieties. Due to the Intensive Research and Development Programme, the crop has now assumed importance as an oil seed in Indian agriculture. With its tremendous advantages over other oil seed crops, sunflower is poised to become a popular oil seed crop in India.

Sunflower is fast becoming popular with our farmers as an oil seed crop in view of its short duration, wide adaptability, photo insensitive nature and ability to with-stand drought. In India, sunflower was cultivated in 0.3 million hectares during 1975, as against an area of about 1.60 million hectares during 2002–2003; thereby recording an increase of 433 per cent and the production has increased nearly four fold times over the years, from a mere 0.2 million tonnes in 1974–1975 to nearly 0.9 million tonnes in 2002–2003.

The sunflower crop can be grown during all the seasons–*kharif*, *rabi* and summer. However, higher yields are obtained during summer. It is observed that sunflower crop has more income potential for the farmers and it is replacing some of the crops like *kharif* groundnut, *rabi* jowar and durum wheat, which are less remunerative. Hence, the area under sunflower is on the increase and it is gaining importance as one of the potential oil seed crops (Sindagi, 1985). Andhra Pradesh stands third in country's sunflower area occupying 29.40 thousand hectares with a production of 11.30 thousand tonnes.

Andhra Pradesh plays an important role in edible oil seed production of the country by contributing 13 per cent of total cultivated edible oil seeds production (1982–1983 to 1984–1985 average) and stands second next only to Gujarat, which contributes 16.6 per cent of total production of edible oil seeds (Anonymous, 1985).

In this context it would be relevant to study the resource productivity and allocation efficiency in the production of sunflower and groundnut. Hence, this study was designed to evaluate empirically the relative profitability of sunflower and groundnut crops in Adoni taluk of Kurnool district and to analyse the constraints, if any, in the production of the sunflower. It is sought to be accomplished by setting up the following objective. Study the resource productivity and allocation efficiency in the production of sunflower and groundnut.

Methodology

Kurnool district was chosen for the study as it ranks first in area under sunflower crop in Andhra Pradesh. The data relating to the area and production in the district were collected from District Agricultural Officer. Of all the taluks in Kurnool district, Adoni taluk was selected because it ranks first in area under sunflower. From the list of villages in Adoni taluk, the top 10 villages with the highest area under sunflower were selected. In each selected village, farmers growing both sunflower and groundnut were listed in consultation with Taluk Statistical Assistant. The farmers were categorized into "small" and "large" based on land size. Those with less than 7.50 acres of dry land were termed "small" and those with 7.50 acres or more of dry land were termed "large" in accordance with the classification followed in Andhra Pradesh.

From each of the 10 villages, 3 small farmers and 3 large farmers growing both sunflower and groundnut were randomly chosen. Thus, 30 small and 30 large farmers constitute the sample for the study.

Data Collection

Data was collected by personally interviewing the sample farmers with the help of pre-tested schedules specially designed for the purpose. Data on some selected socio-economic characteristics of the farmers, land holding, cropping pattern, farm assets were collected. Besides, data on quantity of inputs used and yields obtained with their respective value were recorded both for sunflower and groundnut production from the respondents. Data on marketing also were collected for both sunflower and groundnut crops. The data on opinions of respondents about constraints faced by them in production and marketing of sunflower were also elicited. The data pertaining to the year 1999–2000 *kharif* season were collected during February-March by survey method with assiduity.

Resource Productivity and Allocation Efficiency in Sunflower and Groundnut Production

The main objective of management is to coordinate and utilize the farm resources in the production of different crops in such a manner that they together yield the highest net returns. It is in this context the study of resource productivity assumes paramount importance.

The analysis of resource productivity and allocation efficiency of the involved inputs in sunflower and groundnut production was done to assess the extent to which these factors explained the variability in the yield of the crops, and to determine whether the inputs were optimally used in the production of these crops.

A modified Cobb-Douglas type of production function was fitted to estimate the elasticities (elasticity coefficients) of various factor inputs. Two Cobb-Douglas production functions were estimated, one for sunflower and the other for groundnut. For both the crops the dependent variable was yield (qtls/holding). The explanatory variables that were common to both the crops were:

x_1 =Land area under the sunflower/groundnut crop in acres

x_2 =Human labour in Mondays of eight hours (inclusive of family and hired labour) for the crop

x_3 =Value of seeds in rupees used for the crop

x_4 =Value of farm yard manure in rupees

x_s =Value of chemical fertilizers in rupees.

There was one more explanatory variable namely value of plant protection chemicals (kg) for groundnut. This variable is not included in the function for sunflower because the expenditure on plant protection chemicals in sunflower was negligible.

The equation of the production function for sun-flower was:

$$y = a\, x_1^{b1}\, x_2^{b2}\, x_3^{b3}\, x_4^{b4}\, x_5^{b5} \quad \text{.............(1)}$$

The equation of the production function for groundnut was:

$$y = a\, x^{b1}\, x_2^{b2}\, x_3^{b3}\, x_4^{b4}\, x_5^{b5} \quad \text{.............(2)}$$

where,

y is the physical output (yield in qtls/holding), a is a constant term and x_1, x_2, x_3, x_4, x_5, x_6 are as explained above.

The production function (1) was fitted for:

1. The 30 small farms of sunflower
2. The 30 large farms of sunflower
3. The 60 all farms of sunflower.

Similarly, the production function (2) was fitted for:

1. The 30 small farms of groundnut
2. The 30 large farms of groundnut
3. The 60 all farms of groundnut.

In all six Cobb-Douglas functions were fitted.

Before fitting the function, the zero order correlation coefficients between the explanatory variables were calculated to rule out multi-collinearity.

Return to Scale

The return to scale can be estimated directly by getting the sum of 'b' coefficients. The return will be increasing, constant or diminishing as:

Σb_i is greater, equal or less than unity, respectively.

To see if the sum of 'b_i' coefficients were significantly different from unity, return to scale test (Rao and Miller, 1971) was conducted by using the test statistic of the form.

The resulting 't' computed values are compared with t (n-K)–critical values at 5 per cent level.

The marginal products were calculated at the geometric mean levels of the variables by using the formula:

$$\text{Marginal Product with reference to } x_i^{th} \text{ resource} = b_i \frac{y}{x_i}$$

where,

y = Geometric mean yield from sunflower/groundnut

x_i = Geometric mean of ith independent variable

b_i = The regression coefficient (elasticity coefficient) of ith independent variable.

Marginal value product of each factor was calculated by multiplying the marginal product of each factor with the unit price of the output *i.e.,*

$$MVP = MPP \times P_y$$

where,

P_y = Price per unit of output.

Allocation Efficiency

Allocation efficiency of both sunflower and groundnut crops was studied by comparing the marginal value product of each factor to its marginal cost (Market input price) to determine whether the factors were used optimally in the production of crops and to estimate the possibility of shifting a resource if it can be more economical in production of other competing crop.

Results and Discussions

Resource Productivity and Allocation Efficiency in the Production of Sunflower and Groundnut

To study the productivities of resources involved in the production of sunflower and groundnut, analysis was done separately for small, large and all farms of sunflower and groundnut in order to determine whether or not the factors of production were used optimally. The efficiency of allocation of resources was studied by comparing the marginal value product with the marginal factor cost of each factor of production included in the analysis.

Further, an attempt is made to compare the marginal value products of sunflower with its competing groundnut for the resources in both the size groups and all farms and to estimate the possibility of shifting of resources from one crop to another. For this purpose the ratio between marginal value product and the respective marginal factor cost was calculated for the inputs involved in both the crops grown by different categories of respondents at their geometric mean levels.

A modified Cobb-Douglas type of production function was fitted to the data, since it provides a compromise between adequate fit to the data, computational feasibility and sufficient degrees of freedom to allow for statistical testing (Heady and Dillon, 1964).

Small Farms: Sunflower and Groundnut Crops

The results on elasticity coefficients in the production function of sunflower and groundnut for small farms are presented in Table 4.1. It could be observed from this table that in sunflower and groundnut crops the regression coefficients of seed (0.6606 and 0.4748) and fertilizers (0.0257 and 0.0180), were significant at one per cent level in case of sunflower and at five per cent level in case of groundnut. Similarly, the same for human labour (0.4022 and 0.7511) was significant at five per cent level in both the crops. On the other hand, the coefficients of variables like land (0.0286 and 0.0594), and farmyard manure (0.0123 and 0.0041) were non-significant in both the crops. The regression coefficient of plant protection chemicals (0.0123) was non-significant in case of groundnut.

Table 4.1: Factors Influencing Sunflower and Groundnut Production–Small Farms

Crop	*No. of Observations*	*Intercept*	*Regression Coefficients (Production Elasticities)*						*R^2*	*Sum of Elasticities*
			Area Under Crop	*Human Labour*	*Seed*	*Farm-yard Manure*	*Fertilizer*	*Plant Protection Chemicals*		
Sunflower	30	– 3.0038	0.0286 NS (0.090)	0.4022* (1.732)	0.6606** (4.548)	0.0123 NS (1.67)	0.0257** (4.329)		0.8697**	1.294
Groundnut	30	– 4.9121	0.0594 KB (0.409)	0.7511* (2.241)	0.4748* (2.086)	0.0041 NS (0.778)	0.0180* (2.76)	0.0123 NS (2.76)	0.8821** (1.145)	1.3197

Note: * and ** indicate significant at five and one per cent level of significance respectively.

NS: Not Significant.

Figures in parentheses indicate t-values.

R^2: Significance of this was tested by 'F' test.

The results on allocation efficiency of important variables included in the analysis for sunflower and groundnut in small farms are presented in Table 4.2. In this table, the marginal factor cost of land was nothing but an imputed rental value of land. The marginal factor cost of human labour was the wage rate paid by the farmers per man-day. For the remaining resources such as seed, farmyard manure, fertilizers, plant protection chemicals the marginal factor cost of Re. 1 was considered.

As indicated earlier seed, fertilizer and human labour were significant variables, which would contribute in the production of both the crops. The marginal product of a rupee of seed in sunflower was 0.0161 qtls valued at Rs. 9.77 as compared to 0.0054 qtls valued at Rs. 3.01 in groundnut. Similarly the marginal product of a rupee worth of fertilizer was 0.0180 qtls (MVP = 10.93) and 0.0113 qtls (MVP = 6.29) in sunflower and groundnut respectively. The marginal product of one man-day of human labour was 0.0493 qtls (MVP = Rs. 29.92) in sunflower and 0.0935 qtls (MVP = Rs. 52.08) in groundnut. The marginal product of land however did not cover the cost of renting in additional land.

The ratio of marginal value product to marginal factor cost was less than unity in case of land in both the crops, indicating it's over use. The ratio for all other resources *viz.* human labour, seed, farm yard manure, fertilizer were greater than unity in both the crops indicating the possibilities of increasing the yield with increased doses of these resources baring farm yard manure where regression coefficient was not significant. The ratios were 2.99, 9.77, 2.73 and 10.93 in case of human labour, seed, farmyard manure and fertilizer, respectively, in a case of sunflower. The corresponding ratios for groundnut were 5.21, 3.01, 19.77 and 6.29. The ratio for plant protection chemicals, the variable included only for groundnut was 3.18. The sum of elasticities was 1.1294 for sunflower crop and this was not statistically different from unity indicating constant returns to scale. The sum of elasticities was 1.3197 for groundnut crop and this was statistically different from unity indicating increasing returns to scale

Table 4.2: Allocation Efficiency in Sunflower and Groundnut Production–Small Farms

Variables	*Sunflower*					*Groundnut*				
	GM	*MP (Qtls.)*	*MVP (Rs.)*	*MFC (Rs.)*	*MVP/ MFC*	*GM*	*MP (Qtls.)*	*MVP (Rs.)*	*MFC (Rs.)*	*MVP/ MFC*
Area under crop (Acres)	2.34	0.0717	43.52	133.33	0.3263	2.58	0.1795	99.98	133.33	0.75
Human labour (man-days)	47.80	0.0493	29.92	10.00	2.992	70.88	0.0935	52.08	10.00	5.21
Seeds (Rs.)	240.09	0.0161	9.77	1.00	9.77	683.35	0.0054	3.01	1.00	3.01
Farmyard manure (Rs.)	15.94	0.0045	2.73	1.00	2.73	8.98	0.0036	2.00	1.00	2.00
Fertilizer (Rs.)	8.34	0.0180	10.93	1.00	10.93	12.39	0.0113	6.29	1.00	6.29
Plant protection chemicals* (Rs.)						16.74	0.0057	3.18	1.00	3.18
Yield (Quintals)	5.86					7.79				
Price per quintal**	(Rs.) 606.92					557.00				

Note: * Variable included only for Groundnut.

** Estimated at arithmetic mean level.

The adjusted Coefficient of Multiple Determination (R) was 0.8697 and 0.8821 in case of sunflower and groundnut crops, respectively which was significant at one per cent level. This implied that the variables selected in the function, together explained 86.97 per cent and 88.21 per cent of variation in yield significantly.

Large Farms: Sunflower and Groundnut Crops

The results on production elasticities of production functions for both sunflower crop (with five independent variables) and groundnut crop (with six independent variables) are presented in Table 4.3. It is observed from the table that regression coefficients of labour in sunflower and groundnut crops were 0.7104 and 0.6924 respectively and they were significant at one per cent level. The variable 'seed' was significant at one per cent level in case of sunflower (0.4992) and at five per cent level in case of groundnut (0.3442). In sunflower crop the variable farmyard manure (0.0241) was significant at one per cent level and the same in groundnut (0.0065) was non-significant. The use of 'fertilizer' (0.0186 and 0.0189) was significant at five per cent level in sunflower (0.0186) and at one per cent level in groundnut (0.0189). The plant protection chemicals (0.160), which are included only for groundnut was significant at five per cent level. The coefficient of land (0.0676 and 0.0493) though positive was not significant both in sunflower and groundnut.

Table 4.3: Factors Influencing Sunflower and Groundnut Production–Large Farms

Crop	*No. of Obser-vations*	*Intercept*	*Area Under Crop*	*Regression Coefficients (Production Elasticities)*					*R^2*	*Sum of Elasticities (Return to Scale)*
				Human Labour	*Seed*	*Farm-yard Manure*	*Fertilizer*	*Plant Protection Chemicals*		
Sunflower	30	– 4.2691	0.0676 NS	0.7104**	0.4992**	0.0241**	0.0186*		0.9494**	1.3199
			(0.749)	(3.379)	(4.066)	(3.484)	(2.742)			
Groundnut	30	– 2.934	0.0493 NS	0.6924**	0.3484*	0.0065 NS	0.0189**	0.0160*	0.9372**	1.1315
			(0.877)	(4.299)	(2.692)	(1.718)	(2.922)	(2.591)		

Note: * and ** indicate significant at five and one per cent levels of significance respectively.

NS: Not Significant

Figures in parentheses indicate t-values

R^2: Significance of this was tested by 'F' test.

The results on allocation efficiency of variable included in the analysis for sunflower and groundnut crops in large farms are presented in Table 4.4. We have seen earlier that human labour was highly significant in both the crops. The marginal product of a man-day of human labour was 0.1023 qtls valued at Rs. 62.22 in case of sunflower production and in groundnut production it was 0.0759 qtls valued at Rs. 42.54. The marginal product of a unit (Re.) of seed in sunflower was 0.0137 qtls (MVP = Rs. 8.33) whereas it was 0.0039 qtls (MVP = Rs. 2.19) in groundnut. Regarding farmyard manure and fertilizer the marginal products per unit (Re.) were 0.0032 qtls (MVP = Rs. 1.95) and 0.0031 qtls (MVP = Rs. 1.89) in sunflower and 0.0041 qtls (MVP = Rs. 2.30) and 0.0018 qtls (MVP = Rs. 1.01) on groundnut production. As in the small sized farms land did not significantly influence production.

Table 4.4: Allocation Efficiency in Sunflower and Groundnut Production–Large Farms

Variables	*Sunflower*					*Groundnut*				
	GM	*MP (Qtls.)*	*MVP (Rs.)*	*MFC (Rs.)*	*MVP/ MFC*	*GM*	*MP (Qtls.)*	*MVP (Rs.)*	*MFC (Rs.)*	*MVP/ MFC*
Area under crop (Acres)	9.97	0.2103	127.90	135.40	0.95	8.07	0.1535	86.03	135.40	0.64
Human labour (man-days)	215.29	0.1023	62.22	10.00	6.22	229.06	0.0759	42.54	10.00	4.25
Seed (Rs.)	1125.52	0.0137	8.33	1.00	8.33	2197.33	0.0039	2.19	1.00	2.19
Farmyard manure (Rs.)	237.70	0.0032	1.95	1.00	1.95	39.29	0.0041	2.30	1.00	2.30
Fertilizer (Rs.)	182.55	0.0031	1.89	1.00	1.89	260.60	0.0018	1.01	1.00	1.01
Plant protection chemicals*						79.99	0.0050	2.80	1.00	2.80
Yield (qtls.)	31.03					25.13				
Price per quintal** (Rs.)	608.17					560.48				

Note: * Variable included only for Groundnut; **Estimated at arithmetic mean level.

The ratio of marginal value product to marginal factor cost was greater than unity in case of human labour, seed and farmyard manure in both the crops indicating the possibilities of increasing the yield with increased doses of these resources. The ratio for human labour was Rs. 6.22 and Rs. 4.25 in sunflower and groundnut respectively. The MVP to MFC ratio for seed was Rs. 8.33 in the case of sunflower, while it was only Rs. 2.19 in the case of groundnut. The ratio for farmyard manure in sunflower was Rs. 1.95 and the corresponding value for groundnut was Rs. 2.30. The ratio for fertilizer was Rs. 1.89 in case of sunflower whereas it was Rs. 1.01 in groundnut production. The ratio of marginal value product to marginal factor cost of land was less than unity in both the crops. Since in the regression the land variable was non-significant, the value marginal products have limited significance and only indicate a broad trend.

The sum of elasticities was 1.3199 for sunflower, and this was statistically different from unity indicating increasing returns to scale. The sum of elasticities was 1.1315 for groundnut and this was not statistically different from unity indicating constant returns to scale.

This means that if all the variables were increased together by one per cent the yield would increase by one per cent *Ceterus paribus*.

The adjusted R^2 values were 0.9494 and 0.9372 for sunflower and groundnut crops, respectively, which were highly significant at one per cent level. This meant that the variables included in the production function explained 94.94 per cent and 93.72 per cent of the variation in the yield of sunflower and groundnut crops.

All Farms: Sunflower and Groundnut Crops

Table 4.5 presents the results on regression coefficients of variables included in production functions for sunflower and groundnut crops.

Table 4.5: Factors Influencing Sunflower and Groundnut Production–All Farms

Crop	*No. of Obser-vations*	*Intercept*	*Regression Coefficients (Production Elasticities)*						*R²*	*Sum of Elasticities*
			Area Under Crop	*Human Labour*	*Seed*	*Farm-yard Manure*	*Fertilizer*	*Plant Protection Chemicals*		
Sunflower	60	– 3.5909	0.0966 NS	0.5671**	0.6007	0.0177**	0.0253*#		0.9691**	1.307
			(1.005)	(2.791)	(6.136)	(3.4511)	(5.802)			
Groundnut	60	3.2267	0.0793 NS	0.4746**	0.5322*	0.0074*	0.0124*	0.0203**	0.9658**	1.126
			(1.216)	(2.684)	(2.135)	(2.135)	(2.547)	(3.475)		

Note: * and ** indicate significant at five and one per cent levels of significance respectively.

NS: Not Significant.

Figures in parentheses indicate t-values.

R^2: Significant of this was tested by 'F' test.

It is observed from the table that regression coefficients of seed (0.6007 and 0.5322), farmyard manure (0.0177 and 0.0074), fertilizers (0.0253 and 0.0124) were significant at one per cent level in sunflower production and at five per cent level in case of groundnut production. The elasticity coefficient of human labour (0.5671 and 0.4746) was significant at one per cent level in both the crops. In groundnut crop coefficient of plant protection chemicals (0.0203) was significant at one per cent level. The regression coefficients of land (0.0966 and 0.0793) were not significant in sunflower and groundnut production.

Table 4.6 presents the particulars regarding marginal product, marginal value product, marginal factor cost and the ratio of marginal value product to the marginal factor cost of each variable involved in sunflower and groundnut production in all farms.

Table 4.6: Allocation Efficiency in Sunflower and Groundnut Production–All Farms

Variables	*Sunflower*					*Groundnut*				
	GM	*MP (Qtls.)*	*MVP (Rs.)*	*MFC (Rs.)*	*MVP/ MFC*	*GM*	*MP (Qtls.)*	*MVP (Rs.)*	*MFC (Rs.)*	*MVP/ MFC*
Area under crop (Acres)	4.85	0.2697	163.86	134.37	1.22	4.56	0.2432	135.89	134.37	1.01
Human labour (man-days)	101.39	0.0753	45.80	10.00	4.58	127.49	0.0521	29.11	10.00	2.91
Seed (Rs.)	19.57	0.0155	9.42	1.00	9.42	1225.25	0.0114	6.27	1.00	6.37
Farmyard manure (Rs.)	61.56	0.0038	2.31	1.00	2.31	18.78	0.0055	3.07	1.00	3.07
Fertilizer (Rs.)	27.91	0.0122	7.41	1.00	7.41	56.83	0.0030	1.68	1.00	1.68
Plant protection chemicals*						36.60	0.0077	4.30	1.00	4.30
Yield (Quintals)	13.48					13.99				
Price per quintal** (Rs.)	607.55					558.74				

Note: * Variable included only for Groundnut.

**: Estimated at arithmetic mean level.

In both the crops, all the variables included in the analysis were significant except land. The marginal products of unit (man-days) of human labour were 0.0753 qtls and 0.0521 qtls in sunflower and groundnut production respectively. The marginal products of a rupee worth of seed, farmyard manure and fertilizer were 0.0155, 0.0038 and 0.0122 qtls respectively in sunflower production. The corresponding figures for groundnut were 0.0114, 0.0055 and 0.0030 qtls.

The marginal value product of human labour was Rs. 45.80 in sunflower production and groundnut production it was Rs. 29.11. The MVP of seed in sunflower production was Rs. 9.42 whereas groundnut production it was relatively lower at Rs. 6.37. Similarly, the marginal value product of fertilizer (Rs. 7.41) in sunflower was greater than of groundnut (Rs. 1.68).

The ratio of marginal value product to marginal factor cost was more than unity for all the resources in both the crops indicating the possibilities of increasing the yield with increased doses of these resources. The ratio of marginal value product of seed to its marginal cost was Rs. 9.42 in case of sunflower and Rs. 6.37 in case of groundnut. Similarly, the ratio for fertilizer in sunflower production was Rs. 7.41 and in groundnut production Rs. 1.68. The ratios for land, human labour, and farmyard manure were 1.22, 4.58, 2.31 in sunflower and 1.01, 2.91, and 3.07 in groundnut.

The sum of elasticities was 1.3074 for sunflower crop and this was statistically different from unity indicating increasing returns to scale. In groundnut, the sum of elasticities was 1.1262 and this was not statistically different from unity indicating constant returns to scale.

The values of adjusted Coefficient of Multiple Determination (R^2) were 0.9691 and 0.9658 in case of sunflower and groundnut, respectively, which were highly significant at one per cent level. Thus the variables chosen for the production function, together explained 96.91 per cent and 96.58 per cent of the variation in yield of sunflower and groundnut crops in the study area respectively.

Conclusion

The functional analysis revealed that seeds, labour and fertilizers significantly influence production in both small and large farms. The ratio of MVP to MFC was more than unity for these factors indicating the possibilities for increased use of these resources to increase the yield of sunflower and groundnut. The results revealed that land was not significant in both small and large farms, which indicates that the allocation of additional area to sunflower and groundnut crops is not economical. The study indicated overall higher value of MVP to MFC ratios in sunflower production. It is inferred that the farmers would be better placed if they shift the resources (particularly seed to fertilizers) sunflower production from groundnut to a limited extent. The same holds true in both the size groups.

References

Anonymous. 2000. *Bulletin on Food Statistics*. Directorate of Economics and Statistics, Department of Agriculture and Cooperation, Ministry of Agriculture, Govt. of India, pp. 185.

Anonymous. 2000. *Manorama Year Book*. Malayala Manorama Press, Kottayam, Kerala, pp. 566.

Anonymous. 1987b. *The Europa Year Book*. Europa Publications Ltd., Love lane, Rochester, Kent. J., pp. 1354.

Anonymous. 1987c. *Sunflower Ray of Hope*, National Bank News Review. 2(3): 30.

Heady, E.G. and Dillon, J.L. 1964, *Agricultural Production Functions*. Iowa State Univ. Press, Iowa.

Sindagi, S.S. 1985. The current situation, constraints and prospects of sunflower crop in India. *Oil Seed Production: Constraints and Opportunities*, (Eds). Srivastava, H.C., Bhaskaran, S., Vatsya, B. and Menon, K.K.G. Oxford and IBH Publishing Co., New Delhi, pp. 144.

Chapter 5

Identification of Radish (*Raphanus sativus* L.) and Tomato (*Lycopersicon esculentum* Mill) Varieties by Gel Electrophoresis and Isozyme Analysis

Adarsh Pal Vig and Navneet Kaur*

Department of Botanical and Environmental Sciences,
Guru Nanak Dev University, Amritsar – 143 005, India

ABSTRACT

With the increased number of varieties it becomes difficult to identify the varieties on the basis of seed characteristics alone. This has led to the development of new stable parameters such as use of protein and isozyme analysis as a tool for varietal identification. The present study includes analysis of 5 varieties of radish and 2 varieties of tomato on the basis of their protein profile and presence or absence of two enzymes *i.e.*, esterase and peroxidase. In radish (12 per cent native) the variety Punjab Ageti was showing a clear distinction from other varieties by the presence of a polymorphic band at position 2. Variety White-5 was clearly distinguished from other varieties due to polymorphic band at position 9 in case of 10 per cent SDS-PAGE. In case of tomato both the varieties Rajdhani and Nutech show 42 per cent similarity. In isozyme analysis study, it was found that the peroxidase enzyme was present in only one variety of radish *i.e.*, Punjab Ageti out of 5 varieties and absent in both the varieties of tomato. Esterase enzyme was absent in all the five varieties of radish but present in both the varieties of tomato.

Keywords: *Variety identification, Gel electrophoresis, Seed proteins, Isozyme analysis, Tomato, Radish.*

---

* Corresponding Author: Lecturer, E-mail: adarshpalvig@yahoo.co.in.

Introduction

Identification of these varieties by different parameters plays an important role in seed industry and seed trade. Varieties can be distinguished from the other variety directly by morphological characters. However with the increase in the number of varieties of each crop, it has become difficult to distinguish the varieties on the basis of morphological characters alone. Thus there is a need to develop more rapid and accurate procedures of variety identification to supplement the morphology-based approach. Several biochemical techniques have been developed for this purpose. The main advantage of such methods lies in the fact that these are least affected by the environmental interactions. This has led to the development of the new stable parameters such as use of their genetic material (nucleic acid and proteins) as a tool for varietal identification. Analysis of seed and seedling proteins and enzymes are more preferred for variety identification purpose. Evaluation of protein pattern is still the cheapest and the fastest method among biochemical approaches for screening a large number of clones for varietal differences as this method is independent from the growing season of the plant, there is no need for plant cultivation, availability of material the year round, ease of storing material, the relative speed of the examination, the small sample size needed (Moller *et al.*, 1993). The electrophoresis analysis of total seed proteins or a specific fraction provides the most useful method for variety identification, but presence of too many bands revealing a magnitude of non-polymorphic polypeptides often complicating the reading of the banding patterns (Cooke, 1995). In such cases, isozyme analyses are of great significance. Isozymes are practical and useful genetic and biochemical markers as well as good estimators of genetic variability in plant populations. Gel electrophoresis of soluble seed proteins was used in our laboratory for varietal identification in crops like paddy, moong, brinjal, palak and pea (Singh *et al.*, 2004). The present study includes the identification of 5 varieties of radish (*Raphanus sativus* L.) and 2 varieties of tomato (*Lycopersicom esculentum* Mill) on the basis of their protein profile and isozyme analysis.

Materials and Methods

Procurement of Seeds

Seed samples of five varieties of radish (*Raphanus sativus* L.) *viz.*, Pusa Himani, Punjab Safed, Punjab Ageti, Japanese White and White-5 were procured from Department of Plant Breeding, Punjab Agricultural University (PAU) Ludhiana, India whereas the seeds of two varieties (Rajdhani and Nutech) tomato (*Lycopersicon esculentum* Mill) were procured from Sher-e-Punjab Agro Seed Shop, Batala, Punjab.

Seed Protein Electrophoresis

Proteins from the seeds of 5 varieties of radish and 2 varieties of tomato were extracted in sodium phosphate buffer (0.1 M at pH 7). Protein content in different varieties was estimated by Lowry's method (1951). Gel composition of 12 per cent separating and 5 per cent stacking or 10 per cent separating and 4 per cent stacking was applied for electrophoretic separation of proteins. Commassie Brilliant Blue (CBB) dye was used for staining. Native and SDS-PAGE gel was used for radish varieties to see the differences in results.

Isozyme Analysis/Electrophoresis

For isozyme analysis, enzymes were extracted from seeds in esterase extraction buffer and peroxidase extraction buffer (Sadasivam and Manickam, 1991). Electrophoresis of enzyme sample was done according to the procedure described above. After electrophoresis the gel was incubated in substrate solution (Table 5.1) to initiate the enzyme reaction. The gel was transferred in the stop solution to end the reaction and colour was observed for the presence of the enzymes.

Table 5.1: Preparation of Substrate and Stop Solutions of the Enzymes

Sl.No.	Enzymes	Substrate Solution	Incubation Period	Enzyme Reaction	Stop Solution
1.	Esterase	It was prepared by dissolving 2.8 g sodium dihydrogen phosphate, 1.1 g disodium hydrogen, phosphate, 0.2 g fast blue RR salt and 0.03 g alpha-napthyl acetate in 200 ml of water.	20–30 minutes	Brown coloured band	It was prepared by mixing methanol, water, acetic acid and ethyl alcohol in the ratio 10 : 10 : 2 : 1.
2.	Peroxidase	It was prepared by dissolving 2.08 g benzidine in 18 ml acetic acid, 100 ml hydrogen peroxide (3 per cent and 80 ml of water)	20–30 minutes	Bright blue coloured band	Prepared a large volume of 0.67 per cent NaOH or 7 per cent acetic acid

Data Representation

Electrophoregrams were prepared with the help of computer by plotting the number of bands present in the gel, depending on the staining intensities, the bands were categorized as dark, medium, light and faint. The positions of the bands were numbered from top to the bottom. Varieties were distinguished on the basis of their location, presence, absence and staining intensities of bands.

Results

Protein Analysis

Protein content in different varieties of radish ranged from 4.6–8.6 per cent and in tomato it was ranged from 6.2–8.6 per cent.

Electrophoretic Analysis of Radish (12 per cent Native)

Different banding pattern was observed in all the varieties of radish (Figure 5.1). For study purpose the position of band was represented diagrammatically by numbering and staining was represented by different shading pattern (Figure 5.2). A major point of difference in between the varieties was seen at band no. 2 which was absent in Punjab Ageti and present in all other varieties. Similarly band no.5 was absent in Pusa Himani and present in all the other varieties. These bands were termed as polymorphic bands (those bands which were present in all the varieties but absent in one variety). Presence of medium band at position 4 in variety Pusa Himani made this variety different from other, as this band was not present in any other variety. The band no.7 was present only in two varieties, Japanese White and Pusa Himani.

Electrophoretic Analysis of Radish (10 per cent SDS-PAGE)

Similarly different banding pattern was observed using 10 per cent SDS-PAGE (Figures 5.3 and 5.4). Variety White-5 was clearly distinguished from other varieties due to polymorphic band at position 9. In variety Punjab Ageti, presence of faint band at position 10 and medium band at position 11

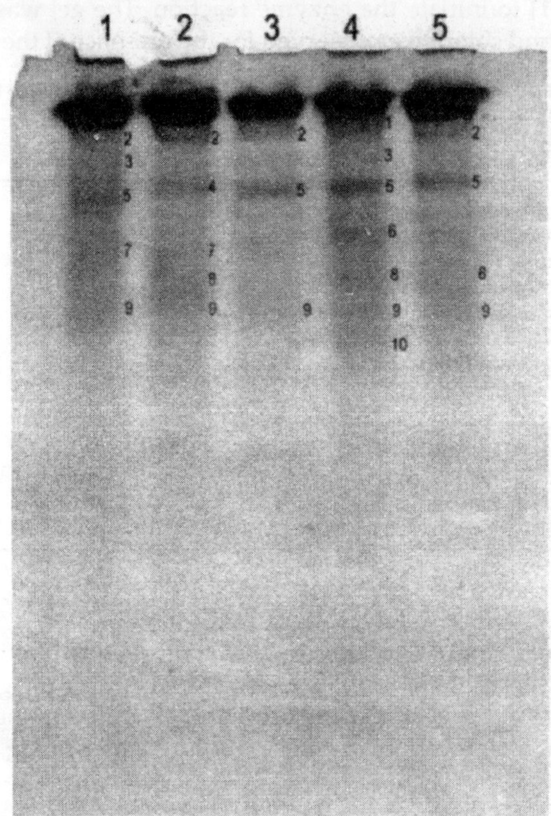

Figure 5.1: Photograph of Seed Protein Banding Pattern of 5 Varieties of Radish Using Native PAGE (12 per cent gel concentration)

1: Japanese White; 2: Pusa Himani; 3: Punjab Safed; 4: Punjab Ageti; 5: White-5

differentiate this variety from the rest of varieties as at these two positions no band was present in any other variety. In all the varieties only a single dark band was present either at position 1 or 2.

Two polymorphic bands were present in 12 per cent native and one in 10 per cent SDS-PAGE. Dark bands were present only in two varieties in case of 12 per cent native, whereas in case of 10 per cent SDS-PAGE in all the varieties only single dark band was present either at position 1 or 2. Punjab Ageti variety had more number of bands in comparison to other varieties in case of both the gels. The SDS present in SDS-PAGE denatures the proteins. In a solution of SDS most multi-subunit proteins are denatured and the individual subunits dissociate (Hayes, 1989).

Electrophoretic Analysis of Tomato (12 per cent Native)

The total numbers of bands observed were 7 (Figures 5.5 and 5.6). Both the varieties were differentiated on the basis that the band at position 1 was present in Nutech variety and absent in Rajdhani variety of tomato. At position 2 and 3, the light and dark bands were at alternate position in both the varieties. Both the varieties showed 42 per cent similarity, which was calculated by the formula given below:

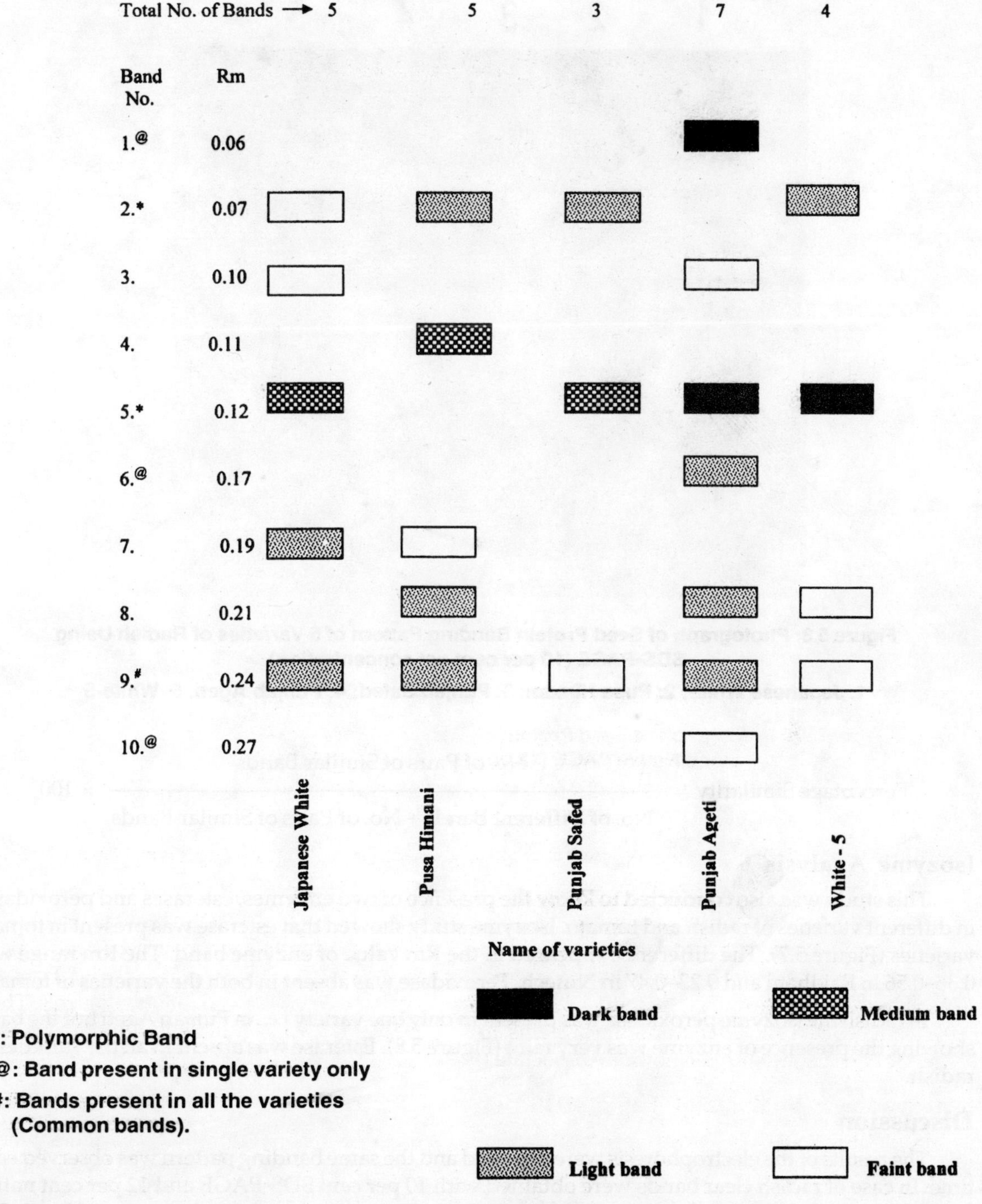

Figure 5.2: Diagrammatic Representation of the Electrophoretic Banding of Soluble Seed Proteins of 5 Radish Varieties in 12 per cent Native Gel Concentration

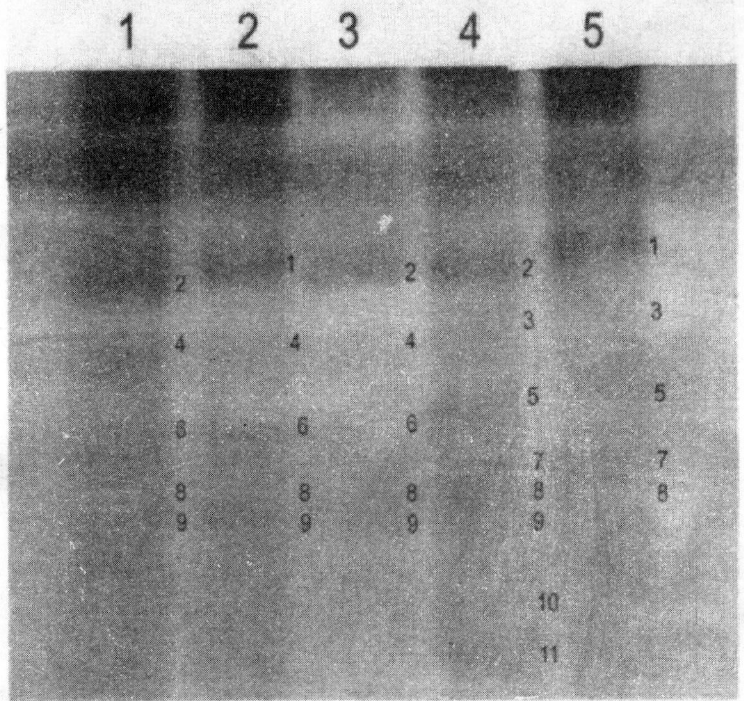

Figure 5.3: Photograph of Seed Protein Banding Pattern of 5 Varieties of Radish Using SDS-PAGE (10 per cent gel concentration)

1: Japanese White; 2: Pusa Himani; 3: Punjab Safed; 4: Punjab Ageti; 5: White-5

$$\text{Percentage Similarity} = \frac{\text{No. of Pairs of Similar Bands}}{\text{No. of Different Bands} + \text{No. of Pairs of Similar Bands}} \times 100$$

Isozyme Analysis

This study was also conducted to know the presence of two enzymes, esterases and peroxidases in different varieties of radish and tomato. Isozyme study showed that esterase was present in tomato varieties (Figure 5.7). The difference appeared in the Rm value of enzyme band. The Rm range was 0.36–0.56 in Rajdhani and 0.23–0.47 in Nutech. Peroxidase was absent in both the varieties of tomato.

In radish the isozyme peroxidase was present in only one variety *i.e.*, in Punjab Ageti but the band showing the presence of enzyme was very faint (Figure 5.8). Esterase was absent in all the varieties of radish.

Discussion

The results of the electrophoresis were repeated and the same banding pattern was observed each time. In case of radish clear bands were obtained with 10 per cent SDS-PAGE and 12 per cent native gels whereas bands obtained with 12 per cent SDS-PAGE and 10 per cent native were diffused. The two varieties of tomato also showed clear bands with 12 per cent native gel.

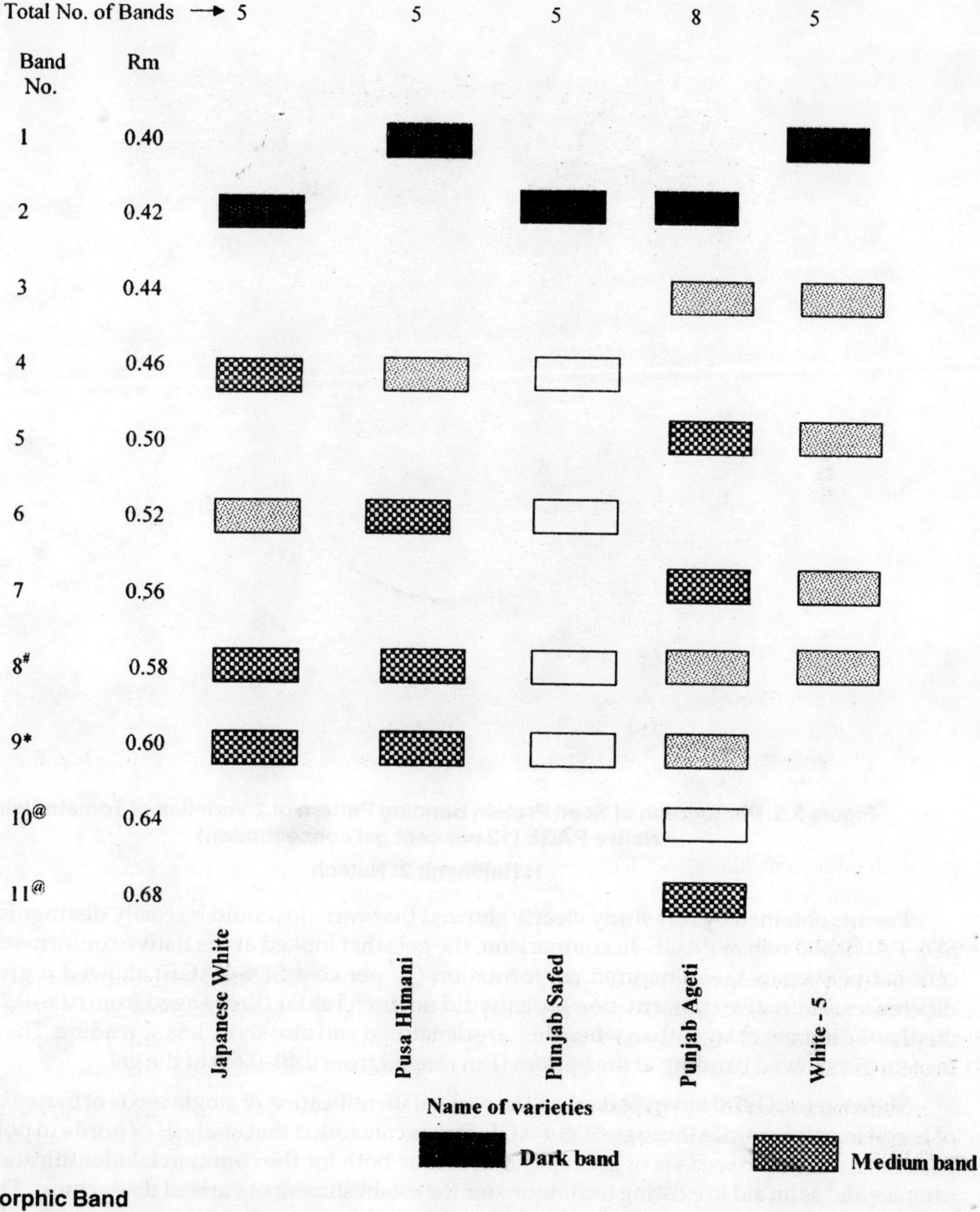

*: Polymorphic Band

@: Band present in single variety only

#: Bands present in all the varieties (Common bands).

Figure 5.4: Diagrammatic Representation of the Electrophoretic Banding of Soluble Seed Proteins of 5 Radish Varieties in 10 per cent SDS-PAGE Gel Concentration

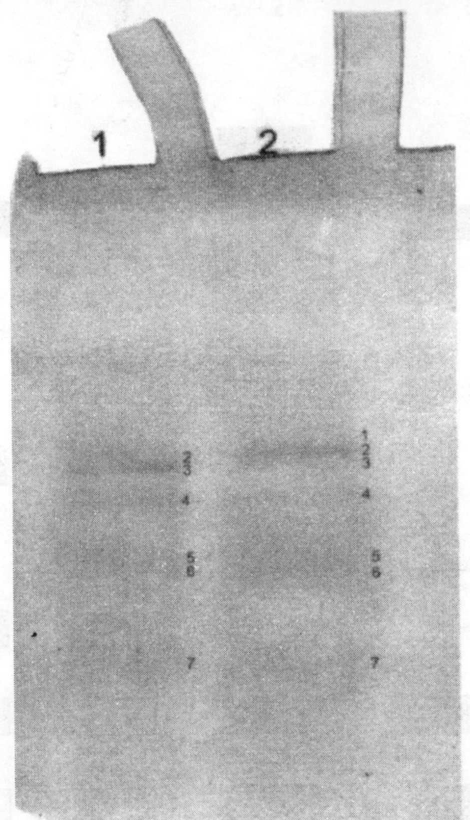

Figure 5.5: Photograph of Seed Protein Banding Pattern of 2 Varieties of Tomato Using Native PAGE (12 per cent gel concentration)

1: Rajdhani; 2: Nutech

Results obtained by this study clearly showed that varieties could be easily distinguished using SDS-PAGE and native PAGE. In comparison, the gels that looked at the native conformation (12 per cent native) versus the denatured conformation (10 per cent SDS-PAGE) showed a great deal of differences. The native conformation proteins did not travel as far (Rm ranged from 0.06–0.27) because they have different charges than when they are denatured and also show less spreading. The denatured protein gel showed banding at the bottom (Rm ranged from 0.40–0.68) of the gel.

Shewry *et al.* (1978) have worked on the varietal identification of single seeds of barley by analysis of horde in polypeptides through SDS-PAGE. It was concluded that analysis of horde in polypeptides patterns from single seeds is of great potential value both for the commercial identification of grain samples and as an aid to existing techniques for the establishment of varietal distinctness. The patterns of protein bands observed on the stained gels obtained from seed of fescue, ryegrass and cocksfoot were used to distinguish the seeds of different cultivars (Gardiner and Forde, 1987). Andersen (1982) used agar gel electrophoresis for examination of 47 barley cultivars of isozymes of α and β amylases in sprouted barley grains and esterase isozymes in the first developed leaf and the results enabled a division of cultivars into 4 groups on the basis of banding pattern. Various procedures for isozyme electrophoresis have been reported for potato (Oliver and Martinez-Zapater, 1985), tomato (Handley *et*

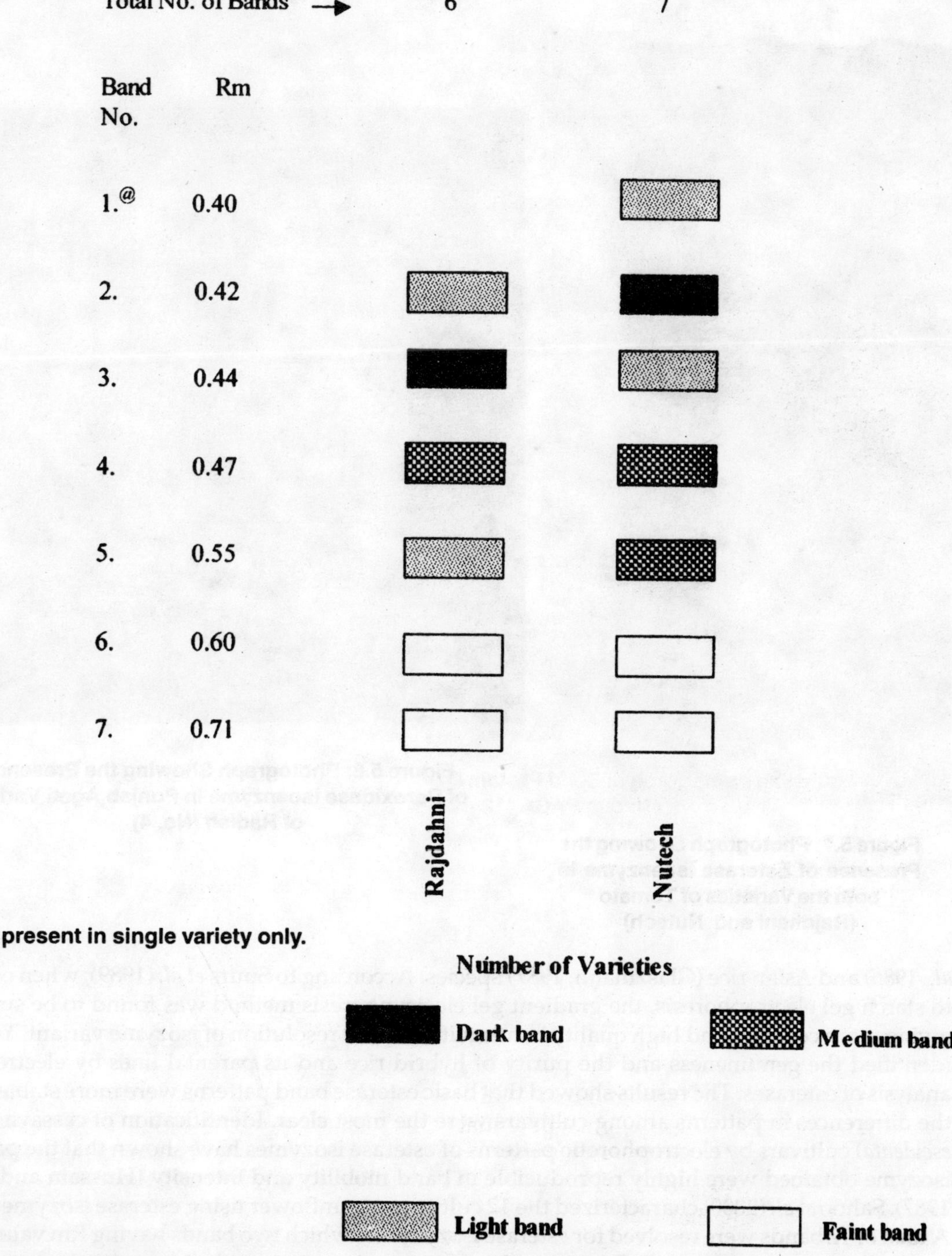

@: Band present in single variety only.

Figure 5.6: Diagrammatic Representation of the Electrophoretic Banding of Soluble Seed Proteins of 2 Tomato Varieties in 12 per cent Gel Concentration

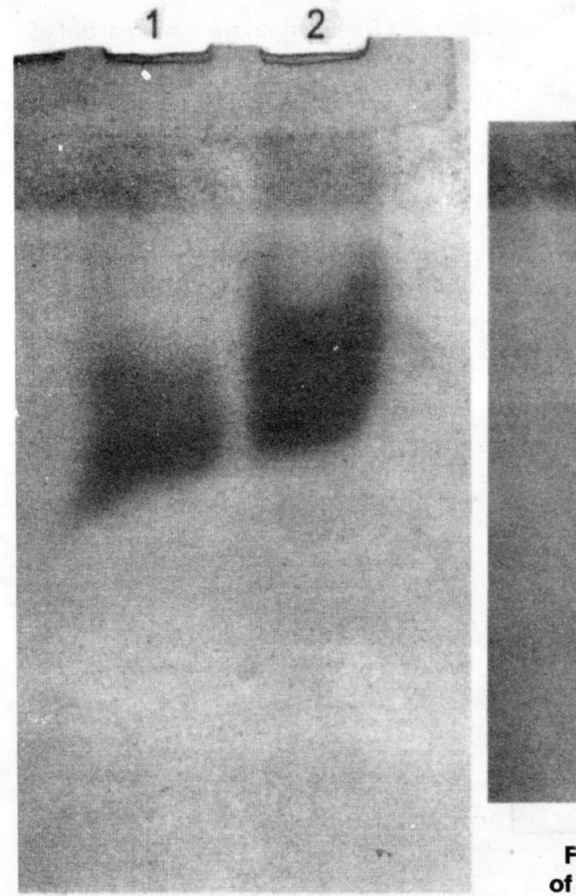

Figure 5.7: Photograph Showing the Presence of Esterase Isoenzyme in both the Varieties of Tomato (Rajdhani and Nutech)

Figure 5.8: Photograph Showing the Presence of Peroxidase Isoenzyme in Punjab Ageti Variety of Radish (No. 4)

al., 1986) and Asian rice (Glaszmann, 1987) species. According to Suurs *et al.* (1989), when compared to starch gel electrophoresis, the gradient gel electrophoresis method was found to be superior in giving more constant and high quality bands with a better resolution of isozyme variant. Yen (1987) identified the genuineness and the purity of hybrid rice and its parental lines by electrophoretic analysis of esterases. The results showed that basic esterase band patterns were more stable and that the differences in patterns among cultivars were the most clear. Identification of cassava (*Manihot esculenta*) cultivars by electrophoretic patterns of esterase isozymes have shown that the patterns of isozyme obtained were highly reproducible in band mobility and intensity (Hussain and Bushuk, 1987). Sahoo *et al.* (2000) characterized the 12 cultivars of sunflower using esterase isozyme analysis. A total of 12 bands were resolved for esterase isozymes of which two bands having Rm values of 0.49 and 0.57 were common to all. Fewer bands of esterase isozymes with characteristic patterns were found to be more reliable and easy to analyze than the protein profile. In a recent study (Vyas *et al.*, 2003), esterase and malate dehydrogenase were found to be the best enzyme systems for characterization of *Juglans regia* cultivars. A total of 13 loci and 24 alleles were observed for five enzyme systems, out of which eight loci were polymorphic. From the present study it was clear that native PAGE, SDS-PAGE of seed proteins and isozyme analysis can be successfully applied to distinguish between varieties of many crops of agriculture importance and these methods continue to be of considerable significance.

References

Andersen, H. J. 1982. Isoenzyme characters of 47 barley cultivars and their application in cultivar identification. *Seed Sci. and Technol.*, 10: 405–413.

Cooke, R.J. 1995. Gel electrophoresis for the identification of plant varieties; *J. Chromat. A.*, 698: 281–299.

Gardiner, S.E. and Forde, M.B. 1987. SDS polyacrylamide gel electrophoresis of grass seed proteins: A method for cultivar identification of pasture grasses. *Seed Sci. and Technol.*, 15: 663–674.

Glaszmann, J.C. 1987. Isozymes and classification of Asian rice varieties. *Appl. Genet.*, 74: 21–30.

Handley, L.W., Nickels, R.L., Cameron, M.W., Moore, P.P. and Sink, K.C. 1986. Somatic hybrid plants between *Lycopersicon esculentum* and *Solanum lycopersicoides. Theor. Appl. Genet.*, 71: 691–697.

Hayes, J.D. and Stockman, P.K. 1989. Electrophoresis of proteins and nucleic acids: II–Techniques and application. *Br. Med. J.*, 299: 907–910.

Hussain, A. and Bushuk, W., Ramirez, H. and Roca, W. 1987. Identification of cassava (*Manihot esculenta* Crantz) cultivars by electrophoresis patterns of esterase isozymes. *Seed Sci. and Technol.*, 15: 19–22.

Lowry, O.H., Rosebrough, M.J., Farr, A.L. and Randall, R.J. 1951. Protein measurement with folin phenol reagent. *J. Biolog. Chem.*, 193: 265–275.

Moller, M. and Spoor, W. 1993. Discrimination and identification of Lolium species and cultivars by rapid SDS–PAGE of seed storage proteins. *Seed Sci. and Technol.*, 21: 213–223.

Oliver, J.L. and Martinez-Zapater, J.M. 1985. A genetic classification of potato cultivars based on allozyme patterns. *Theor. Appl. Genet.*, 69: 305–311.

Sadasivam, S. and Manickam, A. 1991. *Biochemical Methods.* New Age International (P) Limited Publishers, New Delhi, p. 59–67

Sahoo, L., Dadlani, M., Singh, D.P. and Sharma, S.P. 2000. Characterization of sunflower (*Helianthus annus* L.) genotypes using laboratory techniques. *Plant Varieties and Seeds*, 13: 31–43.

Shewry, P.R., Helen, M. and Miflin, B.J. 1978. Varietal identification of single seeds of barley by analysis of hordein polypeptides. *J. Sci. Fd. Agric.*, 29: 587–589.

Singh, G., Vig, A.P. and Velma, R. 2004. Varietal identification in paddy (*Oryza sativa* L.) and moong (*Phaseolus mungo* L.) by gel electrophoresis of soluble seed proteins. *J. New Seeds*, 6: 91–99.

Suurs, L.C.J.M., Jongedijk, E. and Tan, M.M.C. 1989. Polyacrylamide gradient-gel electrophoresis: A routine method for high resolution isozyme electrophoresis of *Solanum* and *Lycopersicon* species. *Euphytica*, 40: 181–186.

Vyas, D., Sharma, S.K. and Sharma, D.R. 2003. Genetic structure of walnut genotype using leaf isozymes as variability measure. *Scientia Horticulturae*, 97: 141–152.

Yen, C.C. 1987. Identification of the genuineness and the purity of hybrid rice (*Oryza sativa* ssp. *indica*) and its parental lines by electrophoretic analysis of esterases. *Seed Sci. and Technol.*, 15: 645–649.

Chapter 6

Effect of Different Levels of NPK on the Growth Behaviour of *Leucaena leucocephala* (Lam.) de Wit and Yield of *Zea mays* L. in Mizoram

U.K. Sahoo, P.C. Vanlalhluna and R. Mohan

Department of Forestry, Mizoram University, Post Box No. 190, Aizawl, Mizoram

ABSTRACT

A study was conducted at Zemabawk in 1999–2000 to compare the effect of three different levels of NPK fertilizers (low, medium and high) on the growth of *Leucaena leucocephala* (Lam.) de wit along with *Zea mays* L. in Mizoram. The various growth attributes *viz.*, plant height, stem girth, number of leaves were recorded at bimonthly interval, however, mean yield of maize, number of grains per cob and the plant biomass were estimated at the time of harvest. Analysis of variance (ANOVA-2 ways, fixed effects model) was used to find the significance of variation in these parameters due to fertilizers. Our results reveal that high level of NPK had an adverse effect on the growth and biomass production of the species compared to other two levels. Similarly, the yield of maize was significantly ($p < 0.05$) higher under medium level of NPK application. In general, leucaena played a positive role in increasing yield and biomass of the crop.

Introduction

Leucaena leucocephala (Lam.) de wit is a multipurpose nitrogen fixing tree belonging to the famil Leguminoseae (Mimosoideae). It is a fast growing species which is widely used for increasing wood fuel wood, green manure, forage (Singh, 1986), conservation of soil moisture (Chirwa *et al.*, 1994 reduction in soil erosion (Lal, 1989) and to enrich the companion crop (Ngambeki and Wilson, 1983 In Mizoram, this species has been relatively recently introduced in the agroforestry systems as an alle crop to protect soil erosion and to increase crop productivity (Mohan *et al.*, 2004). Among the different crops, maize (*Zea mays* L.) is grown widely next to paddy, occupying an area of 8360 hectare (*i.e.* 8.61 per cent of total cultivation area accounting 9.12 per cent of total crop production) in the state. However, this crop is practiced either as mono or mixed crop with others like cereals, vegetables in jhum lands resulting in its low production. The present investigation therefore seeks to assess the role played by leucaena and different levels of NPK on the growth and productivity of maize in Mizoram.

Materials and Methods

An experiment was carried out at Zemabawk (altitude 1132 m asl), which lies at the out skirt of Aizawl, the capital city of Mizoram (Latitude 21°58′–24°35′ N, longitude 92°15′–93°29′ E). The climate of the area is humid tropical characterized by short and dry winter, long summer with heavy annual rainfall (2030 mm), the temperatures in summer is between 21°C to 31°C and in winter temperature between 11°C to 23°C. The site is moderately sloppy. The soil of the site is lateritic, sandy clay, redish brownish and acidic (pH 5.4) in nature.

The experiment had four treatments in RBD with three replications, totaling 12 sub-plots (each measuring 9 m × 8 m and a space of 1.5 m between two sub-plots) thus covering 1300 sqm. One month old poly plot raised seedlings of *Leucaena lecocephala* (Lam.) de wit Var. K-28 were transplanted from a nursery to the filed at a spacing of 3 m × 2 m in the month of September, 1999. After 7 month of establishment of the seedlings of leucaena, *Zea mays* L. Var. Vijaya seeds were in the interspaces at a spacing of 60 cm × 40 cm with 2.3 seeds per hill. Three NPK treatments such as low ($N_{100}P_{50}K_{60}$), medium ($N_{125}P_{70}K_{80}$) and high ($N_{150}P_{90}K_{100}$) were given in the form of urea, super-phosphate and muriate of potash respectively. Fertilizers were given in two equal splits. First half of fertilizers dose was applied to the crops in September, 1999 after 15 days of transplanting of leucaena seedlings. The remaining half dose was applied in the next year in May after 30 days of maize germination. Observation were recorded in respect of vegetative growth of maize, such as plant height stem girth, number of leaves, number of cob at monthly interval and the average grain yield at the harvest. Observations were also recorded in respect of biomass production and vegetative growths of the crops. Fertility of soil (0–10 cm depth) under different treatments was evaluated twice, once before transplanting of leucaena and second, after maize harvest. The data were subjected to ANOVA (2-ways) to see the significance of variation in these parameters due to fertilizers and months.

Results and Discussion

Growth Parameters

Plant height, girth and biomass production of leucaena increased with time. The rate of change in all these was related to the characteristic sigmoid growth curve. The growth of the plant decreased with increase level of fertilizers. Similar was the result for biomass production. With time, a better growth in plant height was observed under control than that of all other treatments. Similar was the case in the girth. The decrease in plant height and girth associated with increase dose of fertilizer indicate that fertilizer had an adverse effect on the growth of plant (Table 6.1). *Leucaena* being a leguminous plant can fix atmospheric nitrogen and enrich soil through nodule formation. The nitrogen accumulated by its root nodule help to meet its nitrogen requirement. The foliage harvested from one hectare after one year growth contained 44 kg of Phosphorus and 187 kg of Potassium as well as calcium and micronutrients (Anon, 1977), the high turnover foliage and efficient nitrogen fixation (Lulandala, 1985) can contribute highly to soil conservation and improvement planted as a fertilizer crop one hectare of leucaena can provide an amount of nitrogen equivalent to a ton of ammonium sulphate fertilizer each year especially when the foliage is incorporated in the soil (NAS, 1989). Moreover, the litter and humus layer on the soil surface act as a cushion against erosion (Nair, 1984). An additional dose of fertilizers therefore became limiting for its growth and development. Szott and Kass (1993) mentioned that such adverse effect of fertilizer on the growth of leucaena may be due to vegetative relationship between nitrogen fixation and quantity of inorganic fertilizer applied.

Table 6.1: Vegetative Growth of *Leucaena leucocephala* Under Different Levels of NPK

Parameters	*Treatment*	*Sep. 99*	*Nov. 99*	*Jan. 2000*	*Mar. 2000*	*May 2000*	*Jul. 2000*	*Sep. 2000*
Height	Control	15.6	25.8	29.8	39.2	55.7	94.3	22.4
		(0.102)	(0.295)	(0.345)	(0.389)	(0.640)	(0.394)	(0.510)
	$N_{100}P_{50}K_{60}$	15.9	22.5	27.9	37.4	54.6	91.0	115.0
	(Low)	(0.189)	(0.078)	(0.128)	(0.256)	(0.412)	(0.415)	(0.684)
	$N_{125}P_{70}K_{80}$	15.2	24.2	29.0	36.4	50.9	83.1	108.5
	(Medium)	(0.119)	(0.178)	(0.273)	(0.335)	(0.242)	(0.605)	(0.623)
	$N_{150}P_{90}K_{100}$	15.2	22.8	27.5	34.3	45.3	74.4	95.0
	(High)	(0.077)	(0.253)	(0.145)	(0.157)	(0.239)	(0.906)	(1.125)
Diameter	Control	1.8	2.2	2.9	4.0	5.3	7.0	9.0
		(0.015)	(0.017)	(0.023)	(0.017)	(0.008)	(0.017)	(0.017)
	$N_{100}P_{50}K_{60}$	1.8	2.3	2.9	3.9	5.3	6.9	8.3
	(Low)	(0.027)	(0.015)	(0.000)	(0.015)	(0.008)	(0.023)	(0.023)
	$N_{125}P_{70}K_{80}$	1.7	2.2	2.8	3.7	4.9	6.2	7.6
	(Medium)	(0.008)	(0.015)	(0.017)	(0.031)	(0.015)	(0.045)	(0.065)
	$N_{150}P_{90}K_{100}$	1.7	2.2	2.7	3.6	4.9	6.1	7.3
	(High)	(0.008)	(0.017)	(0.017)	(0.017)	(0.031)	(0.023)	(0.03)

Values in paratheses are ± SE.

A gradual decrease in leucaena biomass with increase in fertilizer dose can also be related to the above causes. As the plant height and girth were affected adversely due to increase application of fertilizer, obviously the biomass decreased in highly fertilizer dose (Table 6.2). Our present findings are in line with those of Mohan (1992) and Totey *et al.* (1988) who reported excess NPK fertilizers have adverse effect on the growth of *Terminalia myriocarpa* and *Dendrocalamus strictus*, respectively. The plant height, girth and number of leaves of maize varied with fertilizer doses (Table 6.2). The growth in maize height increased with in time, however its rate of growth was not uniform. Under low dose of NPK, the growth performance was higher compared to other doses.

Yield

The yield of maize cob and its number were related to the growth performance of the species is general, which in turn were linked to their nutritional requirement. Since high NPK dose caused stunted growth, it too affected the yield of maize cob and its number thereby reducing the yield. This is in accordance with Kanwar (1976) who reported similar cases in cereals receiving excess nitrogen exhibited poor growth thereby reducing both quality and quantity of yield.

Table 6.2: Vegetative Growth of *Zea mays* Under Different levels of NPK

Parameters	*Treatments*	*April*	*May*	*June*	*July*
Height	Control	28.9 (0.263)	80.6 (0.594)	186.4 (0.38)	190.6 (0.239)
	$N_{100}P_{50}K_{60}$ (Low)	27.2 (0.163)	85.0 (0.471)	190.9 (0.311)	195 (0.181)
	$N_{125}P_{70}K_{80}$ (Medium)	26.2 (0.113)	76.9 (0.259)	180.8 (0.163)	186.6 (0.508)
	$N_{150}P_{90}K_{100}$ (High)	2.7 (0.471)	73.9 (0.433)	169.3 (1.258)	179.9 (0.222)
Diameter	Control	2.03 (0.008)	2.9 (0.015)	5.03 (0.008)	5.23 (0.008)
	$N_{100}P_{50}K_{60}$ (Low)	1.93 (0.008)	3.1 (0.015)	5.13(0.032)	5.4 (0.027)
	$N_{125}P_{70}K_{80}$ (Medium)	2.0 (0.015)	2.9 (0.017)	4.9 (0.023)	5.13 (0.023)
	$N_{150}P_{90}K_{100}$ (High)	2.1 (0.015)	2.7 (0.008)	4.7 (0.017)	4.93 (0.023)
No. Leaves	Control	4.0 (0.000)	7.7 (0.086)	9.7 (0.086)	10.6 (0.178)
	$N_{100}P_{50}K_{60}$ (Low)	4.7 (0.086)	8.0 (0.155)	10.3 (0.086)	11.0 (0.155)
	$N_{125}P_{70}K_{80}$ (Medium)	3.7 (0.086)	7.3 (0.086)	9.0 (0.155)	10.6 (0.086)
	$N_{150}P_{90}K_{100}$ (High)	3.7 (0.086)	7.0 (0.155)	9.0 (0.155)	11.6 (0.086)

Values in paratheses are ± SE.

Changes in Soil Fertility

The soil pH was comparatively higher before intercropping compared to the time of harvest. Conversely, other soil nutrients such as organic carbon, total nitrogen, available phosphorus and potassium were higher after harvesting. This may presumably, due to the addition of fertilizers to the system and their less utility by the crops. As the nutrient applied to the system became surplus which can be indicated by vegetative growth of the plant, the nutrient could not be uptake by the plants properly, thus there was hardly any loss of nutrients through the plants. However, there might have been some loss of nutrients through run-off, leaching, percolation which unfortunately could not be measured in the present study.

References

Anon. 1977. *Leucaena Promising Forage and Tree Crop for the Tropics*. National Academy of Science. Washington D.C., National Academy Press.

Chirwa, P.K., Nair, P.K.R. and Kizza-Nkedi. 1994. Pattern of soil moisture depletion in alley cropping under semi-arid conditions in Zambia. *Agroforestry System*, 26: 89–99.

Kanwar, J.S. 1976. *Soil Fertility and Practices*. ICAR, New Delhi.

Lulandala, L.L.L. 1985. Intercropping *Leucaena leucocephala* with maize and beans. *Ph.D Thesis*, Sorkanic University of Agr. Morogoro, Tanzania.

Lal, R. 1989. Agroforestry systems and soil surface management of tropical alfisol. 2. Water runoff soil erosion and nutrient loss. *Agroforestry Systems*, 8: 97–111.

Mohan, S. 1992. Fertilizer response on the growth and biomass of a common timber species of Arunachal Pradesh, *Terminalia myriocarpa* (hillock) seedlings. *Indian Forester*, 118(11): 822–831.

Mohan, R., Sanglura, C. and Sahoo, U.K. 2004. Effect of different mulches on soil moisture and productivity of maize in agroforestry system of Mizoram. *Science Vision*, 4(4): 14–17.

Ngambeki, D.S and Wilson, G.F. 1983. Economic and on-farm evaluation. Alley cropping with *Leticaena leucocephala*. IITA 1980–1983. Activity consolidated report.

NAS. 1983. *Leucaena: A Promosing Forage and Tree Crop for the Tropics*. National Academy Press, Washington, D.C.

Nair, P.K.R. 1984. *Soil Productivity Aspects of Agroforestry*. Science of practices of agroforestry series No. 1, ICRAF.

Singh, L.P. 1986. Alley cropping in semi-arid tropics of India–CRIDA experience. In: *Proceeding of ICRAF Agroforestry Training-cum-Workshop*, CRIDA, Hyderabad.

Szott, L.T. and Kass, D.C.L. 1993. Fertilizers in agroforestry systems. *Agroferestry Systems*, 23: 157–176.

Totey, N.G., Arun Prasad, K.S., Kapoor, P.K., Chauhan, J.S., Bhownik, K.S. and Dahia, V.K. 1988. Effect of Nitrogen and Phosphorus fertilization on *Dendrocalamus strictus* seedlings. *Indian Forester*, 114(9): 594–598.

Chapter 7

Correlation of Physiological and Biochemical Characters in Seed Conservation Studies of Cucumber and Watermelon

C. Vanniarajan*, Sanjeev Saxena** and T. Nepolean*

**Centre for Plant Breeding and Genetics, Tamil Nadu Agricultural University, Coimbatore – 3, T.N., India*

***National Bureau of Plant Genetic Resources, PUSA, New Delhi – 110 012, India*

ABSTRACT

Correlation coefficient is one of the important biometrics tools for formulating the selection index as it giving the strength of relationship between the group of characteristics. Correlation coefficient of germination percentage with seed vigour, shoot length, root length, seedling length and electrical conductivity and of seed vigour with electrical conductivity over the storage under different storage conditions in cucumber and watermelon was carried out. The electrical conductivity of leachates from stored seeds of cucumber and watermelon showed a significant negative correlation with seed germination percentage. A positive correlation was noticed between germination percentage and seedling vigour in the treatments of high moisture content (15 per cent and 10 per cent) at all the storage temperatures.

Keywords: Correlation, Seed conservation, Cucumber, Watermelon.

Introduction

Ex-situ conservation of germplasm in seed storage facilities necessitates techniques that prolong seed longevity. For many seed species, this is achieved by optimizing the moisture content and reducing the temperature at which seeds are stored. The international standards recommend that for long term conservation in gene banks seeds are dried to 5–2 per cent moisture content and stored – 20°C. The moisture level for maintaining the life spans of seeds needs to be defined for each species. Experiments using a number of different species have shown that longevity may be improved of seeds are dried to less than the recommended mean value. Experiments have also shown that there are limits to the beneficial affects of drying, such that drying below an optimum or critical moisture content will not improve longevity and may have detrimental effects on seed longevity during storage. Thus optimum moisture content for storage may exist at which longevity is maximum. The reports of the behaviour of seeds for the crops like cucumber and watermelon stored at different temperatures with different moisture contents are scanty. Hence an attempt was made to study the seed deterioration pattern in cucumber and watermelon.

Materials and Methods

The variety "Pusa Safed" of cucumber and "Sugar Baby" of watermelon were procured from the National Seed Corporation, New Delhi. This experiment was conducted in the Physiology and Biochemistry Lab of National Bureau of Plant Genetic Resources, New Delhi during February–April 2000.

The cucumber and watermelon seeds were divided into four lots and each lot was conditioned to acquire moisture content of 15, 10, 5 and 3 per cent. The sealed seed lots with known moisture contents were placed at four different storage temperatures of – 20°C, 4°C, ambient and 35°C. The seeds were retrieved periodically of one week interval from the respective stored temperatures and the characters germination percentage, root length, shoot length, seedling length, seedling vigour and electrical conductivity test were recorded. To investigate the relationship between different observations, correlation analysis was carried out using the mean values over different storage periods. The analysis was carried out using MSTAT : C package.

Results and Discussion

The correlation coefficients of germination percentage with seedling vigour, shoot length, root length, seedling length and electrical conductivity and of seedling vigour index with electrical conductivity over the storage period under different storage conditions in cucumber and watermelon are given in Tables 7.1 and 7.2 respectively.

In the present study, the electrical conductivity of leachates from stored seeds of cucumber and watermelon showed a significant negative correlation with seed germination percentage which was in agreement with the results on mustard (Dubey *et al.*, 1989) and on rice (Ray and Gupta, 1980). Deteriorated seeds of muskmelon failed to register any significant increase in electrical conductivity and thus were not a suitable-index of seed quality (Pesis and Ng, 1983). A positive significant correlation is noticed between germination percentage and vigour in the treatments of high moisture contents (15 per cent and 10 per cent) at all the storage temperatures.

A positive significant correlation is also observed at low moisture contents (5 per cent and 3 per cent) only in seeds stored at 35°C. These parameters have no significant correlation at moisture contents of 5 per cent and 3 per cent at other storage temperatures (– 20°C, 4°C and ambient).

In watermelon, all the treatments of high moisture contents (15 per cent and 10 per cent) stored at different storage temperatures and seeds with low moisture contents (5 per cent and 3 per cent) stored at 35°C showed negative significant correlation between electrical conductivity and vigour. In cucumber, all the treatments except seeds with moisture content 3 per cent stored at 35°C showed non significant values indicating that electrical conductivity and vigour are not correlated with each other. Guruswamy *et al.* (1998) stated that biochemical deterioration is faster than the physiological indicators of vigour. However, in the present study the correlation between electrical conductivity and vigour is not following the same trend for all the treatments of the same crop and also between the crops. It can be concluded that, electrical conductivity has the potential of being developed as criteria for assessing seed viability in cucumber and watermelon.

Table 7.1: Correlation Coefficient of Germination Percentage with Seedling Vigour, Shoot Length, Root Length, Seedling Length and Electrical Conductivity and of Seedling Vigour Index with Electrical Conductivity Over 80 Days of Storage Period Under Different Storage Conditions in Cucumber

Storage Condition		*Correlation*					
		Germination (%)					*Vigour*
Seed Moisture (%)	*Temperature (°C)*	*Vigour*	*Electrical Conductivity*	*Shoot Length*	*Root Length*	*Seedling Length*	*Electrical Conductivity*
15	– 20	0.97*	– 0.76*	0.94*	0.88*	0.96*	– 0.14
15	4	0.92*	– 0.94*	0.89*	0.88*	0.91*	– 0.31
15	Ambient	0.96*	– 0.95*	0.94*	0.92*	0.96*	– 0.54
15	35	0.97*	– 0.96*	0.96*	0.97*	0.99*	– 0.55
10	– 20	0.94*	– 0.80	0.85*	0.87*	0.92*	– 0.15
10	4	0.89*	– 0.96*	0.82*	0.94*	0.85*	– 0.23
10	Ambient	0.94*	– 0.96*	0.96*	0.76*	0.94*	– 0.42
10	35	0.92*	– 0.76*	0.75*	0.95*	0.85*	– 0.40
5	– 20	– 0.47	– 0.84*	– 0.78*	0.73*	– 0.73*	0.55
5	4	– 0.47	– 0.71*	– 0.57	0.55	– 0.51	0.52
5	Ambient	– 0.02	– 0.93*	0.37	0.67*	– 0.35	0.48
5	35	0.98*	– 0.90*	0.96*	0.81*	0.98*	– 0.22
3	– 20	– 0.65	– 0.84*	– 0.71*	– 0.85*	– 0.75*	0.49
3	4	– 0.66	– 0.90*	– 0.77*	0.04	– 0.74*	0.79*
3	Ambient	– 0.60	– 0.34*	– 0.77*	0.44	– 0.76*	0.21
3	35	0.98*	– 0.94*	0.97*	0.92*	0.97*	– 0.18

*: Significant at 5 per cent level.

Table 7.2: Correlation Coefficient of Germination Percentage with Seedling Vigour, Shoot Length, Root Length, Seedling Length and Electrical Conductivity and of Seedling Vigour Index with Electrical Conductivity over 80 Days of Storage Period Under Different Storage Conditions in Watermelon

Storage Condition		*Correlation*					
		Germination (%)					*Vigour*
Seed Moisture (%)	*Temperature (°C)*	*Vigour*	*Electrical Conductivity*	*Shoot Length*	*Root Length*	*Seedling Length*	*Electrical Conductivity*
15	– 20	0.95*	– 0.94*	0.96*	0.96*	0.9*	– 0.97*
15	4	0.95*	– 0.92*	0.89*	0.92*	0.92*	– 0.97*
15							
	Ambient	0.99*	– 0.86*	0.96*	0.98*	0.97*	– 0.81*
15	35	0.92*	– 0.94*	0.97*	0.94*	0.97*	– 0.81*

Contd...

Table 7.2–Contd...

Storage Condition		*Correlation*					
		Germination (%)					*Vigour*
Seed Moisture (%)	*Temperature (°C)*	*Vigour*	*Electrical Conductivity*	*Shoot Length*	*Root Length*	*Seedling Length*	*Electrical Conductivity*
10	– 20	0.94*	– 0.92*	0.93*	0.92*	0.94*	0.94*
10	4	0.93*	– 0.95*	0.91*	0.91*	0.92*	– 0.94*
10	Ambient	0.86*	– 0.90*	0.81*	0.88*	0.85*	– 0.97*
10	35	0.87*	– 0.88*	0.91*	0.88*	0.90*	– 0.32
5	– 20	0.64	– 0.88*	0.24	0.76*.	0.50	– 0.61
5	4	– 0.76*	– 0.96*	0.28	0.67*	0.55	– 0.43
5	Ambient	0.15	– 0.91*	– 0.42	0.29	– 0.15	– 0.43
5	35	0.97*	– 0.95*	0.96*	– 0.35	0.95*	– 0.96*
3	– 20	0.44	– 0.95*	– 0.48	0.53	0.16	– 0.46
3	4	0.32	– 0.97*	– 0.45	0.40	– 0.17	– 0.32
3	Ambient	0.57	– 0.93*	0.42	0.27	0.48	– 0.39
3	35	0.91*	– 0.96*	0.83*	0.94*	0.88*	– 0.95*

*: Significant at 5% level.

References

Dubey, A.K., Singh, P., Goyal, R.D. and Katiyar, P. 1989. A comparison between large and small seeds in mustard. *Seed Res.*, 17: 204–207.

Guruswamy, C.S., Pitchammal, J. and Thiagarajan, C.P. 1998. The influence of storage container on the shelf life of seeds of cauliflower. *Orissa. J. Horti.*, 26: 31–33.

Pesis, E. and Ng, T.J. 1983. Viability, vigour and electrolyte leakage of muskmelon seeds-subjected to accelerated ageing. *Hort. Sci.*, 18: 242–244.

Ray, M.B. and Gupta, K. 1980. Effect of storage of early and late varieties of rice seeds in different relative humidity on viability, leaching and dehydrogenase activity. *Seed Tech. News.*, 10: 14.

Chapter 8

Integrated Weed Management in Soybean (*Glycine max.*)

Pardeep Kumar and Sat Paul Mehra

Department of Agronomy and Agrometereology, Punjab Agricultural University, Ludhiana

ABSTRACT

Soybean Cv. PK 416 was sown in six and nine rows of 45 cm and 30 cm and plant spacing of 5 cm and 7.5 cm, respectively using recommended seed rate of 87.5 kg/ha, keeping seed rate per plot constant and applying recommended dose of fertilizers. Sixteen treatment combinations of row spacings (30 cm and 45 cm) and eight weed control treatments *i.e.*, pendimethalin (pre-em.) and trifluralin (pre-plant) each at 0.5 and 0.75 kg/ha, both pendimethalin and trifluralin at 0.5 kg/ha (f.b.) one hand weeding (7 WAS) and hand weeding twice (3 and 6 WAS) and weedy check (control) were tested in a randomized block design with four replications. Both herbicides at applied doses proved safe to the crop. Application of pendimethalin/trifluralin each at 0.5 kg/ha f.b. one handing weeding 7 WAS effectively controlled weeds and yielded statistically at par with each other. Both herbicides applied at higher dose (0.75 kg/ha) gave better control of weeds than their lower dose (0.5 kg/ha). Alone application of both herbicides at lower dose proved less effective in controlling weeds and yielded significantly lower than hand weeding twice. Narrow row spacing (30 cm) proved significantly superior to wider row spacing (45 cm) due to higher growth parameter and yield attributes. The interactions between row spacing and weed control treatments in respect of growth parameters, yield attributes and seed yield were non-significant.

Introduction

Soybean is a high value pulse and oilseed crop with multiple food, feed and industrial uses. Soybean grown during the rainy season faces tremendous competition with weeds. Due to congenial temperature and moisture conditions, soybean is infested by a wide variety of both grass and broadleaf weeds. Wider row spacing and slow growth of the crop during its early growth period further aggravate the problem of weeds. Weeds cause significant reduction in the yield of this crop. Losses in yield to the tune of 50 to 71 per cent due to weed competition are quite usual (Bhan *et al.*, 1972; Muniyappa *et al.*, 1986 and Singh and Singh, 1987). Integrated weed management of soybean is important because no single method of weed control proves effective on long term basis. Through present study, an attempt was made to test the efficacy of trifluralin vis-à-vis pendimethalin and also the possibilities of doing away with requirement of one hand weeding by increasing the dose of herbicide integrated with closer row spacing to exploit the smothering potential of the crop.

Material and Methods

The field experiment was conducted at the Research Farm of the Department of Agronomy, Punjab Agricultural University, Ludhiana during *kharif* 1997 to test the efficacy of trifluralin vis-à-vis pendimethalin for weed control in soybean and to study the effect of soybean planting pattern integrated with herbicides and mechanical methods on weeds and yield of the crop. The soil was sandy loam, normal in pH (8.1), EC (0.30 mm hos/cm) low in organic carbon (0.12 per cent) and available nitrogen (61.4 kg/ha), high in available phosphorus (22.8 kg/ha) and potassium (165.0 kg/ha). A soybean Cv. PK 416 was sown in six and nine rows of 45 cm and 30 cm spacing at a plant spacing of 5 cm and 7.5 cm respectively using recommended seed rate of 87.5 kg/ha keeping seed rate per plot constant. Crop was given recommended dose of N (30 kg/ha) and P_2O_5 (80 kg/ha) in the form of urea and SSP. Sixteen treatment combinations of two row spacings (30 cm and 45 cm) and eight weed control treatments *i.e.*, both pendimethalin (pre-emergence) and trifluralin pre-plant incorporation (PPI) at 0.5 and 0.75 kg/ha, both pendimethalin and trifluralin at 0.5 kg/ha followed by (f.b.) one hand weeding 7 WAS, hand weeding twice (3 and 6 WAS) and weedy check (control) were tested in a randomized block design with four replications.

Results and Discussion

The weed flora of the experimental field comprised mainly of *Digitaria sanguinalis, Eleusine aegyptiacum, Trianthema portulacastrum* and *Euphorbia hirta*.

Effect on Crop and Weeds

The data showed that the effect of various weed control treatments on seedling emergence was non-significant indicating that at 0.5 and 0.75 kg/ha both pendimethalin and trifluralin proved safe to the crop as none of the herbicides showed any toxicity on seedling emergence. Pre-emergence application of pendimethalin 0.5 kg/ha combined with one hand weeding 7 WAS and PPI of trifluralin 0.5 kg/ha f.b. one hand weeding 7 WAS effectively controlled weeds and yielded statistically at par with each other (14.45 and 13.97 q/ha, respectively) (Table 8.2). These herbicides reduced weed biomass and provided conditions in favour of crop growth. Pendimethalin 0.5 kg/ha coupled with one hand weeding 7 WAS gave complete control of *Trianthema portulacastrum, Euphorbia hirta* and nearly complete control of *Eleusine aegyptiacum* and *Digitaria sanguinalis.* Similar effects of herbicides on weed in soybean crop were reported by Singh and Mani, 1977 and Jain *et al.*, 1985. Both herbicides applied at higher dose (0.75 kg/ha) gave better control of weeds than their lower dose (0.5 kg/ha) (Table 8.1). Pendimethalin 0.75 kg/ha (12.89 q/ha) and trifluralin 0.75 kg/ha (12.13 q/ha) though statistically at par with each other, but proved significantly inferior to hand weeding twice 3 and 6 WAS (13.59 q/ha). Alone application of pendimethalin and trifluralin at 0.5 kg/ha proved less effective in controlling weeds (Table 8.1) and yielded significantly lower than hand weeding twice 3 and 6 WAS. The comparatively lower yield was due to partial weed kill by herbicide at low dose (0.5 kg/ha).

Table 8.1: Effect of Row Spacing and Weed Control Treatments on Periodic Dry Matter Accumulation by Weeds (g m^{-2})

Treatment	*30 DAS*			*60 DAS*			*120 DAS*		
	S_1	S_2	*Mean*	S_1	S_2	*Mean*	S_1	S_2	*Mean*
Pendimethalin 0.5 kg/ha (PE)	65.96	79.10	72.52	51.68	69.52	60.60	31.85	42.49	37.17
	(8.12)	(8.82)	(8.47)	(7.12)	(8.35)	(7.74)	(5.64)	(6.53)	(6.09)
Pendimethalin 0.5 kg/ha (PE) f.b. 1 hoeing 7 WAS	61.86	76.85	69.36	10.93	21.43	16.18	7.12	15.10	11.11
	(7.86)	(8.69)	(8.28)	(3.34)	(4.63)	(3.98)	(2.79)	(3.98)	(3.38)
Pendimethalin 0.75 kg/ha (PE)	52.26	60.05	56.16	31.32	47.0	39.11	18.89	24.0	21.45
	(7.23)	(7.76)	(7.50)	(5.60)	(6.88)	(6.24)	(4.15)	(4.95)	(4.55)
Trifluralin 0.5 kg/ha (PPI)	91.97	101.12	96.55	66.86	75.23	71.03	41.26	47.24	44.25
	(9.64)	(10.10)	(9.87)	(8.20)	(8.71)	(8.45)	(6.46)	(6.92)	(6.69)
Trifluralin 0.5 kg/ha (PPI) f.b. 1 hoeing 7 WAS	89.01	98.95	93.98	16.28	29.73	23.01	7.92	18.34	13.13
	(9.49)	(9.99)	(9.74)	(3.39)	(4.95)	(4.17)	(2.94)	(4.35)	(3.64)
Trifluralin 0.75 kg/ha (PPI)	56.85	66.19	61.52	35.89	56.94	46.42	17.44	28.11	22.78
	(7.59)	(8.18)	(7.88)	(5.87)	(7.52)	(6.69)	(4.20)	(5.17)	(4.68)
Hand weeding twice (3 and 6 WAS)	38.48	49.10	43.49	21.68	28.90	25.29	11.98	17.35	14.67
	(6.26)	(7.06)	(6.66)	(4.29)	(5.41)	(4.85)	(3.39)	(4.15)	(3.77)
Unweeded Control	237.09	301.35	269.23	141.0	176.72	158.86	98.16	125.58	111.90
	(15.42)	(17.37)	(16.40)	(11.88)	(13.30)	(11.90)	(13.32)	(12.61)	(10.60)
Mean	86.69	104.10	–	47.0	63.18	–	29.19	39.77	–
	(8.95)	(9.75)		(6.21)	(7.47)		(4.94)	(5.91)	
CD (P = 0.5) for									
Row spacing		0.36			0.72			0.42	
Weed control treatment		0.72			1.44			0.82	
Interaction		NS			NS			NS	

S_1: 30 cm; S_2: 45 cm.

Table 8.2: Seedling Emergence, No. of Pods/Plant, 100 Seed Weight and Seed Yield as Influenced by Various Treatments

Treatment	*Seedling Emergence per Metre Row Length*			*No. of Pods/ Plant (Mean of S_1 and S_2)*	*100 Seed Wt (g) (Mean of S_1 and S_2)*	*Seed Yield (q/ha) (Mean of S_1 and S_2)*
	S_1	*S_2*	*Mean*			
Pendimethalin 0.5 kg/ha (PE)	19.2	26.0	22.6	54.7	12.21	12.21
Pendimethalin 0.5 kg/ha (PE) f.b. 1 hoeing 7 WAS	18.8	25.9	22.4	63.6	12.79	14.45
Pendimethalin 0.75 kg/ha (PE)	17.6	27.0	22.3	57.5	12.64	12.89
Trifluralin 0.5 kg/ha (PPI)	18.5	25.3	21.9	51.6	11.82	11.87
Trifluralin 0.5 kg/ha (PPI) f.b. 1 hoeing 7 WAS	17.8	26.5	22.1	60.7	12.48	13.97
Trifluralin 0.75 kg/ha (PPI)	18.9	24.5	21.7	55.9	12.40	12.13
Hand weeding twice (3 and 6 WAS)	17.5	25.0	21.2	57.5	12.26	13.59
Unweeded Control	18.5	25.1	21.8	38.3	11.36	4.99
Mean	18.4	25.7	–			
CD (P = 0.5) for						
Row spacing		1.1		1.59	0.20	0.75
Weed control treatment		NS		3.18	0.41	1.51
Interaction		NS		NS	NS	NS

PE: Pre-emergence; PPI: Pre Plant Incorporation; WAS: Weeks After Sowing; DAS: Days After Sowing.

Narrow row spacing (30 cm) proved significantly superior to wider row spacing (45 cm) due to higher growth parameter and yield attributes (No. of pods per plant and hundred seed weight) and seed yield (Table 8.1) and weed suppression (Table 8.2).

Pendimethalin or trifluralin 0.5 kg f.b. one hand weeding 7 WAS gave effective control of weeds like *Trianthema portulacastrum, Euphorbia hirta, Eleusine aegyptiacum* and *Digitaria sanguinalis* and yielded significantly higher than weedy check, respectively. The integration of both pendimethalin and trifluralin 0.5 kg/ha with one hand weeding 7 WAS proved superior to pendimethalin or trifluralin alone (0.5 kg/ha). Herbicides applied at higher dose (0.75 kg/ha) provided better weed control than their lower dose (0.5 kg/ha) (Table 8.1). The interactions between row spacing and weed control treatments in respect of growth parameters, yield attributes and seed yield were non-significant.

References

Bhan, V.M., Singh, M. and Maurya, R.A. 1972. Studies on chemical weed control in soybean (*Glycine max* L. Merill). *Weed Sci.*, 4: 16–22.

Jain, H.C., Dubey, S.K., Koshta, L.D. and Raghu, J.S. 1985. Integrated weed management in soybean (*Glycine max.* L. Merill). *Ann. Conf. Indian Soc. Weed Sci.*, 67 (Abstr).

Muniyappa, T.V., Ramchandra, P.T.V. and Krishnamurthy. 1986. Critical stages of weed competition and chemical weed control in soybean. *Indian J. Weed Sci.*, 18: 34–38.

Singh, B.P. and Maini, V.S. 1977. *Investigations on Chemical Weed Control in Soybean*. Div. Agron., IARI New Delhi.

Singh, G. and Singh, D. 1987. Weed control efficiency of pendimethalin and methabenzthiazuron in soybean (*Glycine max* L. Merill). *Indian J. Weed Sci.*, 19: 230–232.

Chapter 9

Response of *Eugenia bracteata* (Wild.) Roxb. to VAM, *Rhizobium* and Molybdenum Source

Ramesh H. Ratageri and H.C. Lakshman

Post Graduate Department and Research in Botany, Karnatak University, Dharwad – 580 003

ABSTRACT

Growth response and biomass production was established on *Eugenia bracteata* (Wild.) Roxb. seedlings by the inoculation of VAM, *Rhizobium* and Molybdenum spray. The results revealed in the increased shoot and root length, per cent of VAM colonization, number of leaves, dry weight of shoots, recorded significantly higher over the control plants. Indigenous V AM fungi, *Rhizobium* and Molybdenum spray improving the agronomic performance through the increased uptake of nutrients like N, P, K, Mg and Molybdenum.

Introduction

Eugenia bracteata (Wild.) Roxb. is a sub tropical tree, well adopted to mix deciduous forests, both dry and moist type. Trees grow up to 70 feet height, its leaves used as fodder and yield good timber used in wheel works and furniture. It is one the promising drought tolerant perennial grows well in varied soils. The main advantage of cultivation, of this plant grows well with a minimum requirement of water and without much agriculture management.

Steinberg (1953) reported that molybdenum is indispensable for; variety of plant species. Croft (1954) reported the profitable response of 8 oz of Sodium molybdate per acre to peas. No reports on tree species, though the role of VA mycorrhizal fungi and *Rhizobium* has been well documented on different plants (Manjunath *et al.*, Rahangdate *et al.*, 1998; Lakshman and Geeta Patil, 2004). However, in recent days, a few research reports are available to demonstrate that biofertilizers stimulate the growth of many tree species in forest nursery, the growth of many tree species (Niranjan *et al.*, 1990; Singh, 2001) there by survival rate of planted seedlings are increased.

Materials and Methods

A total of 5 × 25 seeds were collected from 10 years old plant *Eugenia bracteata* (Wild.) Roxb. Seeds were surface sterilized in 2 per cent sodium hypochlorite, later on the seeds were washed in water and sown in earthen pots contained 3 kg of sterilized sandy loam soil in (1 : 1) ratio, (loam soil and pure sand), Chemical analysis of soil as shown in (Table 9.1). The experiments were set up in a completely randomized design and replicate thrice. The earthen pots were treated with biofertilizers and molybdenum *viz.*, (1) VAM fungi, (2) *Rhizobium*, (3) Molybdenum, (4) VAM + *Rhizobium*, (5) VAM + Molybdenum, (6) VAM + *Rhizobium* + Molybdenum, (7) Uninoculated seedlings were maintained as control.

Table 9.1: Physico-chemical Properties of Soil Used for Pot Experiments

Soil Characteristics	*Values*
Soil type	Sandy loamy
Soil moisture	24.10
pH	6.7
Electric conductivity (µmhos/cm at 25°C)	0.54
Organic Carbon (per cent)	0.27
Available Phosphorus (per cent)	0.19
Available Potassium (per cent)	22.72
Available Nitrogen (per cent)	19.16
Iron (per cent)	13.18
Zink (per cent)	4.91
Copper (per cent)	2.92
Mangesium (per cent)	0.04
Molybdenum (per cent)	

Each value is mean of 14 samples.

Rhizobium was applied through seeds. Inoculant containing (1.0 × 10^{-5} cells/ml) with jaggary based culture (Varma and Subba Rao, 1975). 20 g mixed inoculum of *G. fasiculatum* were applied to each pots at 5 cm depth near the root zone. The VAM inoculum consists of 215 chlamydospores/50 g, hyphae from the pot culture of pennistytum tyhoidenum Roem. (Pearl millet), which was infected by *G. fasiculatum*. A different level of (3, 6 and 8 g molybdenum/L water) sodium molybdate foliar spray was applied with the help of sprayer when plants were 5 cm tall. After inoculation plants were watered on alternate days. Once in three days. 5 ml of minus P Hoagland solution was given per each pot. The parameters such as plant height, root length dry weight of shoot, number of leaves, leaf length was recorded. Per cent V AM colonization in roots following the procedure (Phillips and Hayman, 1970). Chemical analysis of the plant was carried as per the procedure of (Jackson, 1973). The results were subjected to analysis of variance and tested for significant difference (P < 0.05) after Panes and Sukhatme (1967).

Results and Discussion

The used soil for the pot experiments had phosphorus deficiency (Table 9.1). A significant plant height 56.2 per cent was achieved. When the seedling were treated with VAM fungi (*G. fasiculatum*) plus *Rhizobium* and molybdenum 8 g/liter water (Table 9.2). However no favorable increased plant height and per cent root colonization was observed in plants treated with *Rhizobium* or molybdenum alone. Three fold increased leaf length, fourfold increased leaves and two and half fold increased dry weight was recorded over the control plants, when the plants inoculated with VAM fungi, *Rhizobium* and molybdenum treatment (Table 9.3). In contrast to these results not much favorable result were observed in those plants treated with *Rhizobium* alone. Nutrients such as N, P, K, Mg and Mo content in shoots was achieved after dual inoculation of VAM fungi, *Rhizobium* with molybdenum treatment (Table 9.4). A similar trend was observed by earlier workers (Shiva shankar, 1987; Shiva shankar and Hagstrom, 1992) on soybeans and green gram. Several scientists have reported the positive and significant enhancement of nutrients after the inoculation of biofertilizers and the uptake of Mg and cadmium in shoots of Green gram and Tamarind (Ghonskar and Saxena, 1973; Reena and Bagyaraj 2000). A very significant increase in the number of nodules was noticed *Rhizobium* treatment and Mo sprayed solution (0.01 per cent). The effect was greatest with (0.02 per cent) solution. The effect of Mo in increasing the nodulation can very well be realised in view of its position as a component element of nitrogenese proteins (Boswell and Anderson, 1972). They reported similar results of increased nodulation in garden pea with foliar application of Mo. They reported significantly more root nodulation with 0.10 per cent molybedenum. But in the present study, *Eugenia bracteata* (Wild.). Roxb. being a tree species it seems requires optimum (0.20 per cent) foliar application of molybdenum. The influence of Mo may be attributed to its conversion in to IAA. The IAA thus formed appears to influence the lengthening, branching and curling of root hairs, which favors infection process during nodule formation. The results of the present study are in strong support of combined inoculation to *Eugenia bracteata* (Wild.) Roxb. seedlings with VAM plus *Rhizobium* with (0.20 per cent) Mo spray aids in minimizing our dependence on inorganic fertilizers, which are very wide application, and also is assured a pollution free environment (Lakshman 1996). In this context there is a greater need to have a proper survey of forest soils to understand the nutrients status.

Table 9.2: Effect of VAM, *Rhizobium*, and Molybdenum on Shoot Length, Root Length per cent Colonization of *Eugenia bracteata* (Wild.) Roxb. in the Nursery at 120 Days Seedlings

Inoculants/Treatment	*Shoot Length*	*Root Length*	*% of VAM Colonization*
Control	10.2	1.6	53.2
VAM	15.2	1.8	42.4
Rhizobium	12.9	2.4	48.3
Molybdenum	13.50	2.7	47.5
VAM + *Rhozobium*	31.30	3.1	51.5
VAM + Molybdenum	5.52	3.3	55.3
VAM + *Rhizobium* + Molybdenum	69.1	4.1	57.2
Mean	**39.16**	**2.50**	**39.12**
	C.D. (P = 0.05)	C.D. (P = 0.05)	C.D. (P = 0.05)
Treatment (T)	0.51	NS	0.27
Growth stages (GS)	0.29	0.09	0.41
T × GS	0.72	0.06	0.12

Table 9.3: Effect of VAM, *Rhizobium* and Molybdenum on Number of Leaves, Leaf Length and Shoot Dry Weight of *Eugenia bracteata* (Wild). Roxb. in the Nursery at 120 Days Seedlings

Inoculants/Treatments	*Number Leaves/Plant*	*Leaf Length*	*Shoot Dry wt* g^{-1}
Control	16.7	7.4	8.1
VAM	43.2	9.1	11.9
Rhizobium	35.3	11.2	10.2
Molybdenum	45.5	11.7	12.2
VAM + *Rhizobium*	49.3	16.4	12.5
VAM + Molybdenum	49.8	16.9	14.8
VAM + *Rhizobium* + Molybdenum	48.6	18.2	16.8
Mean	42.40	12.37	14.02
	C.D. (P = 0.05)	C.D. (P = 0.05)	C.D. (P = 0.05)
Treatment (T)	2.25	0.22	1.401
Growth stages (GS)	1.46	0.53	1.022
T × GS	3.59	0.74	1.703

Table 9.4: Effect of VAM, *Rhizobium* and Molybdenum NPK and mg Content in Shoots of *Eugenia bracteata* (Wild.) Roxb. at 120 Days Seedlings

Inoculants/Treatments	*Nutrients*			*N, P, K and Mo (%) in Shoots*	
	N	*P*	*K*	*Mg*	*Mo*
Control	1.9	0.05	1.02	0.11	0.03
VAM	2.10	0.13	1.15	1.12	0.03
Rhizobium	3.15	0.14	1.16	0.13	0.03
Molybdenum	3.19	0.14	1.17	0.14	0.04
VAM + *Rhizobium*	4.19	0.19	2.14	0.15	0.04
VAM + Molybdenum	4.23	0.22	2.19	0.16	0.05
VAM + *Rhizobium* + Molybdenum	4.16	0.27	2.72	0.17	0.05
Mean	2.19	0.14	1.62	0.27	0.01
	C.D. (P =0.05)	C.D. (P = 0.05)	C.D. (P = 0.05)	C.D. (P = 0.05)	C.D. (P = 0.05)
Treatment (T)	0.24	0.021	0.07	0.01	NS
Growth stages (GS)	0.26	0.033	0.11	0.01	0.002
T × GS	0.58	0.060	1.20	0.02	0.002

Conclusion

By the present study with the above mentioned results, it can be concluded that VAM *Rhizobium* and Molybdenum influenced greatly in the biomass production of *Eugenia bracteata* (Wild.) Roxb. compare to non-mycorrhizal plants. The information of mycorrhizal with molybdenum treatment in tropical forests, however skill remains scanty and disjoined. Further research should be oriented towards the stability of agro forestry seedlings at nursery stage on nutrient poor soils.

Acknowledgement

Second author indebted to U.G.C. New Delhi for awarding major project with Financial support. Ramesh H. Ratageri thank the U.G.C. for awarding the Teacher Fellowship under F.I.P programme.

References

Boswell, F.C. and Anderson, O.E. 1972. Effect of time of application on soybean yield and nodulation and molybdenum contents. *Agronomy J.*, 61: 58–60.

Croft, P.C. 1954. Superphosphate and molybdenum response in peas at wollonghar. *Agri. Goz. N.S.W.*, 65(4): 185–186.

Ghonsikar, C.D. and Saxena, S.N. 1973. Influence of Molybdenum and tryptophane on nodualtion in cluster bean *Cyamopsis tetragonoloba* (L.) Tamb. *J. Agronomy*, 43(10): 938–941.

Jackson, J.L. 1973. *Soil and Plant Chemical Analysis*. Prentice Hall of India (Pvt.) Ltd., New Delhi.

Lakshman, H.C. 1996. Development and response of *Terminalia bellarica* to VA mycorrhizae and trace elements. *Tropical Forestry*, 79: 109–114.

Niranjan, R.S. and Banwarilal, M. Rao. 1990. Studies on the effect of Rhizobium and Endomycorrhizal interaction in *Dalbergia sisso*. In: *Proceedings of National Conference of Mycorrhiza*, H.A.U., Hissar, India, p. 205–207.

Panse, V.G. and Sukhatme, P.V. 1967. *Statistical Methods for Agricultural Workers*, Indian Council of Agricultural Research, New Delhi.

Phillips, J.M. and Hayman, D.S. 1970. Improved procedure for clearing and staining parasitic and vesicular arbuscular mycorrhizal fungi for rapid assessment of infection. *Trans. Brit, Mycol. Soc.*, 55: 158–161.

Shivashankar, K. 1987. Molybdenum role in leguminous plants. In: *Proc. National Seminar an Micronutrients in Crop Production*, FAT, Bangalore, p. 35–40.

Shivashankar, K. and Hagstrom, G.R. 1992. Molybdenum fertilizers sources and their use in crop production. In: *Proceeding on the Role of S, Mn and Micronutrients in Balanced Plant Nutrition*, Held at Sichan, China, p. 297–305.

Steinberg, R.A. 1953. Symptoms of molybdenum deficiency in Tobacco. *Plant Physiol.*, 28: 319–322.

Varma, A.K. and Subha Rao, N.S. 1970. Effect of different levels of moister on growth, yield and some physiological aspects of nodulation in green granna. *Indian J. Agri. Sci.*, 45: 11–21.

Chapter 10

Seasonal Variation in Population of *Microtermes obesi* Holmgren in Rose Plantations of M.J.P. Rohilkhand University, Bareilly (Uttar Pradesh) (Insecta: Isoptera)

Sachin Srivastava*, Sudhakar Gupta, Pushpa Singh and Kamna Singh

Department of Animal Science, Faculty of Applied Sciences, M.J.P. Rohilkhand University, Bareilly – 243 006, U.P., India

ABSTRACT

The survey of rose plants (*Rosa* spp.) of Mahatma Jyotiba Phule Rohilkhand University Campus, Bareilly was carried out in different seasons and revealed the infestation of *Microtermes obesi* Holmgren. The fluctuating range of ambient temperature and moisture contents of the soils of infested rose plants in winter (10.12–15.22°C, 2.10–2.83 per cent), summer (28.56–30.66°C, 1.94–2.59 per cent) and rainy season (20.19–28.58°C, 3.40–4.35 per cent) were recorded. The population of *M. obesi* Holmgren in winter was higher (332.66±6.56–396.66±8.05 individual/plant) in comparison to that of summer and rainy season (175.33±28.18, 209.33±19.30) plant and 200±1.61–241.66±1.13 individual/plant). The lowest proportion (1.5–2.4 per cent) of solidiers in population of termites was recorded in winter (December–January) in comparison to that of summer (April–May) and rainy season (August–September) respectively. These observations suggested that the fluctuation in total population and proportion of soldiers in termites was inversely related to the climatic variations.

eywords: *Microtermes obsei Population, Soldier, Temperature, Moisture.*

Corresponding Author.

Introduction

Termites (Isopterans) are one of the major destroyers of wood and they are generally known as pest of agricultural, forestry and horticultural crops. Termites occur almost at all states of India and thriving both tropical and subtropical regions of the world. The colonies of subterranean termites are cryptic-polymorphic and comprising different castes such as, alates, soldiers and workers/ pseudoworkers. The main function of soldiers in colonies, however varies among different species of termites and at different seasons.

In case of foraging colonies of termites the proportion of soldiers largely depend on the surface area of the infested wood. Luscher (1961) had stated that soldier production is currently held to be regulated by externally secreted honnones or pheromones. Hardy (1971), Wanyoni and Luscher (1973), French (1974) and Chu *et al.* (1974) had observed that the proportion of soldiers can be produced in the termites colonies by administration of juvenile hormone. In another study, Sen-Sarma and Mishra (1982) observed the fluctuations in soldiers population in three species of termites, such, as *Neotermes bosei* Snyder, *Microcerotermes beesoni* Snyder and *Nasutitermes dunensis* Chatterjee and Thakur. Because, to date no-investigation has been carried out on tennite infestation in rose plantations of university campus of Bareilly. Therefore, present study has drawn up to explore the effect of ambient temperature on population distribution and proportion of soliders in *M. obesi* Holmgren in relation to seasonal variations.

Materials and Methods

The survey of rose plantations in the campus of Mahatma Jyotiba Phule Rohilkhand University, Bareilly district of Uttar Pradesh was carried out to explore the infestation of termites vis-à-vis ambient temperature and moisture content of the infested root soil of the rose plants. The sampling of winter was carried out in December to January and summer sampling was carried out in April to May while the sampling of rainy season was carried out in August to September. The sampling in both seasons followed the same patterns. Termites along with the soil rhizosphere from rose plants were collected and brought to laboratory for the estimation of population/plant and species identification according to the methods of Chatterjee and Thakur (1964), Singh and Singh (2000) and Srivastava and Singh (2004a). The ambient temperature was recorded with a thermometer and soil moisture was determined by ovendry method (Srivastava and Singh, 2004 b).

Results and Discussion

The results on the influence of seasonal changes in relation to the proportion of soldiers in observed population (both workers and soldiers) from rose plantations in the campus of Mahatma Jyotiba Phule Rohilkhand University, Bareilly are presented in Table 10.1. The infestation of *M. obesi* Holmgren in experimental rose plants was recorded.

The atmospheric temperature fluctuating range in winter (10.12–15.22°C), Summer (28.56–31.29°C) and rainy season (20.19–28.58°C) were recorded. The moisture contents of infested root soil of the experimental rose plants Group I (Winter: 2.83 per cent; summer: 2.59 per cent and rainy season: 4.35 per cent), Group II (Winter: 2.80 per cent; Summer: 1.94 per cent and rainy season: 3.73 per cent) and Group III (Winter: 2.10 per cent; Summer: 2.28 per cent and rainy season: 3.40 per cent) were also recorded. The records on proportion of soldiers in winter (1.5–2.4 per cent), summer (6.28–8.6 per cent) and rainy season (5.38–9.11 per cent) were recorded. The records on population of *M. obesi* in winter (Group I: 396.66±5.86 individual/plant; Group II: 332.66±6.56 individual/plant and Group III: 396.66±8.05 individual/plant), summer (Group I: 175±28.18 individual/plant; Group II: 209-66±19.3

individual/plant and Group III: 177±11.98 individual/plant) and rainy season (Group I: 241.66±1.13 individual/plant; Group II: 200±1.61 individual/plant and Group III: 204.33±6.05 individual/plant) were also variable (Table 10.1). In the case of soldier proportion in total population (both workers and soldiers) progressively decreases in winter (2.0–1.5 per cent). It is interesting note that after winter the proportion of soldiers suddenly increases (5.38–9.11 per cent) in summer and rainy season respectively. In these observations proportion of soldiers in summer and rainy seasons marginally variable and higher in comparison to that of winter (Table 10.1).

Table 10.1: Population Fluctuation of *M. obesi* Holmgren in Relation to Ambient Temperature and Moisture Content of the Soils Under Infested Rose Plants of MJP Rohilkhand University, Bareilly

Season	*Experimental Plant Group*	*Ambient Temperature (Mean°C)*	*Moisture % of Infested Root Soil (Mean ± SE)*	*Population Individual/Plant (Mean ± SE)*	*Soldier Proportion in Population (%)*
Winter	I	10.12	2.83±0.37	396.66±5.85	2.0
	II	15.22	2.80±0.75	332.66±6.56	2.4
	III	12.10	2.10±0.61	396.66±8.05	1.5
Summer	I	30.66	2.59±0.25	175.33±28.18	6.28
	II	28.56	1.94±0.65	209.66±19.30	8.61
	III	31.29	2.28±0.03	177±11.98	7.90
Rainy Season	I	28.58	4.35±0.98	241.66±1.13	9.12
	II	20.19	3.73±1.13	200±1.13	9.5
	III	26.17	3.40±0.77	204.33±6.05	5.38

Similar observations in case of *Zootermopsis* spp. and *Microcerotermes beesoni* were made by Castle (1934) and Sen-Sarma and Mishra (1982). This sudden increasement in the proportion soldiers can be expJained by the fact that increased number of soldiers is required to guard the numerous exist holes through which alates swarm out.

The population of termite infesting rose plantations *M. obesi* Holmgren is closely related to fluctuation in ambient temperature and moisture levels of the soil. A reference to Table 10.1, shows that the population of *M. obesi* is higher (332–396.66 individual/plant) during winter (December–January) and lowest (175–209.33 individual/plant) during summer (April–May). These observations suggested that lowest temperature fluctuating range through year (10.12–15.22°C) provided most favourable conditions for tennites. Similar records were made by El-Sebay (1993), Evans *et al.* (1998) Evans and Gleeson (2001) and Srivastava and Singh (2004a). In contrast, high temperature (but not extreme) provided suitable conditions for the survival of termites and their foraging activity (Mishra and Sen-Sarma, 1979 and Haverty *et al.*, 1999). A weak correlation between temperature, relative humidity and activity of termites in Pakistan, reported by Akhtar and Sarwar (1995).

From the above studies concluded that the proportion of soliders and population (both worker and soldiers) of termites varies and depend upon the agroclimatic variations.

Acknowledgement

The authors are thankful to Dr. N.B. Singh, Professor and Head, Department of Animal Science, Faculty of Applied Sciences, M.J.P. Rohilkhand University, Bareilly for his valuable suggestions and providing necessary laboratory facilities for this work.

References

Akhtar, M.S. and Sarwar, G. 1995. The foraging activity of subterranean termites in desert zone of Pakistan. *Pak J. Zool.*, 27: 329–336.

Chatterjee, P.N. and Thakur, M.L. 1964. Revision of the termite genus *Microtermes* Wasmann. *Ind. For. Rec.*, 10(11): 219–260.

Chu, Hsang-Hsoung; Tai, Tchi-Dar Chen; Tse-fue and Ling, Mei-wen. 1974. Induction of soldier differentiation in the termite, *Reticulitermes fleviceps* Oshima with juvenile hormone analogues. *Acta. Ent. Sin.*, 17: 161–165.

Castle, G.B. 1934. The damp-wood termites of western United States, genus *Zootermopsis*. In: *Termites and Termite Control*, (Ed.) C.A. Kofoid. Univ. California, p. 264–282.

El-Sebay, Y. 1993. Ecological studies on the colonies of the harvester termite *Anacanthotermes ochraceus* (Burm.) in Egypt. *Bull. Ent. Soc., Egypt*, 71: 47–54.

Evans, T.A., Lenz, M. and Gleeson, P.V. 1998. Testing assumptions of mark recapture protocols for estimating population size using Australion, mound building, subterranean termites (Isoptera: Rhinotermitidae and Termitidae). *Eco. Ent.*, 23: 139–159.

Evans, Theodore A. and Gleeson, Patrick V. 2001. Seasonal and daily activity patterns of subterranean, wood-eating termite foragers. *Aus. J. Zool.*, 49: 311–321.

French, J.R.J. 1974. A Juvenile hormone analogue inducing caste differentiation in the Australian termite, *Nasutitermes exitiosus* (Hill) (Isoptera: Termitidae). *J. Aust. Ent. Soc.*, 13: 353–355.

Hardy, I. 1972. Der Einfluss von zwei Juvenile Hormon-analogen auf die Differenzierung der soldaten bei *Reticulitermes lucifugus* Santonensis Feyt. (Isoptera: Rhinotennitidae). Z. angew, Ent., 72: 129–134.

Harverty, M.I., Getty, G.M., Copren, K.A. and Lewis, V.R. 1999. Seasonal foraging and feeding behaviour of *Reticulitermes* spp. (Isoptera: Rhinotemitidae) in a wildland and residential location in northern California. *Env. Ent.*, 28: 1077–1084.

Luscher, M. 1961. Social control of polymorphism in termites. In: *Insect Polymorphism*, (Ed.) J.S. Kennedy. *Roy. Ent. Soc., London, Symposia* No. 1: 56–67.

Mishra, S.C. and Sen-Sarma, P.K. 1979. Seasonal fluctuations of colony composition, nest population and foraging in Nasutitermes dunensis Chatterjee and Thakur (Isopetra: Termitidae). *Ind. J. Ent.*, 41: 360–364.

Sen-Sarma, P.K. and Mishra, S.C. 1982. Population of soldiers in termite colonies (Insecta: Isoptera). *Proc. Symp. Ecol. Anim. Popul., Zool. Surv., India*, 4: 23–30.

Singh, M. 2000. Incidences of termite infestation of Sugarcane crops in district Shahjahanpur of U.P. and formulation of effective control measures using certain new synthetic chemicals and plant by products. *Ph.D. Thesis*, M.J.P. Rohilkhand University, Bareilly.

Srivastava, Sachin and Singh, N.B. 2004a. Population fluctuation of termites in Sugarcane fields of Bareilly (U.P.). *Coop. Sug.*, 35(10): 785–792.

Srivastava, Sachin and Singh, N.B. 2004b. Survey of termite infesting mango orchards of Bareilly, Uttar Pradesh. *Bio. Mem.*, 30(1): 48–51.

Wanyoni, K. and Luscher, M. 1973. The action of Juvenile hormone analogues on caste development in *Zootermopsis* (Isoptera). *Proc. 7th Int. Cong. Int. Union Study Social Insects, London*, p. 392–395.

Chapter 11

Management of Brinjal Shoot and Fruit Borer through Insecticidal Application and Varietal Resistance

Rabindra Prasad, Muneshwar Prasad, Rajesh Kumar and Devendra Prasad

Department of Entomology, Birsa Agricultural University, Kanke, Ranchi – 834 006

ABSTRACT

Results of the two years field experiments revealed that fruit damage caused by *Leucinodes orbonalis* Guen to the five promising varieties of brinjal were drastically reduced, by alternate foliar spraying of cypermethrin 10 EC @ 650 ml/ha and *Bacillus thuringiensis* var *Kurstaki*, Btk (Delfin WG) @ 750 g/ha applied at three weeks intervals, to the extent of 4.66 to 23.54 per cent and 5.14 to 25.97 per cent in terms of weight and number of fruits that inturn enhanced yield from 24.87 to 69.72 per cent giving rise to extra production of marketable fruits of brinjal from 50.27 to 91.45 q/ha with net profit of Rs. 14,014 to 28,427 per hectare alongwith benefit cost ratio of 3.91 to 7.85. The overall findings indicated that adoption of protection measures is always profitable and promising, irrespective of brinjal varieties. However, resistant varieties *viz*., Pant Samrat, Mukta Keshi and Pusa Purple Cluster were found to be better yielder on account of lesser pest incidence even in the unprotected condition.

Keywords: Brinjal, Shoot and fruit borer, Bio and chemical insecticides, Economic profits.

Introduction

Brinjal, *Solanum melongena* Linn., an important vegetable crop, is highly endangered in the field by *Leucinodes orbonalis* Guen. The larvae of the insect make injury by boring the growing shoot and developing fruits on large scale, leading to the substantial yield loss, even upto 55 per cent (Tripathi *et al*., 1996). It is difficult to ascertain brinjal varieties that could be absolutely tolerant or resistant to the shoot fruit borer *L. orbonalis* (Sharma *et al.*, 1998). Hence the present study was undertaken to manage the pest through varietal resistance and judicious insecticidal application in the field condition.

Materials and Methods

The field experiment were conducted during spring-summer season in 2003 and 2004 with five popular varieties of brinjal, grown under unprotected and protected conditions in the farmers field in a nearby village of Birsa Agricultural University, Campus. There was ten treatment combinations replicated thrice in the randomized block design, details of which is given in Table 11.1. Plot size was kept 4 × 2.5 m². Row to row and plant to plant distance were maintained 75 and 60 cm for the round/oblong variety and that for long one were kept 60 and 45 cm. Transplanting of seedlings was done in the first week of January, during both of the years. Recommended packages of practices were applied for raising the crop. The protected plots were sprayed with *Bacillus thuringiensis* var. *Kurstaki*, Btk (Delfin WG) @ 0.75 kg/ha and Cypermethrin 10 EC (cilcord 10 EC) @ 650 ml/ha, alternatively at three weeks intervals. The number of spray with each of the two insecticides were three, starting first at the time of pre-flowering stage of the crop with knap sack sprayer with cypermethrin.

Table 11.1: Impact of Incidence of Shoot Borer on Certain Popular Varieties of Brinjal, During 2003 and 2004 Under Protected and Unprotected Conditions

Treatment Combinations	*Shoot Damage Caused by Shoot Borer, Leucinodes orbonalis Guen. (%) During 2003 and 2004*								
Brinjal Varieties (under the protected and unprotected conditions)	*2003*				*2004*				*Pooled Mean of 2003 and 2004*
	*7 DAT**	*14 DAT**	*21 DAT*	*Mean*	*7 DAT**	*14 DAT**	*21 DAT*	*Mean*	
Mukta Keshi (unprotected)	10.38 (18.78)	12.34 (20.57)	14.60 (22.47)	12.45 (20.61)	8.45 (16.85)	9.77 (18.15)	11.70 (20.03)	9.97 (18.34)	11.20 (19.47)
Mukta Keshi (protected)	1.22 (6.29)	1.63 (7.29)	2.06 (8.12)	1.64 (7.23)	1.02 (5.75)	1.08 (5.82)	2.40 (8.92)	1.50 (6.83)	1.57 (7.03)
Pusa purple round (unprotected)	20.36 (26.80)	22.45 (28.26)	23.67 (28.95)	22.16 (28.00)	18.50 (25.50)	19.31 (26.06)	21.50 (27.64)	19.77 (26.40)	20.96 (27.20)
Pusa purple round (protected)	3.21 (10.32)	4.43 (12.17)	4.82 (12.69)	4.15 (11.73)	2.03 (8.15)	3.45 (10.66)	3.74 (11.09)	3.07 (9.97)	3.07 (10.06)
Pant Samrat (unprotected)	8.04 (16.44)	10.12 (18.65)	13.45 (21.48)	10.53 (18.86)	6.77 (15.08)	9.12 (17.58)	10.77 (19.16)	8.89 (17.27)	9.71 (18.06)
Pant Samrat (protected)	0.80 (5.12)	1.18 (6.06)	1.67 (7.35)	1.22 (6.17)	0.42 (3.78)	1.27 (6.41)	1.47 (8.90)	1.12 (6.36)	1.17 (6.26)
Neelum round (unprotected)	22.60 (28.16)	23.76 (29.15)	24.88 (29.89)	23.74 (29.07)	18.76 (25.64)	21.45 (27.58)	23.87 (29.24)	21.36 (27.48)	22.87 (28.27)
Neelum round (protected)	3.88 (11.25)	4.77 (12.62)	5.88 (13.96)	4.84 (12.61)	2.76 (9.56)	3.52 (10.79)	4.70 (12.55)	3.66 (10.97)	4.25 (11.79)
Pusa purple cluster (unprotected)	13.63 (21.66)	14.14 (22.08)	15.03 (22.79)	14.26 (22.17)	13.44 (21.47)	14.74 (22.58)	16.71 (24.16)	14.96 (22.74)	14.61 (22.45)
Pusa purple cluster (protected)	1.66 (7.35)	1.88 (7.85)	2.18 (8.45)	1.91 (7.88)	1.00 (5.75)	1.08 (6.01)	1.83 (7.94)	1.38 (6.57)	1.64 (7.22)
SEm (±)	0.48	0.56	0.83	0.74	0.51	0.63	0.74	0.66	0.70
C.D. (0.05)	1.45	1.65	2.47	2.30	1.53	1.87	2.21	1.96	1.99
C.V. %	7.20	8.30	6.40	10.22	8.22	9.34	6.37	8.11	10.77

Figures under the parentheses are angular transformed values; *DAT: Days after treatments with the insecticidal spray.

Observations were recorded at 7, 14, 21 days after insecticidal treatments (DAT) of each spray by counting total number of shoots and damaged shoots per plant on ten randomly selected and tagged plants in each treatment, replication wise. Percentage of infested shoots were worked out. The means of the observations recorded at 7, 14, 21 days after each of the six sprays application were calculated to work out the pooled mean of the two years in terms of percentage of shoot damage variety wise. Total number of fruits and damaged fruits were counted during each plucking, at proper edible stage to work out pooled mean of percentage of fruit damage of the two years. Weight of total fruits and damaged fruits at the time of each plucking, treatment and replication wise, were taken in order to calculate the pooled mean of percentage fruit infestation in terms of fruit weight as well as the yield of healthy marketable fruits of brinjal of the two years. Accordingly, per cent figures were transformed to angular values by using Sin inverse (Le Clarg, *et al.*, 1962) for the statistical analysis. The data pertaining to shoot and fruit infestation and that of yield of healthy marketable fruits were subjected to the statistical analysis.

Ultimately, economics of the chemical management of shoot and fruit borer (*L. orbonalis*) was worked out on the basis of pooled results of the two years to determine the net return and the benefit cost ratio for each of the five brinjal variety (Table 11.3).

Results and Discussion

The data regarding the average incidence of shoot borer recorded during 2003 and 2004 are presented in Table 11.1 and that of fruit borer incidence as well as yield in Table 11.3, while the economics of the shoot and fruit borer management are shown in Table 11.3.

Shoot Infestation

A perusal of Table 11.1 revealed that the alternate foliar spraying of cypermethrin (Cilcord 10 EC) @ 650 ml/ha and Btk (Delfin WG) @ 0.750 kg/ha brought about a marked decline in the shoot infestation caused by the insect, *L. orbonalis*. The cumulative impact of the insecticidal treatments were significantly superior over the untreated (unprotected) plants of each of the five variety of brinjal at 7, 14 and 21 DAT. It was noted that the shoot infestation ranged from 0.80 to 3.88, 1.18 to 4.77 and 1.67 to 5.88 per cent at 7, 14 and 21 DAT in the protected conditions as against 8.04 to 20.36, 10.12 to 22.45 and 13.45 to 23.67 per cent in the unprotected conditions during 2003. Almost similar trend was observed during 2004 in this regard, though some what lesser incidence of shoot borer was observed during the second year (*i.e.*, 2004) of the experiment. The lowest damage of shoot was recorded in case of Pant Samrat both in protected and unprotected condition, *i.e.*, 0.80, 1.18 and 1.67 and 8.04, 10.12 and 13.45 per cent respectively at 7, 14 and 21 days after the insecticidal treatment. This was closely followed by Pusa purple cluster and Mukta Keshi in this respect during both of the years. Almost similar trend was observed incase of mean of all the three observations as well as that of pooled mean of 2003 and 2004 in this regard. Pant Samrat suffered almost negligible attack of shoot borer *i.e.*, 1.12 per cent only, followed by Mukta Keshi (1.57 per cent) and Pusa purple cluster, when pooled result of both of the year is considered. The highest shoot damage *i.e.*, 22.87 per cent was recorded in case of Neelum round under unprotected conditions as against significantly lower shoot damage (4.25 per cent) was registered under the protected conditions in case of the same variety, Neelum round. On considering the lesser loss of shoot caused by *L. orbonalis* on the basis of pooled results of 2003 and 2004, the varieties ranked in order of Pant Samrat > Mukta Keshi > Pusa purple cluster > Pusa purple round > Neelum round. Some other workers (Sharma *et al.*, 1998) recorded almost similar results.

Table 11.2: Effect of Incidence of Fruit Borer on Some Popular Varieties of Brinjal and their Yield, Recorded During 2003 and 2004 Under Protected and Unprotected Conditions

Treatment Combinations	*Percentage of Fruit Damage Caused by Fruit Borer*						*Yield of Healthy and Marketable Fruits (q/ha)*			
	By Fruit Number			*By Weight of Fruits*						
	2003	*2004*	*Pooled Mean*	*2003*	*2004*	*Pooled Mean*	*2003*	*2004*	*Pooled Mean*	*Increased Yield of 2003 and 2004 Over Unprotected Crop*
Mukta Keshi (unprotected)	27.77 (32.48)	24.48 (29.69)	26.12 (30.08)	26.14 (28.95)	23.43 (29.51)	24.78	164.69	169.50	167.09	–
Mukta Keshi (protected)	2.37 (8.74)	1.70 (7.48)	2.03 (8.11)	1.77 (7.60)	2.12 (8.40)	1.94 (8.00)	235.54	243.60	239.57	72.48 (43.37)**
Pusa purple round (unprotected)	34.47 (35.94)	32.38 (34.64)	33.42 (35.29)	30.44 (33.48)	28.72 (32.38)	29.58 (32.93)	127.88	134.45	131.16	–
Pusa purple round (protected)	8.13 (16.55)	6.45 (14.67)	7.45 (15.61)	8.05 (16.46)	6.98 (15.24)	7.51 (15.85)	219.57	225.66	222.61	91.45 (69.72)
Pant Samrat (unprotected)	8.65 (17.06)	7.75 (16.11)	8.20 (16.58)	7.88 (16.24)	6.67 (14.89)	7.27 (15.56)	203.40	209.70	206.55	–
Pant Samrat (protected)	3.32 (10.48)	2.80 (9.64)	3.06 (10.06)	3.14 (10.18)	2.08 (8.15)	2.16 (9.16)	255.70	260.14	257.92	51.37 (24.87)
Neelum round (unprotected)	28.96 (32.39)	24.38 (29.60)	26.58 (30.99)	26.40 (30.34)	23.79 (29.17)	25.09 (29.75)	132.60	135.75	134.17	–
Neelum round (protected)	1.86 (7.74)	1.56 (7.15)	1.71 (7.44)	1.65 (7.34)	1.46 (6.89)	1.55 (7.11)	221.80	227.80	224.80	90.63 (67.55)
Pusa purple cluster (unprotected)	10.88 (19.23)	8.72 (17.18)	9.80 (18.20)	8.45 (16.86)	7.22 (15.58)	7.83 (16.22)	185.16	191.70	188.43	–
Pusa purple cluster (protected)	1.87 (7.82)	1.22 (6.29)	1.54 (7.05)	1.34 (6.62)	1.03 (5.75)	1.18 (1.18)	235.60	241.80	238.70	50.27 (26.68)
SEm (±)	0.94	0.91	0.86	1.06	1.04	1.02	2.90	3.06	2.72	–
C.D. (0.05)	2.85	2.72	2.56	3.16	3.13	3.04	8.74	9.76	8.23	–
C.V. (%)	7.22	8.66	9.20	8.34	6.72	10.34	10.44	8.36	9.46	–

Figures in the parentheses are angular transformed value.

**: Figures under the parentheses are percentage increase in yield due to the protection measures applied, over the untreated ones.

Table 11.3: Economics of Shoot and Fruit Borer (*Leucinodes orbonalis* Guen.) Control in Some Promising Brinjal Varieties (on the basis of pooled figures of 2003 and 2004) Overall Net Return (Rs/ha)

Sl.No.	Parameters of the Economics	Mukta Keshi		Pusa Purple Round		Pant Samrat		Neelum Round		Pusa Purple Cluster	
		Unprotected	Protected	Unprotected	Protected	Unprotected	Protected	Unprotected	Protected	Unprotected	Protected
1.	Percent shoot damage, caused by shoot borer	11.20	1.57	20.96	3.07	9.71	1.12	22.87	4.25	14.61	1.64
2.	Reduction in shoot damage (%)	–	9.63	–	17.89	–	8.59	–	18.62	–	12.97
3.	Per cent fruit damage										
	– by fruit number	26.12	24.09	33.42	7.45	8.20	3.06	26.58	1.71	9.80	1.54
	– by fruit weight	24.78	1.94	29.58	7.51	7.27	2.61	25.09	1.55	7.83	1.18
4.	Reduction in fruit damage (%)										
	– by fruit number	–	5.82	–	25.97	–	5.14	–	24.87	–	8.26
	– by fruit weight	–	22.84	–	22.07	–	4.66	–	23.54	–	6.65
5.	Yield of healthy fruits of brinjal (q/ha)	167.09	239.57	131.16	222.61	206.55	257.92	134.17	224.80	188.43	238.70
6.	Yield increase over unprotected crop (q/ha)	–	72.48	–	91.45	–	51.37	–	90.63	–	50.27
7.	Per cent increase in yield over the unprotected crop	–	43.37	–	69.72	–	24.87	–	67.55	–	26.68
8.	Value of increased production @ Rs.350 per quintal due to protection measures	–	25368.00	–	32007.50	–	17979.50	–	31720.50	–	17594.50
9.	Cost of control measures, comprising of cost of insecticides plus labour cost	–	3580.40	–	3580.40	–	3580.40	–	3580.40	–	3580.40
10.	Net return (Rs/ha) due to the protection measures applied	–	21787.60	–	28427.10	–	14399.10	–	28140.10	–	14014.10
11.	Benefit Cost Ratio/BCR	–	6.80	–	7.94	–	4.02	–	7.85	–	3.91

Fruit Infestation

The study (Table 11.2) indicated that there were significant differences between protected and unprotected plants in terms of fruit infestation within the same brinjal variety. This was applicable to all the five variety. The lowest fruit damage *i.e.,* 1.87 and 1.22 as well as 1.34 and 1.03 per cent in terms of fruit number and fruit weight under the protected conditions were recorded where as in the case of unprotected conditions, the fruit damage were registered significantly higher *i.e.,* 10.88 and 8.72 as well as 8.45 and 7.22 per cent respectively during 2003 and 2004 in case of Pusa purple cluster.

This was closely followed by Pant Samrat and next to this is Mukta Keshi in the similar conditions. The highest fruit damage was registered in case of Pusa purple round *i.e.,* 34.47 and 32.38 as well as 30.44 and 28.72 per cent, by fruit number and fruit weight basis during 2003 and 2004 respectively as against significantly lower damage of fruit, *i.e.,* 8.13 and 6.45 as well as 8.05 and 6.98 per cent under the protected conditions in this respect in the respective varieties. Some other workers through their field trials reported that Pusa purple cluster was found to be resistant against shoot and fruit borer while Pant Samrat proved to be tolerant against the pest. Mukta Keshi was found to be moderately tolerant against shoot borer and moderately susceptible against fruit borer while Pusa purple round and Neelum round were observed to be highly susceptible to shoot and fruit borer, *L. orbonalis.* (Kumar *et al.,* 1997–98, Sharma *et al.,* 1998).

Yield of Healthy and Marketable Brinjal Fruits

The data on yield (Table 11.2) revealed that the yield varied, significantly between protected and unprotected brinjal plants in case of all the five varieties of brinjal, both on account of degree of protection provided by the protection measures against the pest and by virtue of genetic variation in the yield potentiality of the respective varieties. The yield ranged from 127.88 to 203.40 and 135.75 to 209.70 q/ha under unprotected crop conditions as against 219.57 to 255.70 and 225.66 to 260.14 q/ha under the protected crop conditions during 2003 and 2004 respectively (Table 11.3). So far as, the pooled means of the two years in terms of fruit infestation by fruit weight as well as that of yield of healthy fruit is concerned, fruit damage by weight was found to be significantly more in the highly susceptible varieties like Pusa purple round (29.58 per cent) and Neelum round (25.09 per cent) and hence the proportionate increase in the yield of healthy fruit was noticed also to be more in that case due to control measure that resulted to an extra yield of 91.45 and 90.63 q/ha respectively, in comparison to the tolerant and resistant varieties like Pant Samrat, Pusa purple cluster and moderately tolerant variety, Mukta Keshi giving rise to additional increase in yields, *i.e.,* 51.37, 50.27 and 72.48 q/ha respectively. However, the highest yield was obtained in case of Pant Samrat (257.92 q/ha) followed by Mukta Keshi (239.57 q/ha) which inturn remained at par with Pusa purple cluster. Keeping in view the degree of avoidable loss in yield in the varieties, the least one was found in case of Pusa purple cluster (50.27 q/ha) followed by Pant Samrat (51.37 q/ha) < Mukta Keshi (72.48 q/ha) < Neelum round (90.63 q/ha) < Pusa purple round (91.45 q/ha) that were suffering from the considerable avoidable yield loss to the tune of 26.68, 24.87, 43.37, 67.55 and 69.72 per cent over the protected ones respectively.

Thus, these results clearly indicate that the protection measures are usually promising to minimise the avoidable yield loss irrespective of the varieties of the crop.

Economics of the Protection Measures

A perusal of Table 11.3 indicated that there were yield gains of 50.27 to 91.45 q/ha in the protected crop in comparison to unprotected crop conditions. The reduction in shoot damage to the tune of 8.59,

9.63, 12.96, 18.62 and 17.89 per cent and that of fruit damage by weight *i.e.*, 4.66, 22.84, 6.65, 23.54 and 22.07 per cent were recorded in cases of Pant Samrat, Mukta Keshi, Pusa purple cluster, Neelum round and Pusa purple round that increased their yield of healthy and marketable fruits to the tune of 51.37, 72.48, 50.27, 90.63 and 91.45 q/ha with the benefit cost ratios of 4.02, 6.08, 3.91, 7.85 and 7.94 by enhancing 24.87, 43.37, 26.68, 67.55 and 69.72 per cent in the fruits yield that gave rise to net return of Rs.14,399, 21,787, 14,014, 28,140 and 28,427 Rs./ha in the respective brinjal varieties. Tewari and Moorthy (1983) reported that the rewarding yield with the highest benefit cost ratio could be obtained by protecting brinjal against *L. orbonalis* by spraying cypermethrin at three weeks intervals. Yien (1985) obtained the higher yield, net return and benefit cost ratio by spraying deltamethrin against the pest.

Hence, it is concluded that protection of brinjal against *L. orbonalis* is always beneficial and profitable irrespective of its varieties in order to minimise the avoidable yield loss and maximise net return and benefit cost ratio. However, growing the resistant varieties needs lesser chemical protection to minimize the yield loss.

Acknowledgement

The authors are grateful to the Vice-Chancellor, Dean (Agric.) and Director of Research, B.A.U., Kanke, Ranchi for providing the facilities for conducting this experiment.

References

Kumar, Manoj, Ram, H.H., Singh, Y.V. and Kumar, M. 1997–98. *Recent Horticulture*, 4: 152–155.

Le Clarg, E.L., Leonard, W.H. and Clark, A.G. 1962. *Field Plot Technique.* Burgess Publishing Company, Minnesota, pp. 373.

Sharma, V.K., Roshan, Singh, Arora, R.K., Gupta, Arun, Singh, R. and Gupta, A. 1998. Field response of brinjal cultivars against shoot and fruit borer, *Leucinodes orbonalis* Guen. *Annals of Biology,* Ludhiana, 14(2): 199–201.

Tewari, G.C. and Moorthy, P.N.K. 1983. Effectiveness of synthetic pyrethroids against the pest complex of Brinjal. *Entomon.*, 8(4): 365–368.

Tripathi, M.K., Sengupta, B. and Patra, R. 1996. Seasonal incidence and population fluctuation of *Leucinodes orbonalis* Guen. at Bhubaneshwar, Orissa. *Pest Management and Economic Zoology*, 4(1–2): 15–18.

Yein, B.R. 1985. Field efficacy of some insecticides against shoot and fruit borer (*Leucinodes orbonalis* Guen.) of brinjal. *J. Res. Assam Agric. Univ.*, 6(1): 31–34.

Chapter 12

Medicinal Plants of Ethnobotanical Importance Curing Diabetes from Namakkal District (Tamil Nadu), India

M. Rajasekara Pandian*, G. Sharmila Banu*, G. Kumar** and K.H. Smila*

**Centre for Biotechnology, Muthayammal College of Arts and Sciences, Kakkaveri, Rasipuram, Namakkal, Tamil Nadu – 637 408, India*

***Department of Biochemistry, Selvamm Arts and Science College, Namakkal, Tamil Nadu – 637 003, India*

ABSTRACT

A survey of medicinal plants has been carried out in different locations of Namakkal District, which are used in the treatment of Diabetes. Some of which are less known, but very effective. This paper deals with botanical names, local names and plant parts used in diabetes. During the recent exploration about 100 plants belonging to 53 different families have been recorded.

***Keywords**: Medicinal plants, Diabetes, Namakkal.*

Introduction

Plants have been used as source or medicine throughout the world for more than 5000 years and still continue to occupy an important place in traditional as well as modern systems or medicine. There are about 2000 plant species, which have been found to possess the medicinal value in all these four indigenous systems of medicine *viz.*, Ayurveda, Unani, Siddha and Homeopathy (Farooqi *et al.* 2003). According to the report of the WHO over 80 per cent of the world population rely on traditiona systems of medicines, largely plant based, to meet primary health care needs (Jain, 1967; Tumane *et al.*

2001). Over 7500 species of plants are estimated to be used by 4635 communities for human and veterinary health care (Beulah Jerlin *et al.*, 2004). Efforts have been taken to assess the medicinal plants throughout the country, but a complete inventory is not available so far (Singh *et al.*, 1997).

Most of the supply of drug is obtained from wild plants (Mundappa and Ommen, 1998). Ethnomedicines are receiving great attention all over the world; therefore a great emphasis has been laid to revive the heritable knowledge on the medicinal plants. But some plants have already been extinct and many are facing a danger of extinction. Hence necessity was felt to conduct the survey of medicinal plants carried out from locality of Namakkal district. Especially the plant survey was made to cure diabetes. Because the disease manifesting in the form of silent epidemic is increasing at a very fast rate with estimated 80 million diabetes in the world today (Singh and Sharma, 2001).

Materials and Methods

Ethnobotanical surveys (2003–2004) were conducted in various parts of Namakkal Districts. During field survey certain plants were observed which are used as diabetes. Voucher specimens were collected photographs were taken; plants were properly identified and compared with standard literature and herbarium. The plant species are arranged in alphabetical order according to botanical names followed by family, local name and part used. The ethnobotanical information has been described as collected from the folklore.

The medicinal uses of plants were compiled with the help of earlier publications (Kirtikar and Basu, 1933; Nadkarni, 1954; Jain, 1967; Maheswari *et al.*, 1993; Krishnan, 1992).

Results and Discussion

100 medicinal plants belonging to 53 families recorded in this paper with their botanical and local name and use in the folk medicines.

Scientific studies may elaborate the prospects of growing more and more medicinal plants successively. The distribution of some medicinal plants having good scope for commercial exploitation of pharmaceutically important medicinal plants (Anonymous, 1978), study also revealed that mostly uses of the plants are not recorded in published and compiled literature (Saklani and Jain, 1994; Sharma, 1999; Chandola, 1999; Shriranjini and Acharya, 2001; Chandola and Singh, 2001; Kar *et al.*, 2001; Mohanty and Dwivedi, 2002) which warn for preservation of such valuable knowledge in welfare of mankind.

The present findings indicated that different folk populations inhabiting in the remote villages of Namakkal district are concions for their health. These findings may need phytochemical investigation and proper assessment to establish a crude drug industry for ethnomedicine.

Hence, there is necessity to generate all round awareness in society regarding the conservation of a flora of the plant treasure. For instances, students and local communities must be told that if a particular habitat were destroyed along with the species, what the negative consequences for them would be. Prime importance should be given for multiplication of endangered plant species in botanical garden by *ex situ* conservation or thought tissue culture techniques. For this to be successful, habitat restoration will also have to be a top priority. Several medicinal plants domesticated within the last decenniums but much is to be done in this field in the near future. According to Altieri, 1989; Cunningham, 1991 some of the plants in forest as medicinal plants can also be attributed to increased concern with the relationship of indigenous plants knowledge to genetics evolutional and ecological security.

Table 12.1: Anti-diabetic Plants with Botanical Names, Family and Plant Parts Used in Diabetes

Sl.No.	Name of the Plant	Family	Local Name	Parts Used
1.	*Acacia arabica*	Leguminosae	Karuvelum	Gum
2.	*Acorus calamus*	Araceae	Vashambu	Rhizome
3.	*Achyrunthus aspera*	Amaranthaceae	Naaurivi	Leaf
4.	*Adhatoda vasica* Nees	Acanthaceae	Adadodai	Leaf, Root
5.	*Aegle marmeoles*	Rutaceae	Vilvam	Fruit
6.	*Allium cepa*	Liliaceae	Vengayam	Bulb
7.	*Allium sativum*	Liliaceae	Pundu	Bulb
8.	*Aloe vera*	Liliaceae	Katralai	Leaf
9.	*Alpinia galanga*	Zingiberaceae	Peratattai	Rhizome
10.	*Annona squamosa*	Annonaceae	Sitapalam	Leaf
11.	*Asparagus racemosus*	Liliaceae	Shimi shadavari	Root
12.	*Asteracantha longifolia*	Acanthaceae	Nir mulli	Whole plant
13.	*Azadirachta indica*	Meliaceae	Vembu	Leaf, fruit
14.	*Bauhinia purpurea*	Leguminaceae	Mandarai	Bark
15.	*Bauhinia variegata*	Leguminaceae	Segappu mandaraii	Bark
16.	*Boerhavia diffusa*	Nyctaginaceae	Mukkurattai	Whole plant
17.	*Boswellia serrata*	Burseraceae	Parangisambirani	Gum, Bark, Resin
18.	*Butea monosperma*	Leguminaceae	Parasa	Gum, Leaf, Bark
19.	*Brassica juncea*	Cruciferae	Kadugu	Stem
20.	*Caesalpinia bonducella*	Leguminaceae	Kaliccikkai	Seed
21.	*Cajanus cajan*	Fabaceae	Thuvarai	Seed
22.	*Camellia sinensis*	Theaceae; Fernstroemiaceae	Thayilai	Leaf
23.	*Capsicum annum*	Solanaceae	Milagai	Fruit
24.	*Calophyllum inophyllum*	Guttiferae	Punnagam	Oil of seed
25.	*Cassia auriculata*	Ceasalpiniaceae	Aavarai	Leaves, seed
26.	*Cassia angustifolia*	Ceasalpiniaceae	Nila aavarai	Leaves
27.	*Cassia obovata*	Ceasalpiniaceae	Kattu-nila aavaral	Leaf
28.	*Cassia indica*	Ceasalpiniaceae	Aavarai	Seed
29.	*Cassia fistula*	Leguminaceae	Konnai	Seed
30.	*Casearia esculenta*	Samydaceae	Kottargovai	Root
31.	*Catharanthus roseus*	Apocynaceae	Sudukattu malli	Leaf, Flower
32.	*Cinnamomum tamala* Nees	Lauraceae	Tamalabattirii	Leaf
33.	*Citrus lemon*	Rutaceae	Periya yelumichai	Fruit peel
34.	*Clerodendron phlomidis*	Verbenaceae	Taludalai	Herb
35.	*Clerodendron serratum*	Verbenaceae	Angaravalli	Root
36.	*Coccinia indica*	Cucurbitaceae	Kovaikai	Root, leaf
37.	*Cucumis sativus*	Cucurbitaceae	Vellarikkkay	Fruit

Contd...

Table 12.1–Contd...

Sl.No.	Name of the Plant	Family	Local Name	Parts Used
38.	*Commiphora mukul*	Burseraceae	Gukkulu	Resin
39.	*Curcuma longa*	Zingiberaceae	Viral manjal	Rhizome
40.	*Cyamopsis tetragonoloba*	Fabaceae	Kottavarai	Seed
41.	*Cynadon dactylon*	Poaceae	Arugampillu	Whole plant
42.	*Dolichos lablab*	Fabaceae	Avarai	Seed
43.	*Emblica officinalis*	Euphorbiaceae	Nelli	Fruit bulb
44.	*Enicostemma littorale*	Gentianaceae	Vallari	Whole plant
45.	*Euphorbia hirta*	Euphorbiaceae	Amman pacharchi	Whole plant
46.	*Ficus bengalensis*	Moraceae	Alamaram	Bark
47.	*Ficus religiosa*	Moraceae	Arasu	Bark
48.	*Ficus racemosa*	Moraceae	Atti	Stem bark
49.	*Fumaria parviflora*	Fumaraceae	Tusa	Leaf
50.	*Glycyrrhiza glabra*	Fabaceae	Atimaduram	Root
51.	*Gymnema sylvestre*	Asclepiadaceae	Sirukurunjan	Leaf
52.	*Hemidesmus indicus*	Asclepiadaceae	Nannarai	Root
53.	*Holarrhena antidysentrica*	Apocynaceae	Kudasappalai	Seed
54.	*Indigofera tinctoria*	Fabaceae	Avari	Whole plant
55.	*Mangifera indica*	Anacardiaceae	Mamaram	Tender leaf
56.	*Madhuca latifolia*	Sapotaceae	Kattiluppai	Leaf
57.	*Memecylon umbellatum*	Melastomaceae	Kasai	Leaf
58.	*Mimosa pudica*	Mimosaceae	Thottachinungi	Leaf, Root
59.	*Momordica charantia*	Cucurbitaceae	Pagal	Seed, Fruit, Leaf
60.	*Morus alba*	Moraceae	Musukottai	Root
61.	*Murraya konigii*	Rutaceae	Karuvembu	Leaf
62.	*Musa paradisica*	Musaceae	Kadali valai	Flower
63.	*Nelumbo nucifera*	Nymphaeaceae	Thamarai	Rhizome
64.	*Nymphaea alba*	Nymphaeaceae	Alli thamarai	Plant extract
65.	*Ocimum sanctum*	Lamiaceae	Thulasi	Leaf
66.	*Oroxylum indicum*	Bignoniaceae	Vangam	Root bark
67.	*Oryza sativa*	Poaceae	Nellu, arisi	Root
68.	*Papver somniferum*	Papaveraceae	Kasakasa	Fruit
69.	*Phaseolus aconotifolius*	Fabaceae	Tulkapyrai	Seed
70.	*Phaseolus munga*	Fabaceae	Patchipyaru	Seed
71.	*Phaseolus radiatus*	Fabaceae	Paunippayaru	Seed
72.	*Phaseolus vulgaris*	Fabaceae	Panipayar	Seed
73.	*Phyllanthus emblica*	Euphorbiaceae	Nelli	Fruit

Contd...

Table 12.1–Contd...

Sl.No.	Name of the Plant	Family	Local Name	Parts Used
74.	*Phyllanthus niruri*	Euphorbiaceae	Nelli	Whole plant
75.	*Picrorrhiza kurroa*	Scrophunariaceae	Katukurohini	Rhizome, Root
76.	*Pinus roxburghi*	Pinaceae	Shimaithevari	Bark
77.	*Plumbago zeylanica*	Plumbaginaceae	Kodiveli	Root
78.	*Pongamia glabra*	Fabaceae	Pungamaram	Root, Bark, Leaf
79.	*Polyalthia longifolia*	Annonaceae	Asogu, Nettilingam	Bark
80.	*Prunus amygdalus*	Rosaceae	Vadumai	Seed oil
81.	*Punica granatum*	Puniaceae	Madulai	
82.	*Pterocarpus marsupium*	Fabaceae	Vengai	Wood
83.	*Pterocarpus santalinus*	Fabaceae	Santhana Vengai	Seed
84.	*Rhizophora mangle*	Rhizophoraceae	Kandal	Leaf
85.	*Ricinus communis*	Euphorbiaceae	Amanakku	Leaf
86.	*Saccharum officinarum*	Gramineae	Karumbu	Stem
87.	*Salacia oblongata*	Hippocrateaceae	Ponkoranti	Root bark
88.	*Sesamum indicum*	Pedaliaceae	Ellu	Seed
89.	*Swertia chirata*	Gentianaceae	Nilavembu	Stalk
90.	*Syzigium cumini*	Myrtaceae	Naval	Seed
91.	*Terminalia cebula*	Combretaceae	Kudukkay	Fruit
92.	*Terminalia arjuna*	Combretaceae	Marudu	Fruit
93.	*Tribulus terrestris*	Zygophyllaceae	Neringil	Thorns and Fruit
94.	*Tinospora cardifolia*	Menispermaceae	Seenthil kodi	Stalk, Stem
95.	*Trigonella foenumgraceum*	Fabaceae	Venthayam	Seed
96.	*Trichosanthus dioica*	Cucurbitaceae	Kommu pudalai	Fruit
97.	*Vateria indica*	Dipterocarpaceae	Kundurukkam	Resin, Fruit
98.	*Woodfordia fruitcosa* Kurz	Lythraceae	Velakkai	Flower
99.	*Zingiber officinale*	Zingiberaceae	Inji	Rhizome
100.	*Zizyphus rugosa*	Rhamnaceae	Kattilandai	Bark

Finally, with multiplication, cultivation and selection, it is possible to create an immense number of plant varieties adapted to diverse condition that would eventually sustain our biodiversity and thus preserve a half of our massive habitat loss. If this is achieved in near future, we would go along way in sustaining medicinal plants.

References

Anonymous, 1978. *The Ayurvedic Formulary of Part I*, Edition 1. Ministry of Health and Family. Govt. of India, New Delhi.

Altieri, M. 1989. Rethinking Crop genetic resource conservation: A view from the south. *Conservation Biology*, 3: 77–79.

Beulah Jerlin, S., John De Britto, A., Balasingh, J., Swami doss, D.P. and Prakash, A.A. 2004. Medicinal flora of Koonthakulam bird Sanctuary. *J. Ecobiol.*, 16(1): 7–16.

Chandola, H.M. 1999. Diabetes mellitus in Ayurveda. In: *National Seminar on Medicinal Plants and Traditional Medical Knowledge and Workshop on Traditional Knowledge and Management of Incurable Diseases like AIDS* Organized by Gujarat Ayurved University, Jamnagar.

Chandola, H.M. and Singh, G. 2001. Promotion of mental health of Ayurveda and yoga in diabetes mellitus. In: *Second International Seminar on Ayurveda*, Organized by Jamnagar Ayurved University, Jamnagar.

Cunningham, A.B. 1991. Indigenous knowledge and biodiversity. *Cultural Survival Quarterly*, 15: 4–8.

Farroqi, A.S., Sreeramu, B.S. and Srinivasappa, K.N. 2003. Cultivation of medicinal plants in South India. *Agrobios Newsletter*, 1: 21–22.

Jain, S.K. 1967. Ethanobotany, its scope and study. *Indu. Mus. Bull.*, 2(1): 39–43.

Kar, A.K., Upadhyay, B.N. and Ojha, D. 2001. A study on the prognosis of Prameha (Diabetes mellitus) on the basis of Insulin level of the body. In: *Second International Seminar on Ayurveda*, Organized by Jamnagar Ayurved University, Jamnagar.

Kirtikar, K.R. and Basu, D.D. 1933. *Indian Medicinal Plants*, 2nd ed. Lalit Mohan Basu Co., Allahabad, India.

Krishnan Marg, K.S. 1992. *The Useful Plants of India*. Publications and Information Directorate, CSIR, New Delhi.

Maheswari, J.K., Kunkel, G., Bhandari, M.M. and Duke, J.H. 1993 *Ethnobotany in India*. Scientific Publishers.

Mohanty, B. and Dwivedi, R.B. 2002. Antioxidant mechanism. Away of Biopurification in diabetes mellitus. In: *Second International Seminar on Ayurveda*, Organized by JamnagarAyurved University, Jamnagar.

Mundappa, A. and Ommen, S. 1988. Amruth 2:3 Med. Plan. Conservation society, Bangalore, India.

Nadkarni, K.M. 1954. *Indian Materia Medica*, Vol. I and II, 3rd Ed. A.K. Nadkarni, Popular Book Depot, Bombay, India.

Saklani, Arvind and Jain, S.K. 1994. *Cross-cultural Ethnobotany*, Deep Publication, New Delhi.

Sharma, P.P. 1999. Importance of Dravyaguna Vijnana and list of drugs useful in Madhemeha, Amavata and AIDS. In: *National Seminar on Medicinal Plants and Traditional Medical Knowledge and Workshop on Traditional Knowledge and Management of Incurable Disease like AIDS*, Organized by Jamnagar Ayurved University, Jamnagar.

Shriranjini, S. and Acharya, G.S. 2001. Clinical evaluation of the hypoglycaemic effect of Asanad Kwatha in diabetes mellitus. In: *Second International Seminar on Ayurveda*, Organized by Jamnagar Ayurved University, Jamnagar.

Singh, D.K. and A.K. Sharma. 2001. Clinical evaluation of the hypoglycaemic effect of Asanad Kwatha in diabetes mellitus. In: *Second International Seminar on Ayurveda*, Organized by Jamnagar Ayurved University, Jamnagar.

Singh, J., Ashok Sharma and Susil Kumar. 1997. Inventorisation of herbal drugs used in the traditional and modern medicines. *J. Med. Aromatic Plant Sci.*, 19: 665.

Tumane, H.H., Mehta, P.J. and Chopra, K.K. 2001. Conservation of medicinal plants. In: *Second International Seminar on Ayurveda*, Organized by Jamnagar Ayurved University, Jamnagar.

Chapter 13

Consumptive Use, WUE and Moisture Extraction Pattern by Rapeseed and Greengram as Influenced by Fertility Levels and Straw Mulch

B. Mitra, P. Bose, B.K. Mandal and S.K. Hansda

Department of Agronomy, Faculty of Agriculture, Bidhan Chandra Krishi Viswavidyalaya, Mohanpur, Nadia – 741 252, West Bengal, India

ABSTRACT

An experiment was conducted at 'C' Block Farm of Bidhan Chandra Krishi Viswavidyalaya, Kalyani, Nadia, West Bengal, India to study the effect of levels of fertility and straw mulch on consumptive use, WUE and moisture extraction pattern by winter rapeseed and summer greengram, grown in a rapeseed-greengram sequence under rainfed conditions. The experiment revealed that the application of mulch brought about changes in moisture status of the soil and thereby influenced both consumptive use and WUE in rapeseed and greengram. Mulching treatment gave rise to lower consumptive use and higher WUE for both the crops during both the years of experimentation. Regarding the fertility treatments, the maximum consumptive use and WUE were achieved in both the crops in the treatment where FYM was applied @ 10 t ha^{-1} in *rabi* crop (rapeseed) along with 50 per cent RD through fertilizers to both the crops in the sequence. In mulched treatment, both rapeseed and greengram extracted more moisture from top 0–15 cm layer.

Keywords: *Consumptive use, WUE, Rapeseed, Greengram, Mulch.*

Introduction

A major portion of our cultivated area depends on monsoon rain for raising successful crops. The rainfed agriculture still accounts for over 95.8 m ha or more than 60 per cent of the cropped area.

During *Kharif* season, *aman paddy* is grown almost in every location as rainfed except some HYVs in some pockets with assured irrigation facilities. At present, the rainfed rice areas in eastern part of the country are mostly monocropped and the land remains fallow after the harvest of *kharif* crop. Some oilseed crops especially rapeseed and mustard appeared to be promising under rainfed condition. Inclusion of a legume in the cropping system may replenish the lost nutrients due to continuous mining. So during *pre-kharif* season greengram is a good alternative as it can restore soil fertility. An experiment was, therefore, conducted under rainfed condition to study the effect of straw mulch and fertility levels on consumptive use, WUE and moisture extraction pattern in rapeseed (*rabi*) and greengram (*pre-kharif*).

Materials and Methods

The experiment was conducted at 'C' Block Farm of Bidhan Chandra Krishi Viswavidyalaya, Kalyani, Nadia, West Bengal, India (23°N latitude, 89°E longitude and 9.75 m above mean sea level). It was carried out for two consecutive years (2001–2002 and 2002–2003). The experimental soil was sandy loam in texture with a good drainage facility having a pH of 7.7.

The experiment was laid out in a split-plot design having 14 treatments in 3 replicates. Levels of fertility were randomly allotted in 8 main plots; while mulch treatments were randomly allocated in 2 sub-plots. The treatments were:

Main Plot treatments

T_1: No application of manure and fertilizer.

T_2: RD* to all the crops in the sequence.

T_3: RD to all the crops except N in greengram.

T_4: RD to all the crops except K_2O in rapeseed.

T_5: Organic manure @ 10 t ha^{-1} (applied to rapeseed only).

T_6: 50 per cent RD to all the crops in the sequence.

T_7: Organic manure @ 10 t ha^{-1} (to rapeseed only) + 50 per cent RD to all the crops in the sequence.

Sub-Plot treatments

M_0: Without mulch.

M_1: With rice straw mulch @ 7.0 and 3.5 t ha^{-1} to rapeseed and greengram, respectively.

The varieties used in experiment were YSB 9 for rapeseed and Pusa Baisakhi for greengram.

Soil samples were collected with the help of screw augur from three soil layers: 0–15 cm, 15–30 cm and 30–45 cm. Moisture percentage was computed on oven dry basis. Consumptive use and WUE were calculated accordingly.

Results and Discussion

Among the seven different fertility treatments, the highest consumptive use was recorded with T_7 treatment for both rapeseed and greengram (132 and 160 mm with rapeseed and 342 and 249 mm with greengram during 2001–2002 and 2002–2003, respectively); while consumptive use was recorded to

* RD: Recommended dose.

be lowest when there was no fertilizer and manure application (T_1). For other five fertility treatments, the consumptive use varied between this range; however, there was no definite trend of consumptive use with varying levels of fertility (Table 13.1).

Table 13.1: Effect of Fertility Levels and Mulching on Consumptive Use in Rapeseed and Greengram

Treatments	*Consumptive Use (mm)*			
	Rapeseed		*Greengram*	
	2001–2002	*2002–2003*	*2001–2002*	*2002–2003*
		Fertility levels		
T_1	115	151	283	202
T_2	122	158	321	231
T_3	122	156	310	226
T_4	121	154	319	233
T_5	118	147	298	217
T_6	130	150	320	228
T_7	132	160	342	241
		Mulching		
M_0	137	176	335	232
M_1	120	161	295	209

The treatments receiving paddy straw mulch recorded lower consumptive use (120 and 161 mm for rapeseed and 295 and 209 mm for greengram during 2001–2002 and 2002–2003, respectively) as compared to no mulch treatment (Table 13.1). This was simply because straw mulch checked a significant loss of moisture through evaporation from the surface soil layer. Banik *et al.* (1997) reported similar results in their moisture studies in rapeseed.

Table 13.2: Effect of Fertility Levels and Mulching on WUE in Rapeseed and Greengram

Treatments	*WUE (kg ha^{-1} mm^{-1})*			
	Rapeseed		*Greengram*	
	2001–2002	*2002–2003*	*2001–2002*	*2002–2003*
		Fertility levels		
T_1	3.54	2.79	2.19	2.77
T_2	6.11	5.63	3.26	3.44
T_3	6.13	5.57	2.86	3.27
T_4	4.41	4.56	3.21	3.27
T_5	4.62	4.03	2.54	3.06
T_6	5.10	4.92	3.05	3.25
T_7	6.20	5.11	3.35	4.18
		Mulching		
M_0	4.42	3.94	2.70	3.19
M_1	5.65	5.08	3.19	3.69

Table 13.3: Moisture Extraction Pattern by Rapeseed and Greengram as Influenced by Mulching

Mulching Treatments	*Moisture Extraction Pattern (%)*											
	Rapeseed						*Greengram*					
	0–15 cm		*15–30 cm*		*30–45 cm*		*0–15 cm*		*15–30 cm*		*20–45 cm*	
	2001–02	*2002–03*	*2001–02*	*2002–03*	*2001–02*	*2002–03*	*2001–02*	*2002–03*	*2001–02*	*2002–03*	*2001–02*	*2002–03*
M_0 (without mulch)	34.56	32.15	34.67	36.75	30.77	31.10	31.55	33.27	35.52	36.06	32.93	30.67
M_1 (with mulch)	38.21	39.00	33.27	32.51	28.52	28.49	41.16	38.25	32.00	31.51	26.84	30.24

Water Use Efficiency (WUE) varied with varying levels of fertility as well as mulching. Under various fertility levels, WUE ranged between 3.54 to 6.20 kg ha^{-1} mm^{-1} and 2.79 to 5.11 kg ha^{-1} mm^{-1} in rapeseed as well as 2.19 to 3.35 kg ha^{-1} mm^{-1} and 2.77 to 4.18 kg ha^{-1} mm^{-1} in greengram during 2001–2002 and 2002–2003, respectively (Table 13.2). Higher fertility levels increased seed yield in a greater proportion as compared to its corresponding increase in consumptive use resulting in higher WUE.

Under mulch treatment, WUE was recorded to be higher because of higher seed yield as well as lower consumptive use. The results corroborated with the findings of Uttam *et al.* (1994) and Bhan *et al.* (1997).

Mulching treatment for both rapeseed and greengram influenced moisture extraction pattern. Under mulch treatment, both the crops extracted more moisture from the surface layer of 0–15 cm as compared to the other two layers (Table 13.3). While under no-mulch treatment, the contribution of three different layers was almost equal during both the years of experimentation.

References

Banik, P., Chakraborty, A. and Bagchi, D.K. 1997. Integrated nutrient management in rice and its effect on water use and moisture depletion pattern of following winter crops in rainfed areas. *Indian Journal of Agricultural Sciences*, 67(8): 298–301.

Bhan, S., Uttam, S.K. and Tripathy, R.Y. 1997. Yield, moisture use and root development of rainfed mustard as influenced by sowing methods and moisture conservation practices. *Indian Journal of Soil Conservation*, 25(1): 55–60.

Uttam, S.K., Chand, M. and Tripathy, R.Y. 1994. Effect of different cultural and moisture conservation practices on mustard production under rainfed conditions. *Bharatiya Krishi Anusandhan Patrika*, 10(4): 183–189.

Chapter 14

Relative Salt Tolerance of Selected Perennial Tree Species

B.R. Gurumurthy*, M.B. Shivanna Naryan S. Mavarkar* and H.V. Rudramurthy***

**College of Agriculture, Shimoga – 577 204*
***Department of Applied Botany, Kuvempu University, Jnanasahyadri – 577 451, Karnataka, India*

ABSTRACT

Seedlings of six tree species–neem, jack, acacia, eucalyptus, pongamia and tamarind previously grown in normal soil were transplanted into Polybags containing saline soil and grown in green house for a period of 360 days. The chlorophyll content of all the plant species decreased with increase in the concentration of salts in soil. The effects of EC 4 were less but salinity above EC 8 significantly reduced the pigment content. Similarly, all plants accumulated high proline with in a period of 60 days. The species differences were highly significant both in respect of proline accumulation and chlorophyll content in different growth stages. On the basis of relative chlorophyll and proline contents on 360 DAT at high salinity level of EC 20, the tree species–pongamia could be rated as tolerant based on the fact that its growth was comparable to control plant by 360 DAT.

Keywords: Chlorophyll, Proline, Relative salt tolerance.

Introduction

Growth of plants is adversely affected by soil salinity but plants differ in their ability to grow and yield under saline conditions. Information on effects of water salinity or soil salinity on the growth, yield and quality is very scant in perennial tree species although such effects are apparent and have been noticed in field crops. A few workers have even attempted to relate salt tolerance of tree species to their growth (Yadav and Singh, 1970; Mass, 1986). Soil salinity although a serious problem is not insurmountable; salinity could be redressed using technological and/or biological approaches. Biological approaches are inexpensive (Shannon, 1984) and include the identification of potential salt tolerant plants. In view of the above, present study was undertaken to study changes in chlorophyll and proline contents that might occur in seedlings grown under induced salinity.

Materials and Methods

Experiments were conducted in the greenhouse during June, 2002–May, 2003 and repeated in June, 2003–May, 2004 at the campus of Agriculture College, Shimoga. Six tree species, neem, jack, acacia, eucalyptus, pongamia and tamarind, which are generally preferred for agro/social forestry, were selected for the investigation. A potting mixture was prepared by mixing red earth, farmyard manure and sand in a ratio of 3 : 2 : 1. Seeds of individual tree species were sown in small poly bags (12 × 5 cm) containing potting mixture at the rate of two seeds of jack, tamarind, neem or pongamia or ten seeds of eucalyptus or acacia per poly bag. They were placed in the greenhouse and watered regularly to raise uniform and healthy seedlings. As soon as the seedlings attained a height of 5 cm, thinning was done to retain only one seedling per bag. Saline soils were prepared artificially by mixing sodium chloride and calcium chloride to the potting mixture in a ratio of 2 : 1 (Todd and Reed, 1998) so as to obtain EC of 4, 8, 12, 16 or 20 dsm^{-1}. A control set without any salt addition was maintained for comparison. Artificially prepared saline soils were filled into large poly bags, watered and after a little compaction of soil, a hole was made in the center. Four hundred and fifty-day-old-seedlings raised in potting mixture were placed carefully into these holes and the surrounding soil was slightly compacted. Field capacity of the soil was calculated and each time only measured amount of water was applied to each bag, once in three days. The experiment was laid out in a complete randomized design. Plants were labeled in each treatment, chlorophyll and proline content was estimated at 5 stages (0, 60, 120, 240 and 360 days) by the spectrophotometric (Arnon, 1949) and colorimetric (Bates *et al.*, 1973) methods, respectively. Finally the data of two years were subjected to pooled statistical analysis and significance was tested at P 0.01.

Results and Discussion

Effect of Salinity on Chlorophyll Content (mg g^{-1} fresh weight)

Chlorophyll content was significantly affected by the soil salinity levels (Table 14.1). In neem, salinity of EC 8 and EC >12 caused the lowest chlorophyll production at 60 and 120 DAT, respectively. When jack seedling were grown in saline soils, except in case of treatment EC 4 all other treatments caused gradual reduction in chlorophyll content over time. The salinity of EC 12 did not cause mortality, however, chlorophyll content was drastically reduced. In case of plants treated with EC 16, the chlorophyll content reduced from 0.9 to 0.27 mg and in EC 20 from 0.9 to 0.27 mg within 120 DAT (Table 14.1). Acacia plants treated with EC 8 had decreased chlorophyll content by 60 DAT, which increased slightly by 360 DAT. The plants grown in EC 4 and EC 8 had 45.1 and 52.9 per cent less chlorophyll content compared to control by 360 DAT. The soil salinity of EC 16 and 20 significantly reduced the chlorophyll content at early stages. But soil salinity of EC 12 and above resulted in mortality of all plants after 120 DAT. The soil salinity of EC 4 did not reduce the chlorophyll content considerably as compared to control in eucalyptus. However, salinity above EC 8 reduced the chlorophyll content and plants died after 120 DAT, when the plants were grown in salinities at and above EC 12. Plants grown in EC 8 had 91.6 per cent less chlorophyll than the control. Similarly, in case of pongamia also, salinity caused a reduction in chlorophyll to a greater extent at higher salinities, but it was more tolerant than the other species. The effect of EC 4 was less but salinity of EC 20 significantly reduced the chlorophyll content. The influence of salinity was more severe in tamarind, however, the mortality of plants was not noticed even at the higher salinity of EC 20. Only in case of control and salinity of EC 4, increase in chlorophyll content was noticed on 360 DAT as compared to initial values at zero day.

Table 14.1: Effect of Soil Salinity on Chlorophyll Content (µg g^{-1}) in Selected Tree Species

	Neem					Jack					Acacia				
DAT	0	60	120	240	360	0	60	120	240	360	0	60	120	240	360
C	0.6	0.81	1.17	1.53	1.77	0.9	1.06	1.43	1.8	2.1	0.45	0.61	0.95	1.3	1.5
EC 4	0.6	0.75	0.78	1.17	1.37	0.8	1.10	1.20	1.4	1.7	0.46	0.51	0.51	0.6	0.8
EC 8	0.6	0.39	0.48	0.71	1.25	0.9	0.62	0.57	0.3	0.8	0.46	0.28	0.35	0.6	0.7
EC 12	0.6	0.41	0.29	0.52	0.72	0.9	0.67	0.23	0.1	0.1	0.42	0.40	0.11	–	–
EC 16	0.6	0.45	0.08	0.51	0.69	0.9	0.69	0.27	–	–	0.46	0.51	0.08	–	–
EC 20	0.6	0.29	0.24	0.33	0.39	0.9	0.62	0.27	–	–	0.44	0.38	0.07	–	–
Mean	0.6	0.52	0.50	0.80	1.03	0.9	0.79	0.66	0.9	1.2	0.45	0.45	0.36	0.8	1.0
P 0.01	NS	0.01	0.01	0.02	0.01	NS	0.01	0.01	*0.7*	*0.1*	NS	0.01	0.01	*0.3*	*0.3*
CV (%)	1.5	1.93	2.02	1.25	0.96	1.0	0.94	1.23	–	–	1.14	1.95	2.89	–	–
	Eucalyptus					Pongamia					Tamarind				
DAT	0	60	120	240	360	0	60	120	240	360	0	60	120	240	360
C	0.6	0.91	1.37	2.27	2.73	1.1	1.33	2.25	2.8	3.1	0.49	0.56	0.76	0.9	1.0
EC 4	0.6	1.07	1.17	1.77	2.27	1.2	1.43	2.53	3.1	3.7	0.49	0.55	0.72	0.8	0.9
EC 8	0.6	0.52	0.61	0.37	0.23	1.1	1.44	1.92	2.5	3.4	0.48	0.42	0.31	0.2	0.1
EC 12	0.6	0.81	0.60	–	–	1.2	1.46	1.50	2.0	2.3	0.49	0.41	0.33	0.1	0.1
EC 16	0.6	0.41	0.40	–	–	1.2	1.63	1.58	2.7	2.5	0.47	0.27	0.23	0.1	0.1
EC 20	0.6	0.40	0.38	–	–	1.1	1.19	0.67	1.1	1.1	0.46	0.29	0.18	0.1	0.1
Mean	0.6	0.69	0.76	*1.47*	*1.74*	1.2	1.41	1.74	2.4	2.7	0.48	0.42	0.42	0.4	0.4
P 0.01	NS	0.01	2.62	0.80	1.08	NS	0.01	0.02	0.1	0.1	NS	0.01	0.01	0.1	0.1
CV (%)	1.4	1.37	2.4	–	–	0.8	0.66	1.13	1.0	0.6	2.05	2.11	1.19	2.4	2.6

Note: Standard deviation values are in italics.

From the Table 14.1 it is evident that the chlorophyll content decreased with increase in concentration of salt added to soil. Mumtaz *et al.* (1997) also reported that salinity stress reduced the chlorophyll content and caused yellowing of leaves. In the present study, plant species like acacia, jack and eucalyptus could not survive beyond a certain period when salinity was increased and plants showed symptoms like leaf curling, yellowing, necrosis, early senescence and finally dropping of leaves. In case of acacia, leaves of plants turned whitish and dropped off within few days of transplanting at and above salinity level of EC 8 and leaves of surviving plants remained brownish for the remaining period. In 1997, El Hog and Sidahmed reported that increased salinity in soil caused progressive decrease in plant height, number of leaves, branches and chlorophyll content in case of lime seedlings raised in pots containing soils of various salinities. They also noticed increased tip burn at higher salinity level.

Effect of Salinity on Leaf Proline Accumulation (µg g^{-1} fresh weight)

In response to salinity stress plant species synthesize diverse group of chemicals like sucrose, glycine betaine, proline, aspargine, valine, glutamine and others which was extensively studied in annual crops. But such studies are very few in perennial plants. In the present study, proline content was estimated to assess the osmotic adjustments of tree species in response to soil salinity stress at seedling stage.

Table 14.2: Effect of Soil Salinity by Addition of Salts on Proline Content (µg g^{-1}) in Selected Tree Species

	Neem					*Jack*					*Acacia*				
DAT	0	60	120	240	360	0	60	120	240	360	0	60	120	240	360
C	30	30	28	28	29	28	29	27	28	30	27	27	26	25	28
EC 4	28	240	232	220	222	29	78	74	69	67	29	138	134	133	128
EC 8	30	335	322	305	300	27	138	136	130	125	26	211	205	203	195
EC 12	32	383	300	240	238	28	143	140	135	130	26	338	320	–	–
EC 16	30	500	477	458	455	29	171	163	–	–	28	369	330	–	–
EC 20	32	590	562	540	500	27	175	170	–	–	27	247	360	–	–
Mean	30	346	320	299	391	28	122	118	*91*	*88*	NS	222	229	*120*	*117*
P 0.01	NS	13.3	14.5	15.2	7.3	NS	14.4	9.4	–	–	5.2	14.5	8.2	–	–
CV (%)	3.2	2.6	3.0	3.4	1.7	3.5	8.1	6.3	–	–	27	4.5	2.5	–	–
	Eucalyptus					*Pongamia*					*Tamarind*				
DAT	0	60	120	240	360	0	60	120	240	360	0	60	120	2.40	360
C	36	38	39	38	40	36	37	36	34	35	35	36	35	35	33
EC 4	37	117	110	102	98	35	289	283	271	260	39	160	154	148	138
EC 8	39	216	210	200	194	36	388	365	352	346	36	323	307	300	290
EC 12	38	210	220	–	–	38	531	500	481	470	40	348	331	316	302
EC 16	40	237	226	–	–	37	628	588	563	540	37	360	340	320	315
EC 20	38	240	230	–	–	36	635	594	570	595	40	390	375	360	320
Mean	38	176	173	*113*	*111*	36	418	394	379	374	38	270	257	247	233
P 0.01	NS	6.9	14.8	–	–	NS	12.5	11.8	12.2	21.1	NS	11.3	10.2	12.1	10.9
CV (%)	6.5	2.7	5.8	–	–	2.3	2.1	2.9	3.5	9.9	3.9	2.9	4.5	6.2	5.5

Note: Standard deviation values are in italics.

In neem, mean of saline treatments indicated that due to salinity, proline content significantly increased and reached highest within 60 days. Treatment differences were highly significant (P 0.01) at all the stages of observations (Table 14.2). On 60 DAT, the proline content was 29 µg in untreated jack seedling but the seedling grown at the salinity of EC 20 accumulated increased proline content of 175 µg. Even though initial proline contents were non-significant, further an increase in salinity significantly enhanced the proline content (Table 14.2). The effects of soil salinity in acacia were similar to that of jack. The higher salinity resulted in significantly increased accumulation of proline within 60 days. In case of eucalyptus proline content could not be estimated after 120 DAT at more than EC 12 when the plant mortality occurred. In EC 20, the proline contents were 240 and 230 µg on 60 and 120 DAT and recorded 632 and 605 per cent increase over control. Mean data over treatments indicated that proline contents were lower in control plants but significantly higher in treated plants of pongamia at all the stages of observations. Among the six tree species tested, maximum proline accumulation was observed in pongamia. The proline content of plants grown in EC 20 was 17 times higher than that of the control plants on 360 DAT. In tamarind, salinity of EC 20 induced increase in proline content from 40 to 390 µg within 60 days. On 360 DAT, the proline contents were 33, 138, 290, 302 and 320 µg in control, EC 4, 8, 12, 16 and 20, respectively.

Relative Salt Tolerance

Salt tolerance was studied on the basis of relative accumulation of chlorophyll and proline contents under induced salinity stress conditions. The actual data on the day of treatment imposition was considered as 100 and percentages were worked out at different stages for the subsequent variations in these parameters and these values were considered as relative values. Although 5 levels of saline solution were studied, to arrive at the relative growth value of different parameters, only salinity of EC 20 was considered, because of the reason that, absolute salt tolerance is directly related to the maximum growth potential of a given plant under high salinity regardless of its growth rate under low salinity or non saline conditions (Gupta and Gupta, 1997).

The chlorophyll content was significantly affected in seedlings of all the six tree species as it was observed from the data of chlorophyll contents estimated on 5 different stages. However, differences were noticed among the species. Except pongamia, which showed increased chlorophyll content, in all the other species, decline in chlorophyll was noticed. On 360 DAT, the relative chlorophyll content was 101, 62 and 13 in the surviving plants of pongamia, neem and tamarind at the salinity level of EC 20 (Figure 14.1). By this time, the remaining three tree species–acacia, eucalyptus and jack completely collapsed due to increased salinity stress. Saha and Gupta (1993) also noticed decrease in chlorophyll, free amino acids and soluble carbohydrates in sunflower and bean seedlings in growth chamber.

The tree species accumulated large quantity of proline in response to induced soil salinity stress. By 120 DAT, the relative proline contents were 1650, 1756, 938, 1286, 630 and 605 in pongamia, neem, tamarind, acacia, jack and eucalyptus respectively. The actual proline contents on 360 DAT under the salinity stress of EC 20 were 595, 500 and 320 µg g^{-1} fw in pongamia, neem and tamarind, respectively, where as the relative proline values were 1653, 1563 and 800 in these plants. Similarly, rise in contents of proline and variation between species 'under salt stress was also evidenced by Choudhuri and Choudhuri (1993) wherein, they observed that *Corchorus olitorius* was more susceptible than *C. capsularis* to salinity stress. A marked raise in proline level was noticed in tolerant species in comparison to susceptible ones.

Among the six tree species subjected to salinity stress, only three tree species pongamia, neem and tamarind survived upto 360 DAT and continued to grow under the higher salinity stress of EC 20. Based on the higher relative chlorophyll and proline contents and growth of surviving tree species on 360 DAT pongamia could be rated as tolerant based on the fact that its growth was comparable to control plant by 360 DAT. However, the growth of the latter two species was very much stunted at this time indicating that probably these species needed some more time to recover from salinity stress and hence these species could be rated as less tolerant.

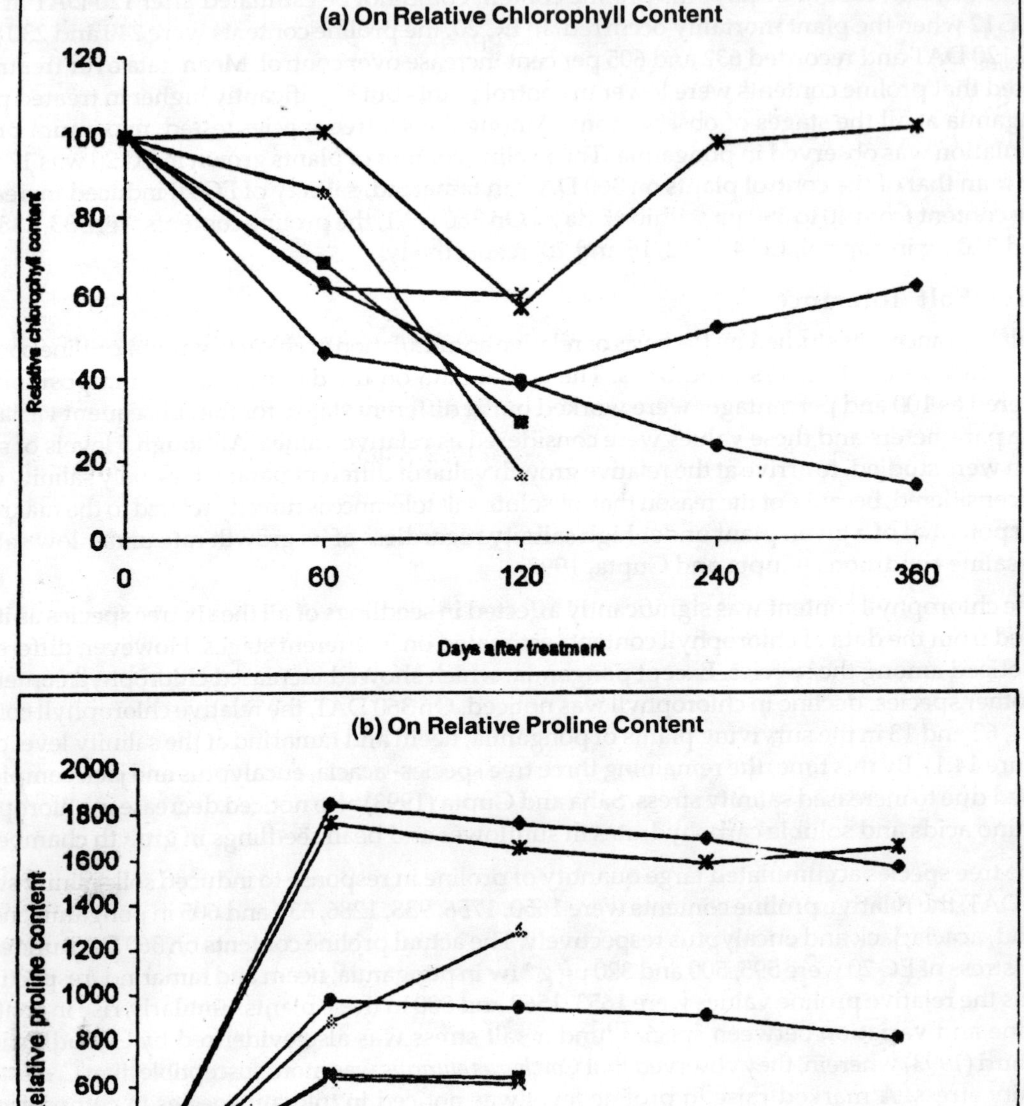

Figure 14.1: Effect of Induced Soil Salinity of EC 20 by Addition of Salts in Different Tree Species

References

Arnon, D.I. 1949. Copper enzyme in isolated chloroplast, polyphenol oxidase in *Beta vulgaris*. *Pl. Physiol.*, 24: 1–15.

Bates, L.S., Waldren, R.P. and Teare, I.D. 1973. Rapid determination of free proline in water stress studies. *Plant and Soil*, 38: 205.

Choudhuri, K. and Chaudhari, M.A. 1993. Differential tolerance of seedling of two jute species to NaCl salinity stress. *Ind. J. Experimental Biol.*, 31: 90–92.

Elhag, E.B. and Sidahmed, O.A. 1997. Response of lime seedling growth and chemical composition to salinity stress. *Soil Science and Plant Analysis*, 28: 13–14.

Gupta, S.K. and Gupta, I.C. 1997. *Management of Saline Soils and Waters*. Scientific Publishers, Jodhpur, p. 98–103.

Mass, E.V. 1986. Salt tolerance of plants. *Applied Agric. Res.*, 1: 12–26.

Mumtaz, S., Naqui, S.S.M., Shereen, A. and Khan, M.A. 1997. Salinity stress and senescence process in wheat. *Pakistan J. Bot.*, 29: 299–303.

Saha, K. and Gupta, K. 1993. Effect of LAB 150978 a plant growth retardant on sunflower and mung bean seedling under salinity stress. *Ind. J. Pl. Physiol.*, 36: 151–154.

Shannon, M.C. 1984. Breeding, selection and the genetics of salt tolerance. In: *Salinity Tolerance of Plants: Strategies for Crop Improvement*, (Eds.) R.C. Staples and G.H. Toennieson. New York.

Yadav, J.S.P. and Singh. 1970. Tolerance of certain forest species to varying degree of salinity and alkalinity. *Ind. Forester*, 96: 587–599.

Chapter 15

Effect of Fly Ash-Soil Amendment on the Cultivation of Root Crops: A Case Study

N. Tripathi, R. S. Singh*, P. K. Mishra** and P.S.M. Tripathi****

**Central Mining Research Institute, Barwa Road, Dhanbad – 826 001, Jharkhand, India*

***V.B. University, Hazaribagh, Jharkhand, India*

****Central Fuel Research Institute, Dhanbad – 828 001, Jharkhand, India*

ABSTRACT

The use of fly ash in agriculture sector is now gaining wide acceptance by the scientists, planners as well as farmers. It has now been established beyond any doubt that the characteristics of Indian fly ash generated from Thermal Power Plants (TPPs) are quite different from that of its foreign counterparts in that Indian fly ashes are alkaline in nature in sharp contrast with the other fly ashes of the world, which are acidic. Apart from this, fly ash is a repository of micro- and macro-elements, which are essential for plant growth. It also advantageously affects the physico-chemical properties of soil, namely water holding capacity (WHC), bulk density (BD), porosity, electrical conductivity, apart from its potentiality to neutralize the acidic soils. As a result of amendment of problematic nutrient-deficient and less fertile soils with fly ash, they become enriched in essential plant nutrients and also more productive and fertile, leading eventually to significant increase in yields of various crops, including the root crops too.

Reported in the present article are the results of investigation on the effect of amendment of red lateritic soil of Dhanbad with different doses of fly ash on the growth and yields of some root vegetable crops, like carrot and radish. The results evince that in the case of radish, the crop yield increased with increasing doses of fly ash compared to control; the maximum yield being at 15 per cent fly ash amendment. This trend has also been found to hold true for carrot; the percentage increase in the yield at the maximum dose being at 15 per cent amendment. The analytical results also show that the physico-chemical properties of the soil get greatly improved with the amendment with fly ash at different doses; the maximum being at 15 per cent amendment of soil with fly ash, which is why the crop yields have increased due to improvement in the fertility status of soil. Most importantly yet that there is no carry-over of trace and heavy elements in these crops. In this way, the amendment of soils with fly ash is quite safe and effective as well for getting increased yields of the root crops.

Introduction

Coal will continue to dominate the energy scenario of the world in the coming several decades (Coal Abstracts, 1992) and electricity will be generated mostly via Pulverized Fuel Combustion (PFC) and also by Fluidized Bed Combustion (FBC) in Thermal Power Plants (TPPs), on a much larger scale to meet the ever-increasing demand of energy. Combustion of coal on a megascopic scale, however, is not without certain serious problems. However, during thermal power generation, of particular concern is the problem of generation of gargantuan quantities of both fly- and bottom-ashes being generated by 85 TPPs in the country (current production of fly ash alone being ca. 110 million tonnes per annum) [C.F.R.I. Report (2002); Central Electricity Generating Board (1973)], the handling, transportation, and storage of which is a big problem. The current generation of fly ashes by the TPPs combined with those already accumulated over the past several years in the ash ponds of all the TPPs in the country is compounding the problem further, resulting in colossal quantity of coal ash, which poses a multitude of serious disposal and associated deleterious environmental problems, apart from covering excessively large areas of fertile and productive land for storage in big ash ponds in the vicinity of TPPs. It has already assumed alarming proportions, which calls for taking urgent remedial measures for proper and bulk gainful utilization of coal ashes, fly ash in particular, in an eco-friendly manner. Particularly, with reference to fly ash utilization in agriculture, it has been found that fly ash can be used as a soil amendment (Davis *et al.*, 2002) and soil-coal ash admixture (20–30 per cent) is optimum in terms of soil health, plant growth, fertility status and yield, where heavy metals in crop produce were within the permissible limits (Mandal, *et al.*, 1997, 98, 99).

In the present article the effect of amendment of red lateritic soil with different doses of fly ash on the growth and yields of some root vegetable crops, like carrot and radish has been discussed.

Materials and Methods

Pot experiments in acidic soil (pH 5.0) at CFRI, Dhanbad for cultivation of root crops with and without different doses of fly ash from DPPL, Durgapur (West Bengal) under Green House conditions were done.

Collection and Characterization of Samples

Sample Characterization

The soil samples from Sudamdih area of Dhanbad and fly ash samples from DPPL, Durgapur were collected and analyzed for various physico-chemical properties, following prescribed standard methods.

The physical properties, *viz.*, colour, texture, BD, porosity, WHC, available water of fly ash from DPPL, Durgapur and soil sample from Dhanbad with and without the additions of fly ash were determined following the prescribed methods (Piper, 1950 and Black, 1965).

Chemical properties, *viz.*, pH, electrical conductivity, total and available elements (N, P, and S) of fly ash and soil were also estimated following the methods as described by Black (1965), Jackson (1973), and Tandon (1995).

Pot Experiment

After proper mixing of fly ash with the soil with pre-decided doses, *i.e.*, 0 (control), 50, 100, 200 and 300 t/ha (equivalent to 0.0, 2.5, 5,0, 10.0, 15.0 per cent fly ash w/w, respectively), 4 kg each of amended soils was added in their respective pots. Recommended doses of NPK fertilizers were applied in both control and fly ash amended soil. The pots were properly lined with the polythene sheets to check the leaching of trace and heavy metals, if any, from fly ash-soil amended system under green house conditions. Two root crops, *i.e.*, radish and carrot were grown in pots with different concentrations of trace/heavy metals present in fly ash samples through the addition of different doses of fly ash to study whether there is any carry-over of toxic metals present in fly ash, such as copper, zinc, manganese, iron, lead, nickel, cobalt, arsenic, mercury, chromium and cadmium by these root crops. The crops grown with the fly ash were harvested and their yield data were recorded both on harvest and oven-dry (60°C) basis. These root crop samples were analyzed for trace and heavy metals.

Analysis of Trace and Heavy Metals

The digestion of root of different crops, soil and fly ash was done in an Analytical Microwave System (Prolabo Microwave Oven) using different "Suprapure" acids for the analysis of micro nutrients (Cu, Zn, Fe, Mn) and trace elements and heavy metals, such as Pb, Cr, Ni, Co, As, Hg, Cd. The trace elements in various digested samples were analyzed by Atomic Absorption Spectrometry technique Schimadzu, Japan; Model AAS-680), but As and Hg were analyzed by another Spectrometer (Unicom make SP-2900 AAS), using hydride vapour generator.

Results and Discussion

The analytical results of soil samples from DPPL, Durgapur, soil samples and roots before the start of experiment and after harvest of each crop are shown, in respect of various physico-chemical parameters, including trace and heavy metals, in Tables 15.1–15.3 while the crop yield data of the crops grown are presented in Table 15.4.

Table 15.1: Initial Physical Characterization of Soil and Fly Ash

Parameter	*Soil*	*Fly Ash*
pH	5.01	8.92
Electrical Conductivity (d s/m)	0.05	0.12
Bulk Density (g/cc)	1.63	0.81
Water Holding Capacity (per cent)	30.92	54.35
Porosity (per cent)	37.31	65.81
Texture		
Sand (per cent)	66.10	33.50
Silt (per cent)	21.60	50.10
Clay (per cent)	12.30	16.40
Textural Class	Sandy loam	Silty loam

Physical Properties of Soil

The comparative physico-chemical characteristics of the fly ash and soil show (Tables 15.1 and 15.2) that the soil is acidic (pH = 5.01), while the fly ash is alkaline (pH = 8.92) in nature. Bulk density (BD), water holding capacity and porosity values of the fly ash are 0.81 g/cc, 54.34 per cent and 65.81 per cent, respectively, while those of the soil are 1.63 g/cc, 30.92 per cent and 37.31 per cent, respectively.

The results of the physical characteristics of soil and fly ash clearly indicate that due to the alkaline nature of fly ash, it can be suitably applied to the acidic soils to neutralize their acidity. Thus, fly ash can be used as a liming agent for the acidic soils. Martens, *et al.* (1976) have reported that the beneficial effects of coal ash on plant growth may result from a combination of pH and nutrient addition. Similarly, the water holding capacity (WHC) and porosity of the soil are seen to be significantly improved by fly ash addition into the soil. Thus fly ash can be potentially used and is quite effective to act as an efficient soil modifier/conditioner.

Table 15.2: Initial Chemical Characterization of Soil and Fly Ash

Parameters	*Soil*		*Fly Ash*
	Total	*Available*	*Total*
Nitrogen (%)	0.0515	0.0120	BDL
Phosphorus (%)	0.0140	0.00095	0.29
Potassium (%)	0.77	0.0098	0.84
Calcium (%)	0.86	0.0024	1.09
Magnesium (%)	0.56	0.00179	0.79
Sulphur (%)	0.052	0.0045	0.026
Zinc (%)	44.10	2.10	71.72
Copper (%)	42.90	2.34	78.20
Iron (%)	229	45.8	293
Manganese (%)	284.40	17.0	212.80
Lead (%)	22.60	3.70	52.0
Nickel (%)	228.30	0.58	134.0
Chromium (%)	179.0	BDL	225.0
Cadmium (%)	6.90	BDL	10.60
Arsenic (%)	0.56	BDL	0.59
Mercury (%)	BDL	BDL	0.01
Cobalt (%)	32.05	BDL	51.20

Major and Secondary Nutrients in Soil

The results (Table 15.3) reveal that the application of maximum dose of fly ash results in the enrichment of the available major and secondary nutrients (N, P, K, S, Ca and Mg) status of the soil, which has contributed to the improvement of soil fertility, most probably owing to higher retention of the fertilizer ions applied through the application of inorganic fertilizers in the presence of fly ash. Further, the results show that at the highest dose of the fly ash and the maximum yield at these doses, the final status of the major nutrients in the soil is proportionately increased in accordance with their contents in the soil and in fly ash. The presence of essential plant nutrients in fly ash has also been reported by several other workers (Ashokan, *et al.*, 1995; Neelima, *et al.*, 1995).

Total and Available Trace and Heavy Metals in Soil

Results indicated that (Table 15.4) the application of maximum dose fly ash does not influence at all the available status of trace elements and heavy metals such as Pb, Ni, Se, Cr, Cd, As, Hg and Co in the soil system after harvest of each crop. This could be ascribable to the fact that the heavy metals are present in alkaline ashes in oxide forms. Application pf dry fly ash in soil thus seemingly does not pose any threat to ground water contamination (Elseewi, *et al.*, 1980).

Table 15.3: Physico-chemical Characteristics of Soil After Fly Ash Amendment

Parameter	*Fly Ash-amended Soil*	*Parameter*	*Fly Ash-amended Soil*
pH	5.48	EC (d s/m)	0.03
BD (g/cc)	1.52	WHC (%)	34.82
Porosity (%)	42.35		
Mechanical composition			
Sand (%)	61.95	Silt (%)	25.30
Clay (%)	12.75	Textural Class	Sandy loam
Nitrogen (%)	0.0156	Phosphorus (%)	0.00108
Potassium (%)	0.01004	Calcium (%)	0.00288
Magnesium (%)	0.00206	Sulphur (%)	0.00420
Zinc (ppm)	2.36	Copper (ppm)	2.70
Iron (ppm)	44.10	Manganese (ppm)	15.62
Lead (ppm)	3.66	Nickel (ppm)	0.61
Chromium (ppm)	BDL	Cadmium (ppm)	BDL
Arsenic (ppm)	BDL	Mercury (ppm)	BDL
Cobalt (ppm)	0.05		

Table 15.4: Root Crop Yield After Amendment with Different Doses of Fly Ash (on harvest and oven dry basis)

Yield (g/pot)	*Radish (after 4 months)*	*Carrot (after 4 months)*
Harvest basis		
Control	124.6	89.6
$^*T_1$	137.4	100.8
$^*T_2$	145.0	108.8
$^*T_3$	156.8	120.9
$^*T_4$	174.6	141.6
CD (5%)	1.6	0.8
Oven dry (60°C) basis		
Control	6.59	8.10
$^*T_1$	7.34	9.22
$^*T_2$	7.55	9.92
$^*T_3$	8.32	10.9
$^*T_4$	9.22	12.5

*: T_1, T_2, T_3, T_4 are the treatments at different doses of fly ash (2.5 per cent, 5 per cent, 10 per cent and 15 per cent, respectively).

Effect of Treatment on Yield of Cropping System

The results of the yield of all the crops grown with different fly ash doses show that the yield of these crops increases from 10 to 30 per cent, without any apparent adverse effect. Further, significant residual effect of fly ash at all doses (one time application before cultivation of first crop) with respect to the increase in yield of the root crops grown in successive years was also observed. The increase in root crop yield can be envisaged due to the improved physico-chemical properties and fertility status of soil after fly ash amendment. More interestingly, the penetration of the roots of such crops was also found to be better in fly ash-amended soil than in control. Taylor *et al.* (1988) and Pichtel *et al.* (1994) have also reported the similar observation as far as root penetration is concerned.

The concentration of major and secondary plant nutrients (N, P, K, S, Ca and Mg) had also been found to increase progressively with the increasing dose of fly ash in comparison to that in the control. The same increasing trend was observed in case of the micronutrients (such as Cu, Zn, Mn, Fe) concentration in all the crops grown with different doses of fly ash application over control.

Nevertheless, the concentration of trace and heavy metals, such as Cu, Zn, Mn, Fe, Cr, Co, Pb, Cd, As, Hg and Ni is non-significant in comparison to control (Table 15.5), albeit the concentrations of trace and heavy metals was increased in all the root crops through the fly ash application. Although the trend of percent increase of some of the heavy metals, such as Cr, Co, Pb, Ni is apparently higher (Table 15.5), yet, it is almost insignificant as the results reported are less than 1 ppm in all the cases, These results are within the normal range generally found in crop produce, as suggested by Srivastava, *et al.* (1996). Sims *et al.* (1994) have also reported that the uptake of trace and heavy metals by the plants grown with fly ash at higher dose (20 per cent by weight of soil) is quite insignificant.

Table 15.5: Trace and Heavy Metal Concentration in Crops Grown on Fly Ash Amended Soil

Elements	*Radish*					*Carrot*				
	Control	*50 t/ha*	*100 t/ha*	*200 t/ha*	*300 t/ha*	*Control t/ha*	*50 t/ha*	*100 t/ha*	*200 t/ha*	*300*
N (%)	2.28	2.25	2.31	2.32	2.34	1.57	1.59	1.61	1.62	1.63
P (%)	0.15	0.18	0.19	0.22	0.25	0.08	0.10	0.12	0.14	0.16
K (%)	2.85	2.87	2.89	2.93	2.96	2.38	2.40	2.42	2.45	2.47
S (%)	0.15	0.18	0.20	0.23	0.23	0.17	0.19	0.21	0.23	0.25
Ca (%)	2.17	2.21	2.25	2.31	2.34	1.18	1.20	1.21	1.23	1.25
Mg (%)	0.19	0.21	0.24	0.25	0.28	0.19	0.21	0.22	0.25	0.26
Cu (ppm)	2.47	2.52	2.56	2.60	2.63	3.75	3.79	3.81	3.83	3.86
Zn (ppm)	10.6	10.8	11.1	11.6	11.9	18.9	19.1	19.5	19.3	19.6
Mn (ppm)	30.8	31.2	30.6	30.8	31.2	52.6	53.1	53.2	53.6	54.0
Fe (ppm)	52.4	52.6	53.0	53.2	53.5	93.7	94.2	94.2	94.3	94.4
Cr (ppm)	0.38	0.35	0.42	0.46	0.49	0.66	0.68	0.71	0.76	0.78
Co (ppm)	0.25	0.27	0.34	0.37	0.40	0.72	0.75	0.77	0.80	0.82
Pb (ppm)	0.15	0.21	0.23	0.28	0.32	0.18	0.16	0.21	0.23	0.25
Cd (ppm)	BDL	BDL	BDL	BDL	BDL	BDL	BDL	BDL	BDL	BDL
As (ppm)	BDL	BDL	BDL	BDL	BDL	BDL	BDL	BDL	BDL	BDL
Hg (ppm)	BDL	BDL	BDL	BDL	BDL	BDL	BDL	BDL	BDL	BDL

Therefore, this study evinces that inspite of the increasing concentration of trace and heavy metals in fly ash, the uptake of these metals by root crops, grown with different doses of fly ash under green house conditions, is insignificant.

Conclusion

The experimental results obtained from these studies have clearly evinced that the bulk application of fly ash in agriculture sector is safe for advantageously amending the soil characteristics and improving upon the acidic soil health and also for increasing yields of various crops to a significant extent (25–30 per cent).

Thus, the soil-fly ash amendment is not only beneficial in making rather less productive lands more fertile via improvement in physico-chemical properties and biological activity, resulting in enhanced crop yields, but is also safe from environmental point of view. It thus simultaneously addresses the problem of environmentally safe disposal and gainful bulk utilization of fly ash as soil amendment in agriculture sector.

Acknowledgement

Ministry of Environment and Forests (MOEF), New Delhi and Council of Scientific and Industrial Research (CSIR), New Delhi are acknowledged.

References

Ashokan, P., Saxena, M., Hose, S.K.J. and Khazenchi, A.C. 1995. *Proceedings of Workshop on Fly Ash Management in the State of Orissa*, RRL, Bhubaneshwar, April 11, p. 64–75.

Black, C.A. (Ed.). 1965. *Methods of Soil Analysis*, Part 1 and 2. American Society of Agronomy Inc. Publisher, Wisconsin, USA.

C.F.R.I. (Central Fuel Research Institute). 2002. Utilization of fly ash in agriculture, at Bakreshwar Thermal Power Project, Bakreshwar (West Bengal. C.F.R.I. Dhanbad (India) Report No. TR/ C.F.R.I.13. 0312001–2002.

Carlson, C.L. and Adriano, D.C.J. 1993. *J. Environ. Qual.*, 22: 227–247.

Central Electricity Generating Board. 1973. The uptake of minerals by plants growing on pulverised fuel ash. CEGB, London, U.K. Report No. SSD. MID./R.2/73, pp. 30.

Coal Abstracts nos. 1992. 9207635 and 9207645, April.

Daniels, W. Lee, Stewart, B., Haering, Kathryn and Zipper, Carl. 2002. Virginia Tech Publication Number 460–134.

Elseewi, A.A. Bingham, F.T. and Page, A.L. 1980. *J. Env. Qual.*, 7: 69–73.

Jackson, M.L. 1967. *Soil Chemical Analysis*. Prentice Hall of India Pvt. Ltd., Delhi.

Mandal, S. and Saxena, M. 1998. Response of Soyabean Grown on Vertisols amended with coal ash. *Chem. Environ. Res.*, 7(3&4): 309.

Mandal, S. and Saxena, M. 1997. Role of fly ash in the improvement of black cotton soil. *J. Env. Studies and Policy*, 2(1): 29–38.

Mandal, S. and Saxena, M. 1999. Fly ash: A boon for clay and sandy soil. *J. Env. Res.*, 9(1): 19–24.

Martens, D.C. 1971. Availability of plant nutrients in fly ash. *Bumines report of Inv.*, p. 15–18

Martens, D.C. and Beham, B.R. 1976. *Proceedings of 4th Intl. Symp. on Ash Utilisation*, St. Lois, MO, p. 24–25 March, p. 657–664.

Neelima, M. Reddy, Khandaul, S. Tripathy, A. and Sahu, P.K. 1995. *Proceedings of Workshop on Fly Ash Management in the State of Orissa*, RRL, Bhubaneswar, April 11, p.76–87.

Pichtel, J.R., Dick, W.A. and Sutton, P. 1994. *J. Environ. Quality*, 23: 766.

Piper, C.S. 1950. *Soil and Plant Analysis*. The University of Adelaide, Adelaide, Australia.

Sims, J.T., Vasilas, B.L. and M.M. Ghodrati. 1994. Soil Hydraulic properties and elemental leaching due to amendment of soil by fly ash addition. *Water Air and Soil Pollution*, 81: 349–361.

Srivastava, P.C. and Gupta, U.C. 1996. *Trace Elements in Crop Production*, Oxford and IBH Publishing Co. Pvt. Ltd., New Delhi.

Tandon, L. (Ed.). 1995. *Methods of Analysis of Soils, Plants, Water and Fertilizers*. Fertilizer Development and Consultation Organization, New Delhi.

Taylor E.M. Jr. and Schuam G.E. 1988. Fly ash and lime amendment of acidic coal spoil aid revegetation. *J. Env. Qual.*, 17(1): 120–124.

Chapter 16

Distribution and Uses of *Euryale ferox* Salisbury in Manipur

P.K. Singh

Ethnobotany and Plant Physiology Laboratory, Department of Life Sciences, Manipur University Canchipur, Imphal – 795 003, India

ABSTRACT

Euryale ferox Salisbury is one of the rare and threatened aquatic plants of Manipur State. From time immemorial it is witnessed by the *Meitei* communities in folklore, edible as special delicacies as vegetable. Distribution in valley districts of the state is discussed. Symbiotic relationship of some aquatic plants in various lakes and their habitat is being disturbed due to urbanization, land reclamation, encroachment, siltation, pollution, etc. of the fresh water lakes and swamps.

Keywords: Euryle ferox, Thangjing, Folklore, Distribution, Uses.

Introduction

Euryale ferox Salisbury is described as a perennial aquatic herb (Haines, 1925; Biswas and Calder, 1936; Subramaniam, 1962) and is a prickly water plant with gigantic floating leaves which are rugulose especially while young, prominently nerved and reticulate beneath. The plant grows in standing shallow water (0.2–2 in depth) with a rhizomatous stem deeply rooted in the sediments or mud by fleshly, thick root clusters. The leaves, green on the upper side and purple red on the lower, are thorny, oval or rounded having a diameter of 0.2–1.3 m.

Results and Discussion

Folklore of Thangjing

Records showing the existence of *E. ferox* plant in Manipur are fragmentary due to the unavailability of written documents. However, the people of Manipur have known the plant. A

Manipuri proverb refers to the *Thangjing* fruit of Kakwapat (a major wetland of the state) is very big in size. Sometimes, people cut jokes like this–"Kakwapat's *Thangjing* is of this much size, if you decrease the size while measuring with your hands you will feel the pain or scratch of the prickles of the *Thangjing* fruit". The exact Manipuri goes like this–"*Kakwagi Thangjigdi yamna chaoyee khutna tambadagi hanjillurabadi khutta tingkhang yurakkani*". In Manipur the *Thangjing* fruits attain their maturity during August–November. One more popular saying goes like this–"*Koroklaobi Khonglabadi Thangjing Loray*" which means that when the very bird called *Koroklaobi* cries, the *Thangjing* reaches its maturity and the mature seeds burst out from the fruit and are found floating on the surface of water. People think that the eternal power of the bird's cry may even break the fruit of *Thangjing*.

Distribution

Euryale ferox grows naturally in Loktak and other lakes. The plant grows along with other aquatic plants like *Trapa natans, Nelumbo nucifera, Nymphaea lotus, Ipomoea aquatica, Marsilea* sp., *Pistia* sp., *Lemna* sp., *Eichhornia* sp., and other aquatic plants. *E. ferox* grows wild in Manipur but in rare case it is cultivated in unused water bodies of the Manipur Valley The plant is restricted mainly in Imphal, Thoubal and Bishenpur district. Local people of Manipur, especially the rural population engaged in agriculture, practice the cultivation of *E. ferox*.

The percentage of frequency of the plant in the lake was 5, the density and abundance expressed, as plant/m were 2 and 15 respectively. The percentage of frequency the ponds were 2, whereas the value of density and abundance of the plant expressed in plant/m were 0.5 and 7 respectively (Singh, 1984). It also grows at Sugnu about 690 m above the mean sea level in Tengnoupal district of Manipur. The district has an area of 3,375 km^2 and has an elevation varying from 550 in to 2000 m from the sea level. In this district, *E. ferox* grows along with other aquatic herbs like *Nymphaea pubescens* (*Tharo*). *N. nouchali* (*Thariktha*) *and Nelumbo nucifera* (*Thambal*). *E. ferox* plant might have been introduced in the district and has been rapidly spreading as part and parcel of its flora (Singh, 1987).

Swamps and Lakes

The swamps and lakes are the main still water bodies, which are the natural home of *E. ferox*. There are numerous lakes and swamps in the plain area especially in the southern part of this state. Altogether, there are about 23 shallow lakes and swan1ps in the valley covering an area of about 600 km^2 out of the total valley area of about 1800 km^2, which constitutes almost one-third of the total plain area. As a result, a considerable portion of the valley is water-logged or marshy or can be claimed as wetlands. The Loktak Lake is the largest lake and is also the most striking landmark of Manipur. It covers an area of about 64 km^2 in the dry season and about 94 km^2 during monsoon rains. Among the other lakes some are sufficiently large and can retain water all the year round. The Lamjaokhong with the Phumlenpat covering an approximate area of about 62.68 km^2 is the second largest lake of the valley. Some other important lakes in the valley are: Kharungpat and Ikopat (59.90 km^2); Poiroupat (0.79 Km^2); Sanapat (1.97 km^2); etc. These lakes play an important role as the source of bio-energy for the entire state. Varieties of aquatic flora including *E. ferox* grow abundantly in these lakes. A large section of the rural population living in this vicinity depends on them for their livelihood (Singh, 1981). All these lakes are of immense utility as fishery reservoir providing a state occupation to the fishing communities of the numerous villages (Singh, 1991).

Swamps are confined only in the valley areas of Imphal, Bishnupur and Thoubal districts. The swams are usually located in the valley areas. The total estimated area under swamp is 290 km^2, which is 1.30 per cent of the total geographical area of the state. Out of this 94.92 km^2 are in Bishnupur, 75.12 km^2 in Imphal (east and west) and 120.25 km^2 in Thoubal districts. All the swampy or marshy areas are heavily covered with aquatic plants like *Nelumbo* sp., *Trapa* sp., *Eichhonia crassipes*, *Pistia* sp. *Ipomoea* sp. and *Euryale ferox*, etc. However, several marshes or swamps have disappeared due to heavy siltation and encroachment for settlement and agricultural activities (Singh, 1991). Marshes are the third category of water bodies of the state, which dry up partially during winter season. Marshes like Utrapat, Sanapal, Laphupat, Leingangpat, I.amphelpal, Porompat, Kakwapat, etc. are filled up with stagnant water during monsoon and the environment is suitable for germination and continuation of *E. ferox* seeds which were at the bottom of water during the previous year. The reduction of wetlands in the state adversely affects the natural aquatic life of *E. ferox*. At present, villagers cultivate the plant by transplanting the seedlings and sowing seeds by following traditional cultivation practices.

Uses of Thanjing

Euryale ferox is used as a delicious vegetable by the Manipuris during autumn and summer seasons. *E. ferox* is an important food plant of the people of the valley (Singh, 1986). The leaf petiole (*Makhok*) and seeds (young–*Lolang*, mature–*Aroba*) are taken raw in salad (*Singju*) and *Chatni* (*iromba*) forms and also as curry in boiled form. The use of the seeds along with the other vegetables in die preparation of vegetable dish is worth mentioning. However, in some other areas, people also use tender leaves (*Kheiyom*), seed aril (*Maparam*) and fruit skin (*Maku*) in the preparation of *Chatni* after removing the prickles by means of fire or some other means. The mature and full-grown leaf petioles (*Makhok*) are taken afresh after removing the spiny cover and are commonly sold in Moirang bazar about 40 kms from Imphal, the capital of Manipur, during July–November at. Rs. 2–5/bundle, each bundle having 8–10 leaf stalks of 30–60 cm in length. The tender leaves are not sold in the market. However, the leaf stalks are also found selling along with fruits in nearby markets of Moirang like Kumbi, Churachandpur and other small bazars of the locality. The leaf petiole and tender leaves are not sold in other market of the state. The seeds of *E. ferox* and some other aquatic herbs like–*Nelumbo nucifera*, *Nymphaca noucliali* are eaten raw or roasted mainly by the tribes of the valleys in Manipur and Tripura (Arora, 1981). Seeds of *E. ferox* after roasting in sand are dehusked and eaten as one of the unconventional cultivars or wild plants used as cereals (Millre, 1981). *E, ferox* is also one of the wild edible plants sold in the markets of Manipur along with some other plants which are being used by different tribes or castes for different purposes. This shows that each tribe has its own peculiarities on the usage of wild plants (Singh and Singh, 1985; Singh *et al.*, 1988).

E. ferox fruits are also used as one of the important and precious offers to the deities mainly by the plain dwellers (*Meiteis*). During June–October, the Meitei ethnic groups perform many traditional functions for their deities and the fruit is used as one amongst the vegetables. Majority of the *Meiteis* are Hindus and as such they use the fruit as an ingredient of Hindu ritualistic purposes also. The use of *E. ferox* as medicine against rheumatism (Me Clure, 1970) and as a traditional Ayurvedic system of Indian medicine as an expectorant and emetic (Nadkarni, 1978) is not practised by the local people of Manipur.

The production of *E. ferox* is self-sufficient for Manipur. However, in recent years, the market price of this crop has been raised due to ever increasing urbanization, land reclamation, encroachment, siltation, pollution of water by domestic refuse, sewage, field outwash, indiscriminate use of a number of pesticides, insecticides and herbicides, etc. Some of the rapidly growing aquatic weeds like *Eichhornia*, *Salvinia*, *Lemma*, etc., adversely affect the suitable environment of the plant. Due to the rapid drying up of wetlands, a number of swamps, marshes and lakes are converted to cultivable fields, settlement area, grazing ground, etc. Lamphelpat, Kakwapat, Porompat, etc. are some or the dried up lakes in the valley.

Schanz (1974) and Hejny (1978) opined that conservation of aquatic plant resources including the protection of nymphaeids has attracted the attention of scientist worldwide. The view is also supported by Jha *et al.* (1991). Kaul (1984) and Kak (1985) stated that *E. ferox* has become rare in the wetlands of Kashmir and they have stressed the need for conservation or plant resources in Indian aquatic ecosystems with reference to some aquatic habitats of north-western India.

Schneider and Kadono (1984) stated that rapidly increasing eutrophication and reclamation of water in Japan has brought about both decline and sometimes the total disappearance of the plant. Siddiqui and Jha (1988) reported that the area under Euryale plantations is receding due to silation as a result of frequent floods in Mithila (north Bihar).

In Bihar *Euryale* cultivation is recognized as an industry along with other major cash crops like banana, litchi, mango, betel etc. and various agencies have agreed to include the *Euryale* industry in their funding programmes. In Manipur also steps should be taken up for the reclamation of waterlogged wastelands for *Euryale*-fish cultivation to produce healthy and better yielding varieties and to draw attention of the scientists for protection of this rare and endangered plant.

References

Anonymous. 1973. *25 Years of Manipur*. Publicity Department, Government of Manipur, Imphal, pp. 60.

Arora, R.K. 1981. Native food plant of the North-eastern tribal. In: *Glimpses of Indian Ethnobotany*, (Ed.) S.K. Jain. Oxford and IBH Publishing Co., New Delhi.

Biswas, K. and Calder, C.C. 1936. *Handbook of Common Water and Marsh Plants of India and Burma*. Government of India. New Delhi.

Choudhary, B.P. and Patnaik, S.N. 1985. Aquatic angiosperm of Bhuwaneshwar. *Journal of Economic and Taxonomic Botany*, 7: 527–536.

Haines, H.H. 1925. *The Botany of Bihar and Orissa*, Part-II. Botanical Survey of India, Kolkata.

Hejny, S. 1978. Conservation of plant communities in fish pond littorals. In: *Pond Eco System*, (Eds.) Dykyjova, D. and J. Kvet. Springer, New York, p. 429–433.

Jha, V., Kargupta, R.N., Jha, A.N., Dutta, U.N., Mishra, R.K. and Saraswati, K.C. 1991. Utilization and conservation of *Euryale ferox* Salisbury in Mithila (North Bihar), India. *Aquatic Botany*, 39: 295–314.

Kaul, V. 1985. Appeal to protect *Euryale ferox* Salisb. getting extinct from Kashmir lakes, Abstract. *National Symposium on Pure and Applied Limnology*, Sagar, pp. 33.

Kaul, V. 1984. Conservation of plant resources in aquatic ecosystems with reference to some aquatic habitat of Kashmir. In: *Conservation of Tropical Plant Resources*, (Ed.) S.K.Jain. B.S.I. Publications, Kolkata, p. 118–131.

Mani, M.S. 1974. *Ecology and Biogeography in India*. Dr. W Junk. The Hague.

McClure, J.W. 1970. Secondary constituent of aquatic angiosperms. In: *Phytochemical Phylogeny*, (Ed.) Harborne, J.B. Academic Press, London, p. 233–268.

Mittre, V. 1981. Wild plants in Indian folk life: A historical perspective. In: *Glimpes of Indian Ethnobotany*, (Ed.) S.K. Jain. Oxford IBH publishing Co., New Delhi.

Nadkarni, K.M. 1976. *The Indian Materia Medica*. Popular Prakashan, Mumbai.

Okada, Y. 1935. Study of *Euryale ferox* Salisb. VIII. Miscellany. Science Reports of the Tohoku Imperial University Ser. IV(2): 455–459.

Schanz, F. 1974. Are there still aquatic plants worthy of protection along the bank of lake of Zurich (Swithzertand)? Viertel Jahrssichr rift. Nat. Forch.Ges. Zurich, 119: 349–359.

Schneider, E.L. and Kadono, Y. 1984. Floral biology, development and structure of *Euryale ferox*. *American Journal of Botany*, 71: 131.

Siddiqui, E.N. and Jha, V. 1988. Conservation of *Euryalc ferox* Salisb. (Makhana) with reference to wetlands of Darbhanga. Paper presented at the Seminar on 'Wetlands of Bihar' BNHS and Govt. of Bihar (Dept. of Forest and Env.) at Purnea 28–29 January, (Unpubl.), pp. 4.

Simpson, J.B. 1936. Fossil pollen in Scottish tertiary coals. In: *Proceedings of Royal Society of Edinburgh*, 56: 90–108.

Singh, H.B. 1987. Flora of Tengnoupal district, Manipur. *Ph.D. Thesis*, Manipur University, Imphal, Manipur, India.

Singh, P.K. 1984. Studies on the physiological and biochemical aspects of *Zizania latifolia goll* induced by *Melanopsichium esculentum. Ph. D. Thesis*, Manipur University, Imphal, Manipur, India.

Singh, P.K., Singh, N.I. and Singh, L.J. 1988. Ethnobotanical studies on wild edible plants in the markets of Manipur II. *Journal of Economic and Taxonomic Botany*, 12(1): 113–119.

Singh, R.K.R. 1981. The pattern of land use in the Manipur valley: A geographical analysis. *Ph.D. Thesis*, Guwahati University, Guwahati, University, Assam, India.

Singh, R.K.S. 1991. Special distribution of diseases in Manipur: A geo-ecological analysis. *Ph. D. Thesis*, Manipur University, Imphal, Manipur, India

Singh, R.S. and Singh, N.I. 1985. A preliminary ethnobotanical study on wild edible plants in the markets of Manipur–I. *Journal of Economic and Taxonomic Botany*, 6(3): 699–703.

Sinha, S.C. 1986. Ethnobotanical study of Manipur. *Ph.D. Thesis*, Manipur University, Imphal, Manipur, India.

Sobolewska, M. 1970. *Euryale ferox* Salisb. in the Elesitocene of Poland. *Acta Palaeobotanica*, 11: 13–20.

Subramaniam, K. 1962. *Aquatic Angiosperms*. CSIR, New Delhi.

Syamananda, R.K. 1991. Study of nutrient enrichment in Loktak lake with reference to biological indices. *Ph.D. Thesis*, Manipur University, Imphal, Manipur, India.

Chapter 17

Evaluation of Basmati Rice Based Crop Sequences for Productivity, Economic Returns and Soil Health

M.S. Sidhu and R. Sikka***

**Senior Agronomist and **Soil Chemist*

Department of Agronomy and Agrometerology, Punjab Agricultural University, Ludhiana – 141 004, Punjab

ABSTRACT

A field experiment was conducted consecutively for three years at research farm of Punjab Agricultural University, Ludhiana, to evaluate the productivity of different basmati rice based crop sequences and their effect on soil health. Maximum mean grain yield of basmati rice was recorded in fallow-basmati rice sequence (22.9 q/ha) and minimum grain yield was recorded in green manure-basmati rice sequence (14.3 q/ha). However, the yield of subsequent crops, wheat, celery, field peas and sunflower increased substantially where green manure was practiced. Maximum basmati rice equivalent yield was recorded in *bajra* (f)-basmati rice-sunflower crop sequence (53.8 q/ha) and minimum was obtained in green manure-basmati rice-celery sequence (25.7 q/ha). The gross returns also followed the same trend. The mean net returns were, however, maximum in *bajra*-basmati rice crop sequence (Rs. 19900 /ha) and minimum in green manure-basmati rice sequence (Rs. 8800 /ha) but there was not much variation in the fallow-basmati and *bajra (f)*-basmati rice crop sequence. Incorporation of *Sesbania aculeata* increased the organic carbon from 0.33 per cent as its initial value to 0.49 per cent and available N from 110 kg/ha to 127 kg/ha after three crop cycles. The content of available P increased than its initial status under all the treatments. However, a significant decrease in available K content of the soil was observed after three crop cycles under all treatments.

Keywords: Diversification, Green manure, Basmati rice, Productivity, economic returns, soil fertility

Introduction

Growing demand for food in the country due to burgeoning human population on one hand and progressive shrinking of agricultural land available per capita on the other hand led to intensification of time and space dimensions in Indian agriculture. This has been made possible through adoption of multiple cropping where more than two crops are grown one after another during a year on the same piece of land.

Rice-wheat crop sequence is the backbone of India's food security. This system occupies an area of 10.5 m ha in the Indo-Gangetic plains of India. There are reports of grain yield stagnation and decline in soil productivity (Yadav *et al.*, 1998), due to continuous adoption of this system. Rapid spread of rice-wheat system in the post-green revolution era in this region caused an eclipse on sustainability of soil productivity (Rattan and Singh, 1997) and low economic returns.

The decline in crop yields and economic returns could be minimized by introduction of alternate crops and crop sequences and inclusion of green manures (Chauhan, 2000). Basmati rice is one such alternative known for its aroma and cooking quality and has great export potential. Generally wheat is grown in sequence with basmati rice but wheat gives poor production due to its delayed sowing. Moreover, both the crops are cereals, which create problem of mining of nutrients from soil. Sunflower another such crop has also been found remunerative than late sown wheat

Therefore, a need is felt to diversify the rice-wheat system by introducing celery, field peas sunflower and *bajra* in rotation with basmati rice and including organic source of fertilizer like green manure crop May and June before planting basmati rice. Accordingly, different basmati rice based crop sequences were tried to evaluate their productivity, economic returns and their effect on soil health.

Materials and Methods

A field experiment was conducted at the research farm of Punjab Agricultural University, Ludhiana for 3 consecutive years. The experiment was conducted on a sandy loam soil, low in organic carbon (0.33 per cent) and available N (110.2 kg/ha), high in available P (42.8 kg/ha) and low in available K (132.6 kg/ha) content. The pH and EC of the soil (1 : 2) were 7.62 and 0.23 dS/m, respectively. The experiment was laid out in split plot design. There were three main treatments *viz.*, fallow (where no crop was grown before basmati), *bajra* (a fodder grown before basmati) and green manure (*Sesbania aculeata* grown before basmati) which were kept as main plots and four crop sequences followed thereafter *viz.*, basmati rice (*Oryza sativa*)-wheat (*Triticum aestivum*), basmati rice-field peas (*Pisum sativum*) basmati rice-celery (*Apium graveolens*) and basmati-sunflower (*Helianthus annuus*) which were kept as sub plots. *Bajra* was grown as fodder and harvested after 8 to 7 weeks and green manure dhaincha (*Sesbania aculeata*) was grown for about 60 days and incorporated in soil a day before transplanting basmati rice in the respective treatments.

In all there were twelve treatments, which were replicated thrice. All recommended package of practices were followed to grow different crops. The varieties and fertilizer applied to different crops are given in Table 17.1. The crops were harvested at proper maturity and yield data were recorded and the yield of basmati rice, wheat, celery, field peas and sunflower were adjusted at 18, 10, 10, 10 and 14 per cent moisture respectively.

The productivity of different crop sequences were compared by converting the yield of all the component crops of a crop sequence into basmati equivalent yield on the basis of minimum support price and gross and net returns were also worked out. After the completion of three cycles of crop rotations surface soil samples (0-15 cm) were taken and analyzed for pH (1 : 2), EC (1 : 2), OC (Walkley and Black 1934), available N (Subbiah and Asija 1956) available P (Olsen *et al.*, 1956) and available K (Meernin and Peach 1950) to determine the changes in soil fertility.

Table 17.1: Varieties of Crops and Nutrients Applied to Different Crops

Crop	*Variety*	*Nutrient Applied (kg/ha)*		
		N	P_2O_5	K_2O
Sesbania aculeate (GM)	Local	0	0	0
Bajra	Local	50	0	0
Basmati rice	Basmati–370/386	60	30	0
Field peas	Field peas-48	30	40	0
Wheat	PBW–373	120	60	30
Celery	Local	100	40	0
Sunflower	MSFH–8	60	30	30

Results and Discussion

Crop Productivity

The data on yield of component crops of different crop sequences are given in Table 17.2. The grain yield of basmati rice ranged from 22.0 to 24.1 q/ha. in fallow-basmati rice sequence and from 18.2 to 19.4 q/ha in different *bajra (f)*-basmati rice sequence and the corresponding values ranged from 14.2 to 14.4 q/ha in green manure-basmati rice crop sequences. When the mean values were compared it was found that maximum mean grain yield of basmati was recorded in fallow-basmati rice sequence (22.9 q/ha) and minimum mean grain yield of basmati was recorded in green manure-basmati rice sequence (14.3 q/ha).

The reduction in grain yield in green manure-basmati rice sequence may be due to severe lodging which occurred in two out of the three years of experimentation. As basmati rice grown after green manured crop also received recommended N which might have made the crop more succulent and, therefore, resulted in lodging. In the *bajra (f)*-basmati rice crop sequences the mining of nutrients by *bajra* (fodder) might have resulted in the low yield of following basmati rice crop as is also evident from nutrient status of soils particularly Nand K after harvest of the crop (Table 17.4).

Earlier also it has been reported by Sidhu *et al.* (2004) that where green manuring was practiced along with application of recommended level of N the basmati crop suffered from severe lodging which resulted in low grain yield. On the other hand the yield of wheat, celery, field peas and sunflower succeeding basmati rice increased substantially where green manure was practiced. The mean grain yield of wheat was 44.9, 43.9 and 47.1 q/ha, respectively, in fallow-basmati rice-wheat, *bajra (f)*-basmati rice-wheat and green manure basmati rice-wheat crop sequence. The seed yield of celery, field peas and sunflower were also higher in crop sequences with green manuring, indicating there by that there was considerable residual effect of green manure in the form of increasing the available nutrients in the soil (Table 17.4) and the other associated beneficial effects.

There was not much variation in the seed yield of celery and sunflower in fallow-basmati rice and *bajra* (fodder)-basmati rice crop sequences. However, the seed yield of field peas was high in fallow-basmati rice sequence as compared to *bajra (f)*-basmati rice-field pea sequence. Green manure of dhaincha not only benefit the crop before which incorporated but also leave sufficient residual effect to benefit the succeeding crop.

Table 17.2: Grain Yield of Component Crops and BREY of Basmati Rice Based Crop Sequences (3 year mean)

Crop Sequence	*Grain Yield (q/ha)*		*BREY (q/ha)*
	I	*II*	
Fallow			
Basmati rice-wheat	22.0	44.9	41.4
Basmati rice-celery	24.1	5.6	34.2
Basmati rice-field peas	22.2	14.6	35.5
Basmati rice-sunflower	23.4	27.0	46.7
Mean	22.9		39.5
Basmati rice-wheat	(472) 18.5	43.9	50.3
Basmati rice-celery	(455) 19.4	5.7	42.1
Basmati rice-field peas	(455) 18.7	11.7	41.7
Basmati rice-sunflower	(443) 18.2	27.2	53.8
Mean	18.7		47.0
Green manure			
Basmati rice-wheat	14.2	47.1	34.6
Basmati rice-celery	14.2	6.3	25.7
Basmati rice-field peas	14.3	16.2	29.1
Basmati rice-sunflower	14.4	28.2	38.8
Mean	14.3		32.1

Figures in parentheses are fresh biomass of *bajra* (fodder) q/ha.

Basmati Rice Equivalent Yield (BREY) and Economic Analysis

The grain yield of component crops of different crop sequences was converted into basmati rice equivalent yield (BREY) and the data are presented in Table 17.2. The maximum BREY was obtained in *bajra (f)*-basmati rice-sunflower crop sequence (53.8 q/ha) and minimum was obtained in green manure-basmati rice-celery system (25.7 q/ha) which could be due to inclusion of *bajra* crop in the former and in the later crop sequence it could be because of low grain yield of basmati rice. Maximum mean BREY was obtained in *bajra (f)* based crop sequence (47.0 q/ha) followed by fallow (39.0) and minimum was obtained in green manure based crop sequence (32.1).

As expected gross returns (Table 17.3) also followed the same trend and the mean gross returns were maximum (Rs. 51,700 /ha) in *bajra*-basmati rice sequence and minimum in green manure-basmati rice sequence, which could also be due to low yields of component crops in these sequence (Table 17.2).

On the other hand when net returns (Table 17.4) were compared it was found that the mean net returns obtained from green manure-basmati rice were significantly below as compared to fallow-basmati rice and *bajra (f)*-basmati rice which may be due to poor yield of basmati rice obtained in this crop sequence. However there was not much variation in the mean net returns obtained from fallow-basmati rice and *bajra* (*f*)-basmati rice (Rs. 19200 and 19900/ha). But when sub crop sequences were compared fallow-basmati rice-sunflower and *bajra* (*f*)-basmati rice-sunflower gave the highest net returns, which were significantly better than fallow-basmati rice-wheat and *bajra* (*f*)-basmati rice-wheat sequence.

Table 17.3: Gross Returns from Basmati Rice Based Crop Sequences (mean of 3 years)

Crop Sequence	*Fallow*	*Bajra (fodder)*	*Green Manure*	*Mean*
Basmati rice-wheat	45.6	55.3	38.0	46.6
Basmati rice-celery	37.6	46.3	28.3	37.4
Basmati rice-field peas	39.0	45.9	31.9	38.9
Basmati rice-sunflower	51.5	59.2	42.7	51.1
Mean	43.4	51.7	35.2	

CD at 5% Main plot: NS, Sub plot: 5.7, Interaction: NS.

Table 17.4: Net Returns from Basmati Rice Based Crop Sequences (mean of 3 years)

Crop Sequence	*Fallow*	*Bajra (fodder)*	*Green Manure*	*Mean*
Basmati rice-wheat	20.8	23.0	11.0	18.3
Basmati rice-celery	15.2	16.4	3.7	11.8
Basmati rice-field peas	14.4	13.8	5.1	11.1
Basmati rice-sunflower	26.3	26.5	15.3	22.7
Mean	19.2	19.9	8.8	

CD at 5% Main plot: 4.7, Sub plot: 2.8, Interaction: NS.

Nitrogen Addition by Incorporating Green Manure

About 60 days old dhaincha was incorporated for green manure during all the three years of cropping which produced 51.6, 54.5 and 62.0 q/ha of dry weight during 1996, 1997 and 1998 respectively. The N content of green manure during the three years was 2.2, 1.97 and 1,87 per cent, respectively and that added 113.5,107.4 and 115.9 kg N/ha during different years. Earlier, Meelu *et al.* (1994) had also reported that incorporation of green manure resulted in addition of 75–100 kg N/ha. Therefore, the improvement in soil fertility and BREY was observed in green manure compared to other treatments due to addition of biomass, nitrogen and other associated beneficial effects.

Soil Health

After three cycles of cropping there was a reduction in the soil pH than the initial value (Table 17.5). The mean soil pH decreased from 7.62 to 7.45, 7.48 and 7.46 in fallow-basmati rice, *bajra* (*f*)-basmati rice and green manure-basmati rice crop sequence. Although within the mean values there were marginal changes but when compared with initial pH (7.62) there was considerable decline in soil pH. This could be due to addition of urea which is known to be acid forming fertilizer and release of some organic acids produced during the decomposition of roots of different crops remaining in the soil.

Table 17.5: Influence of Basmati Rice Based Crop Sequences on Soil Health

Crop Sequence		*pH (1 : 2)*	*OC (%)*	*Available N*	*Available P (kg/ha)*	*Available K*
Fallow						
Basmati rice-wheat		7.52	0.34	114.4	47.6	119.0
Basmati rice-celery		7.56	0.43	125.4	55.0	108.9
Basmati rice-field peas		7.36	0.36	122.3	55.0	119.7
Basmati rice-sunflower		7.36	0.35	101.8	52.0	88.6
Mean		7.45	0.37	115.9	52.4	109.0
Bajra (fodder)						
Basmati rice-wheat		7.50	0.36	112.9	47.6	89.9
Basmati rice-celery		7.46	0.38	109.7	51.7	99.7
Basmati rice-field peas		7.49	0.39	106.6	55.0	59.3
Basmati rice-sunflower		7.49	0.41	103.4	54.6	106.1
Mean		7.48	0.38	108.1	52.0	95.0
Green manure						
Basmati rice-wheat		7.54	0.51	130.1	64.4	119.7
Basmati rice-celery		7.53	0.52	126.9	57.2	115.2
Basmati rice-field peas		7.39	0.49	128.5	59.5	123.2
Basmati rice-sunflower		7.38	0.45	122.2	58.2	116.6
Mean		7.46	0.49	127.0	59.2	118.7
CD at 5%	Main plot	0.05	0.06	NS	5.6	7.9
	Sub plot	NS	NS	NS	NS	NS
	Interaction	NS	NS	NS	NS	NS
Initial value		7.62	0.33	110.2	42.8	132.6

The mean organic carbon content of the soil increased over its initial content but the extent of increase was higher where green manure was incorporated and it increased from its initial status of 0.33 per cent to 0.49. Mahapatra *et al.* (2002) reported beneficial effects of *Sesbania aculeate* green manuring in rice-wheat sequence on soil organic carbon and available N. There was not much variation in the available N content of the soil than its initial content except in green manuring treatment where it increased to 127 kg/ha over its initial content of 110 kg/ha.

On the other hand mean available P of the soil increased than its initial status under all the treatments but again the extent of increase was maximum under green manure incorporation. This may be due to addition of P fertilizers to crops and because of lesser P uptake by the crops than the amounts applied to different crops. Earlier Sikka and Sidhu (1997) had reported that lesser P uptake in spite of much higher P application to different crop sequences resulted in build up of available P in soils.

However, the available K content of the soil decreased significantly than its initial status in all the crop sequences but the magnitude of reduction was little less in green manure incorporation. The decrease in available K status may be due to exceedingly higher K uptake by the crops and no addition of K fertilizer to the crops as is evident from the data in Table 17.1. The mean available K content was minimum under *bajra*-basmati rice sequence, which may be due to greater removal by *bajra* crop, which received no K application.

It may be concluded that fallow basmati rice-sunflower sequence was more viable both economically and biologically as it produced substantially higher rice equivalent yield and higher returns over variable cost than the prevalent basmati rice-wheat sequence and also sustained the nutrient availability in the soil except K. It may also be mentioned that green manuring did not prove beneficial for basmati rice as it resulted in severe lodging of basmati causing a reduction in its grain yield.

References

Chauhan, D.S., Sharma, R.K. and Kharub, A.S. 2000. Diversification of rice-wheat system for sustainability and profitability. In: *Extended Summary, International Conference on Managing Natural Resources*, held during 14–19 Feburary at IARI, New Delhi.

Mahapatra, B.S., Kumar, Ajay, Shukla, D.K. and Shukla, R K. 2002. Summer legumes in relation to productivity and fertility in rice-wheat cropping system. In: *Paper Presented at the International Conference on Managing Natural Resources*, held during 14–19 Feburary at IARI, New Delhi, 1: 155–156.

Meelu, O.P., Singh, B. and Singh, Y. 1994. Effect of green manure and crop residue cycling on N economy, organic matter and physical properties of soil in rice-wheat cropping system. In: *Proceeding of Symposium on Sustainability of Rice-wheat System in India*, held during 7–8 May at Regional Station, Karnal, p. 15–24.

Merwin, H.D. and Peech, M. 1950. Exchangeability of soil potassium in the sand, silt and clay fractions as influenced by the nature of the complementary exchangeable cations. *Soil Sci. Soc. Am. Proc.*, 15: 125–128.

Olsen, S.R., Cole, C.Y., Watanabe, F.S. and Dean, L.A. 1954. Estimation of available phosphorus by extraction with sodium bicarbonate. *USDA Circular*, 939: 19.

Rattan, R.K. and Singh, A.K. 1997. Role of balanced fertilization in rice-wheat cropping system. *Fertilizer News*, 42(4): 592–600.

Sidhu, M.S., Thakar Singh and R.K. Sikka. 2002. Response of basmati rice to nitrogen in different cropping systems. In: *Extended Summary of 2nd International Agronomy Congress*, New Delhi, p. 402–403.

Sidhu, M.S., R. Sikka and Thakar Singh. 2004. Performance of transplanted Basmati rice in different cropping systems as affected by nitrogen application. *IRRI Notes*, p. 62–65.

Sikka, R. and M.S. Sidhu. 1997. Evaluation of summer groundnut based cropping systems for nutrient uptake and changes in soil fertility. In: *Proceedings of the 62nd Annual Convention of the Indian Society of Soil Science* held from 18–21 October at Kolkata, pp. 209.

Subbiah, B.V. and Asija, G.L. 1956. A rapid procedure for estimation of available nitrogen in soils. *Curr Sci.*, 25: 259–260.

Walkley, A. and Black, I.A. 1934. An estimation of degtareff method for determining soil organic matter and a proposed modification of the chromic acid titration method. *Soil Sci.*, 4: 51–56.

Yadav, R.L., Yadav, D.S., Singh, R.M. and Kumar, A. 1998. Long term effects of inorganic-inputs and crop productivity in rice-wheat cropping system. *Nutrient Cycling in Agroecosystems*, 51: 193–200.

Chapter 18

Assessment of Phytotoxicity of SO_2 and NH_3 of Poaceae and Fabaceae Crop Plants

Nirmala Babu Rao, Amena and I. Sobha Rani

Department of Botany, University College for Women, Koti Hyderabad – 500 097

ABSTRACT

The present study deals with the evaluation of phytotoxicity of pollutants like SO_2 and NH_3, on two members of Poaceae (*Zea mays* and *Triticum vulgare*) and two members of Fabaceae (*Arachis hypogea* and *Cyamopsis tetragonoloba*) crops. The plants were exposed to, 1 ppm, 0.001 ppm (for SO_2), and 25 ppm and 1.6 ppm (for NH_3). The plants were exposed to these gases at 15th day stage. Chlorophyll, ascorbic acid, biomass, S and N were found to decrease significantly at every stage up to the harvest period depending upon the duration of SO_2 and NH_3 exposure.

Keywords: *Sulphur dioxide, Ammonia, Phytotoxicity, Crop plants.*

Introduction

Plants are reported to be more susceptible to gaseous pollutants like sulphur dioxide and ammonia due to uptake into leaves Guderian (1997). Air pollutants are phytotoxic if their absorption rates exceed detoxification rates in the plant tissues Tingey and Taylor (1983), Tingey and Olszyk (1984) but the real manifestation of gaseous effects depends mostly on the concentration, duration of exposure to gas and genetic constitution of the plant. However, the studies on the phytotoxicity of these gases on crop plants are very few. The present investigation has been carried out to study the effect of these gases at different dosages on the above crop plants.

Materials and Methods

7.6 mg of Na_2SO_3 in dilute H_2SO_4 releases 1 ppm of SO_2 in 1 m^3 air Rao and LeBlance (1966) and 11.5 ml of freshly prepared liquor ammonia is necessary for the evolution of 1 ppm of ammonia gas Leveggi *et al.* (1973). Fifteen days old standing crop of Poaceae and Fabaceae members namely *Zea mays* (Var. 70 EH. starch), *Triticum vulgare* (Var. 1192), *Arachis hypogea* (Var. T.M.V.3) and *Cyamopsis tetragonoloba* (Var. DSF-1-6-J) were exposed to 0.001 ppm and 1 ppm of sulphur dioxide for a duration of 1 hour and 3 hours by enclosing in a chamber. Similarly, plants were exposed to 1.6 ppm and 25 ppm of ammonia for a duration of 5 minutes and 60 minutes. The untreated plants constituted the control. 15, 45, 60, 90 and 120 days old plants were analysed for chlorophyll (Arnon, 1949), Ascorbic acid (Singh, 1977), Nitrogen (Piper, 1966), Sulphur (Patterson, 1958) and Biomass (Puri, 1950).

Results and Discussion

The result of various factors in plants treated with sulphur dioxide and ammonia on chlorophyll are presented in Tables 18.1 and 18.2. Poaceae plants exposed to SO_2 showed a reduction of 80 to 88 per cent at 15th day age *i.e.*, immediately after the exposure to the gas levels, but the trend reduced at lower concentration of gas. However, when the plants have attained old age, recovery from the damage in terms of increase in chlorophyll was remarkable. There was about 60–66 per cent recovery in *Zea mays* and at higher concentrations there was no significant recovery due to early senescence.

Table 18.1: Percentage Reduction and Recovery in Chlorophyll Content in Sulphur Dioxide Treated Graminaceous and Leguminous Plants

Chlorophyll	*Zea mays*				*Triticum vulgare*			
	1ppm × 1 hour	*1ppm × 3 hour*	*0.001ppm × 1 hour*	*0.001ppm × 3 hour*	*1ppm × 1 hour*	*1ppm × 3 hour*	*0.001ppm × 1 hour*	*0.001ppm × 3 hour*
Reduction at 15th day age	80.80	86.67	55.00	66.66	83.44	88.69	37.50	56.25
Reduction at 90th day age	91.79*	98.62*	18.34	26.60	97.68*	98.25*	22.50	41.25
% recovery by 90th day	–	–	66.65	60.09	–	–	40.00	26.66
	Arachis hypogaea				*Cyamopsis tetragonoloba*			
	1ppm × 1 hour	*1ppm × 3 hour*	*0.001ppm × 1 hour*	*0.001ppm × 3 hour*	*1ppm × 1 hour*	*1ppm × 3 hour*	*0.001ppm × 1 hour*	*0.001ppm × 3 hour*
Reduction at 15th day age	80.00	86.00	60.00	70.06	62.16	64.28	52.08	55.57
Reduction at 90th day age	45.00	48.00	19.00	21.00	92.35*	98.04*	42.64	48.09
% recovery by 90th day	43.75	44.18	68.33	70.02	–	–	18.12	13.46

–: No recovery; *: Due to early senescence.

The same trend was noticed in Fabaceae plants as well. *Arachis hypogea* showed very good resistance against the effect of gas than *Cyamopsis tetragonoloba*.

There was good resistance shown by *Zea mays* and *Arachis hypogea*. Chlorophyll concentration increased with age in control plants, whereas, it drastically reduced in treated plants there by delimiting the photosynthetic activity Halgren (1978) and Black (1982). It has also been noticed that plants exposed to pollutants for shorter duration (1 hour) can recover (0.001 ppm of SO_2) Thomas photosynthesis in plants fumigated with lower concentration of SO_2. Such rapid recovery was noticed in *Zea mays* and *Arachis hypogea* showing development of any acute injury in plants Beannett and Hill, (1973), Sij and Swanson, (1974). Thus these two plants were resistant possibily due to high peroxides, high sulphite or nitrate turn over rate; similar observation was also made by Varshney and Garg (1979).

Table 18.2: Percentage Reduction and Recovery in Chlorophyll Content in Ammonia Treated Graminaceous and Leguminous Plants

Chlorophyll	*Zea mays*				*Triticum vulgare*			
	25 ppm × 60 min	*25ppm × 5 min*	*1.6ppm × 60 min*	*1.6ppm × 5 min*	*25ppm × 60 min*	*25ppm × 5 min*	*1.6ppm × 60 min*	*1.6ppm × 5 min*
Reduction at 15th day age	44.66	25.04	16.68	10.00	47.68	43.53	14.65	8.46
Reduction at 90th day age	86.25*	83.95*	8.28	0.91*	95.00*	94.25*	6.25	2.50
% recovery by 90th day	–	–	50.66	90.90	–	–	57.33	70.44
	Arachis hypogaea				*Cyamopsis tetragonoloba*			
	25 ppm × 60 min	*25ppm × 5 min*	*1.6ppm × 60 min*	*1.6ppm × 5 min*	*25ppm × 60 min*	*25ppm × 5 min*	*1.6ppm × 60 min*	*1.6ppm × 5 min*
Reduction at 15th day age	70.06	56.34	28.00	12.06	8.29	53.69	28.57	13.94
Reduction at 90th day age	40.00	32.00	11.00	3.50	55.35	35.28	12.50	5.35
% recovery by 90th day	42.90	43.20	60.71	70.97	31.09	34.28	56.24	61.62

–: No recovery; *: Due to early senescence.

Ascorbic acid is primary to chloroplast. It is also an important electron donor in photosynthesis. It also scavengers free oxygen radicals and may play an important part in overcoming gas toxicity.

At higher concentrations of the gases and at 90th day age the reduction in ascorbic acid concentration was not as much as found in 15th day age (Tables 18.3 and 18.4). There was less decrease of ascorbic acid at lower concentration of the gases. The reduction in ascorbic acid content supports the hypothesis that the tolerant tissue can elevate the endogenous level of ascorbic acid under stress by gaseous pollutants there by offering protective effects by scavenging toxic radicals Pierre and Queiror (1981). The order of tolerance and sensitivity being *Zea mays, Arachis hypogea* were found to be tolerant and *Triticum vulgare* and *Cyamopsis tetragonoloba* as sensitive.

Table 18.3: Percentage Reduction and Recovery in Ascorbic Acid in Sulphur Dioxide Treated Graminaceous and Leguminous Plants

Ascorbic Acid Content	*Zea mays*				*Triticum vulgare*			
	1ppm × 1 hour	*1ppm × 3 hour*	*0.001ppm × 1 hour*	*0.001ppm × 3 hour*	*1ppm × 1 hour*	*1ppm × 3 hour*	*0.001ppm × 1 hour*	*0.001ppm × 3 hour*
Reduction at 15th day age	74.40	80.00	44.00	60.00	60.00	73.00	36.00	50.00
Reduction at 90th day age	63.26	77.55	22.44	44.89	57.77	70.59	28.57	38.00
% recovery by 90th day	14.90	3.06	49.00	25.00	3.43	3.30	20.63	23.88
	Arachis hypogaea				*Cyamopsis tetragonoloba*			
	1ppm × 1 hour	*1ppm × 3 hour*	*0.001ppm × 1 hour*	*0.001ppm × 3 hour*	*1ppm × 1 hour*	*1ppm × 3 hour*	*0.001ppm × 1 hour*	*0.001ppm × 3 hour*
Reduction at 15th day age	65.00	75.00	39.50	66.33	75.00	80.00	51.83	58.33
Reduction at 90th day age	60.00	72.00	26.00	56.00	69.38	78.59	38.77	52.18
% recovery by 90th day	8.30	4.00	34.11	15.57	7.49	1.76	21.19	10.54

Table 18.4: Percentage Reduction and Recovery in Ascorbic Acid in Ammonia Treated Graminaceous and Leguminous Plants

Ascorbic Acid Content	*Zea mays*				*Triticum vulgare*			
	25 ppm × 60 min	*25ppm × 5 min*	*1.6ppm × 60 min*	*1.6ppm × 5 min*	*25ppm × 60 min*	*25ppm × 5 min*	*1.6ppm × 60 min*	*1.6ppm × 5 min*
Reduction at 15th day age	85.30	83.84	26.76	4.84	62.80	48.00	26.00	16.00
Reduction at 90th day age	81.81	78.18	10.09	1.81	61.22	44.89	14.28	4.08
% recovery by 90th day	4.09	6.75	56.42	62.60	2.51	6.47	45.07	70.50
	Arachis hypogaea				*Cyamopsis tetragonoloba*			
	25 ppm × 60 min	*25ppm × 5 min*	*1.6ppm × 60 min*	*1.6ppm × 5 min*	*25ppm × 60 min*	*25ppm × 5 min*	*1.6ppm × 60 min*	*1.6ppm × 5 min*
Reduction at 15th day age	58.33	54.16	31.00	12.30	58.33	44.66	38.33	22.66
Reduction at 90th day age	52.00	40.00	16.00	6.00	53.18	38.97	22.44	12.24
% recovery by 90th day	10.85	26.14	48.38	51.21	8.83	12.74	41.45	45.98

It was revealed that shoot biomass of gas treated plants have decreased in comparison to control with the increase in age. It has also been recorded that there was decline in biomass even in 1 hour treated plants in case of both the gases. Maximum damage was observed in 3 hours and 60 minutes (Tables 18.5 and 18.6).

Table 18.5: Percentage Reduction and Recovery in Dry Weight of Shoot in Sulphur Dioxide Treated Graminaceous and Leguminous Plants

Dry Weight of Shoot	*Zea mays*				*Triticum vulgare*			
	1ppm × 1 hour	*1ppm × 3 hour*	*0.001ppm × 1 hour*	*0.001ppm × 3 hour*	*1ppm × 1 hour*	*1ppm × 3 hour*	*0.001ppm × 1 hour*	*0.001ppm × 3 hour*
Reduction at 15th day age	40.16	43.17	16.74	26.71	41.90	43.33	24.28	33.33
Reduction at 90th day age	20.72	27.18	6.63	11.46	24.66	27.71	11.30	16.21
% recovery by 90th day	48.39	37.03	60.39	57.07	41.14	36.03	54.49	51.39
	Arachis hypogaea				*Cyamopsis tetragonoloba*			
	1ppm × 1 hour	*1ppm × 3 hour*	*0.001ppm × 1 hour*	*0.001ppm × 3 hour*	*1ppm × 1 hour*	*1ppm × 3 hour*	*0.001ppm × 1 hour*	*0.001ppm × 3 hour*
Reduction at 15th day age	32.94	36.82	17.67	23.01	34.75	37.80	25.60	29.87
Reduction at 90th day age	15.14	17.71	5.16	9.01	21.98	25.06	9.08	12.84
% recovery by 90th day	54.04	57.88	70.80	60.84	36.75	33.69	64.33	57.02

The amount of increase in sulphur content (excess sulphur) in SO_2 treated plants and the amount of increase in nitrogen content in NH_3 treated plants (excess nitrogen) are greater than in control (Tables 18.7 and 18.8), the plants with excess sulphur showed a reduction in nitrogen contents and plants with excess nitrogen exhibited decline in sulphur content.

Table 18.6: Percentage Reduction and Recovery in Dry Weight of Shoot in Ammonia Treated Graminaceous and Leguminous Plants

Dry Weight of Shoot	*Zea mays*				*Triticum vulgare*			
	25 ppm × 60 min	*25ppm × 5 min*	*1.6ppm × 60 min*	*1.6ppm × 5 min*	*25ppm × 60 min*	*25ppm × 5 min*	*1.6ppm × 60 min*	*1.6ppm × 5 min*
Reduction at 15th day age	36.98	33.96	21.42	17.46	40.34	38.74	24.34	21.23
Reduction at 90th day age	16.23	14.13	3.71	2.36	20.71	18.88	9.01	6.32
% recovery by 90th day	56.11	58.38	82.68	86.84	48.66	51.28	62.98	7.23
	Arachis hypogaea				*Cyamopsis tetragonoloba*			
	25 ppm × 60 min	*25ppm × 5 min*	*1.6ppm × 60 min*	*1.6ppm × 5 min*	*25ppm × 60 min*	*25ppm × 5 min*	*1.6ppm × 60 min*	*1.6ppm × 5 min*
Reduction at 15th day age	36.34	32.86	20.16	16.85	39.15	38,.10	24.25	19.00
Reduction at 90th day age	16.30	13.66	4.34	2.36	20.21	18.85	9.70	7.10
% recovery by 90th day	55.14	58.42	78.45	85.99	48.27	50.45	60.00	64.32

Table 18.7: Order of Sensitivity of Graminaceous and Leguminous Plants to Sulfur Dioxide on the Basis of Excess Sulfur Percentage and Percent Recovery from the Damage

Name of the Plants	*Dosage*	*Excess Sulfur %*		*Dry Weight of Shoot*		*Chlorophyll*	
		15th Day	*90th Day*	*Initial Reduction*	*% Recovery*	*Initial Reduction*	*% Recovery*
	1 ppm × 1 hour						
Zea mays		11.45	6.32	40.16	41.06	80.80	–
Triticum vulgare		16.20	12.00	35.25	23.87	83.44	–
Arachis hypogaea		20.42	8.32	32.04	52.14	80.26	43.75
Cyamopsis tetragonoloba		29.48	15.03	34.75	36.75	62.16	–
	1 ppm × 3 hour						
Zea mays		15.70	7.32	43.17	34.03	86.67	–
Triticum vulgare		23.50	16.73	38.00	22.22	88.69	–
Arachis hypogaea		22.42	10.32	33.82	47.63	86.00	44.18
Cyamopsis tetragonoloba		32.12	18.14	31.80	33.69	64.28	–
	0.001 ppm × 1 hour						
Zea mays		3.76	1.64	26.74	60.24	55.00	66.65
Triticum vulgare		10.36	3.84	25.33	46.63	37.50	56.25
Arachis hypogaea		11.62	3.09	17.67	70.80	60.00	68.33
Cyamopsis tetragonoloba		17.32	7.65	25.60	64.53	52.08	18.12
	0.001 ppm × 3 hour						
Zea mays		5.68	3.99	28.71	57.07	66.66	60.09
Triticum vulgare		11.60	4.43	26.39	42.36	40.00	26.66
Arachis hypogaea		12.32	4.32	27.01	66.64	70.06	70.02
Cyamopsis tetragonoloba		19.52	8.64	29.87	57.87	55.57	13.46

Table 18.8: Order of Sensitivity of Graminaceous and Leguminous Plants to Ammonia on the Basis of Excess Nitrogen Percentage and Percent Recovery from the Damage

Name of the Plants	Dosage	Excess Sulfur %		Dry Weight of Shoot		Chlorophyll	
		15th Day	90th Day	Initial Reduction	% Recovery	Initial Reduction	% Recovery
	25 ppm × 60 mints.						
Zea mays		37.82	21.55	36.98	56.11	44.66	–
Triticum vulgare		23.23	32.36	35.11	48.47	45.68	–
Arachis hypogaea		26.42	21.55	36.34	55.14	70.06	42.90
Cyamopsis tetragonoloba		21.31	30.52	39.15	48.37	80.29	31.09
	25 ppm × 5 mints.						
Zea mays		28.60	20.43	32.96	58.38	25.04	–
Triticum vulgare		8.99	25.07	33.71	52.56	43.53	
Arachis hypogaea		18.61	14.65	30.71	54.93	50.34	43.20
Cyamopsis tetragonoloba		18.51	19.04	38.10	50.45	53.69	34.28
	1.6 ppm × 60 mints.						
Zea mays		19.40	14.52	21.42	82.68	16.66	50.66
Triticum vulgare		5.92	4.68	14.40	70.62	14.65	57.33
Arachis hypogaea		12.68	9.48	20.16	78.45	28.00	60.71
Cyamopsis tetragonoloba		12.96	8.42	24.25	60.00	28.57	56.24
	1.6 ppm × 5 mints.						
Zea mays		14.60	9.40	17.46	86.84	10.00	90.00
Triticum vulgare		2.68	1.83	21.23	70.23	8.46	70.44
Arachis hypogaea		10.43	5.42	16.85	85.99	12.06	70.97
Cyamopsis tetragonoloba		5.55	2.66	19.90	64.32	13.94	61.62

Arachis hypogea and *Zea mays* treated with SO_2 and NH_3 showed remarkable; damage despite low sulphur content (Table 18.7), but there was appreciable recovery at the later stages which means that toxic sulphite ions might have gradually converted into sulphate where they could have been incorporated into organic sulphur leading to more growth Launorth *et al.* (1979).

In *Triticum vulgare* and *Cyamopsis tetragonoloba* excess nitrogen was observed (Table 18.8). The increase in nitrogen content was found to be varying with the advancement of age of the plants, similar observation of differential absorption and assimilation was made by Hutchinson *et al.* (1972) in soya bean and corn.

While drawing the order of sensitivity or tolerance with respect to SO_2 and NH_3 treated plants *Arachis hypogea* and *Zea mays* are found to be resistant and *Triticum vulgare* and *Cyamopsis tetragonoloba* as sensitive to the pollutants.

References

Arnon, D.I. 1949. Copper enzymes in isolated chloroplasts polyphenol oxidase in *Beta vulgare*. *Plant Physiol.*, 24(1): 1–5.

Bennett, J.H. and Hill, A.C. 1973. Absorption of gaseous air pollutants by standardized plant canopy. *J. Air Pollution Control Assoc.*, 23: 203–206.

Black, V.J. 1982. Effect of sulphur dioxide on physiological process in plants. In *Effects of Gaseous Pollutants in Agriculture and Horticulture*, (Eds.) M.H. Unsworth and D.P. Ormrod. Proc. 32nd School in Agriculture Sciences. Butterworths, London, pp. 67–91.

Guderian, R. 1977. Air pollution, phytotoxicity of acidic gases and its significance. In: *Air Pollution Control.* Springer Verlag, N.Y.

Hallgren, J.E. 1978. In: *Sulphur in the Environment*, Part II: Ecological impacts. (Ed.) J.O. Nriagu. John Wiley and Sons Inc., London, p. 164–209.

Hutchinson, G.L., Willington, R.J. and Peters, D.B. 1972. Atmospheric ammonia by plant leaves. *Science*, 175: 771–772.

Lauenorth, W.K., Black, C.J. and Dodd, J.L. 1979. Sulphure accumulation in western wheat grass exposed to three controlled sulphur dioxide concentration. *Plant and Soil*, 53: 131–136.

Leveggi, D.A., Siw and Feldstein, M. 1973. Nesslerization method for estimation of ammonia. *J. Air Pollution Control Assoc.*, 23–30.

Patterson, G.B. 1958. *Colorimetric Determination of Non-metals.* International Science Public Inc., NY, p. 216–308.

Pierre, L.M., and Queirz, Q. 1981. Enzymic and metabolic changes in bean leaves during continuous pollution by subnecrotic leaves of sulphur dioxide. *Environ. Pollution* (Ser. A). 251: 41–51.

Piper, C.S. 1966. *Soil and Plant Analysis*. International Science Public Inc., New York, p. 368.

Puri, G.S. 1950. The distribution of conifers in the Kulu Himalayas with special reference to geology. *Indian Forest*, 76: 144–153.

Rao, D.N. and Le Blance, F. 1966. Effects of sulphur dioxide on the lichens alga with a special reference to chlorophyll. *Bryologist*. 69: 69–75.

Sij, J.W. and Swanson, C.A. 1974. Short term kinetic studies on inhibition of photosynthesis by sulphur dioxide. *J. Environ. Qual.*, 3: 103–107.

Singh, N. 1981. Responses of certain leguminous plants to sulphur dioxide pollution. *Ph.D. Thesis*, Banaras Hindu University, Varanasi, India.

Tingey, D.T., Olszyk, D.M. 1984. In: *SO_2 and Vegetation, Physiology Ecology and Policy Issues*, (Eds.) Winner, W.E., Mooney, H.M., Gold-stein, R. Standard University Press, Palo Alto (In Press).

Tingey, D.T. and Toylor. 1983. Air pollution damage to vegetation. *Plant Physiol.*, 172: 237–244.

Thomas, M.D. and Hill, G.R. 1937. Relation to sulphur dioxide in the atmosphere to photosynthesis and respiration to alfalfa. *Plant Physiol.*, 12: 309–383.

Varshney, C.K. and Garg, G.R., C.R.C. 1979. Plant responses to sulphur dioxide pollution. *Critical Reviews in Environmental Control*, 9: 27–49.

White, K.L., Hill, A.C. and Bennett, J.H. 1974. Synergistic inhibition of apparent photosynthesis rate of alfalfa by combination of sulphur dioxide and nitrogen dioxide. *Environ. Sci. Technol.*, 8: 574–576.

Chapter 19

A Note on Air Pollution Tolerance Index of Some Plants Growing at Hyderabad

Nirmal Babu Rao, Amena and I. Sobha

Department of Botany, Osmania Univeresity College For Women, Koti, Hyderabad – 500 195, A.P., India

ABSTRACT

The role of plants in the atmosphere acts as pollution sink. In this context, an attempt has been made in the investigation to screen the plants for their Air Pollution Tolerance Index (APTI). Tolerance index were computed on the basis of pH of leaf extract, Total chlorophyll extract, ascorbic acid and (RCW) relative water content. The tree *Albizzia lebbeck, Annona squamosa, Azadirachta indica, Cassia fistula, Eugenia jambolana,* shrub *Hibiscus rosans, Ixora coccina, Hemelia patens, Quisqualis indica, Asperagus racemoses* and climbers *Tinospora cordifolia, Pyrostegia grandusa, Adenocalymma ailiaceum, Thumbergia latifolia, Thumbergia alba,* are highly tolerant to air pollution.

Keywords: Air pollution tolerance index, Hyderabad growing plants.

Introduction

The plants act as pollution sink and they replenish the atmosphere with oxygen. A search of literature reveals that there exist several reports for the trees acting as pollution sink (Rich, 1907; Smith, 1970 and Smith and Dochinger, 1975). For proper management of environment and abatement of pollution a knowledge of their tolerance level of plant species to air pollution is required. Air pollution tolerance index has been calculated with the help of total chlorophyll, relative water content, ascorbic acid, pH of leaf extract. In view of this the present investigation was taken up to evaluate primarily air pollution tolerance index of vegetation of Hyderabad from the sites free from air pollution.

Materials and Methods

The leaf samples of different species of trees, shrubs and climbers from Hyderabad region were collected in morning hours.

The leaf samples collected were analysed for pH of leaf extract (Singh and Rao, 1983), total chlorophyll (Arnon, 1949), ascorbic acid (Singh, 1977) and relative water content (RWC) Singh (1977).

Air Pollution Tolerance Index

An attempt has been made to determine the air pollution tolerance index (APTI) which gives an empirical value for the tolerance level of a plant to air pollution (Singh and Rao, 1983). This has been calculated with the help of the formula:

$$\text{APTI} = \frac{A\,(T + P) + R}{10}$$

where,

A: Ascorbic acid content of leaf in mg/g dry wt.

T: Total chlorophyll of leaf in mg/g dry wt.

P: pH of leaf extract and

R: Relative water content of leaf extract (in per cent).

Results and Discussion

Plant species differ in their sensitivity levels to pollutants. A knowledge of tolerance level of a plants species to air pollution will help in abatement of pollution and proper management of environment. The plants act as pollution sink and they replenish the atmosphere with oxygen. In this investigation some trees, shrubs, and climbers (Table 19.1) have been categorized as tolerant and sensitive plants on the basis of APTI.

To recognize the potentiality of plants for serving as pollution sink (Rich, 1970; Smith, 1974; Smith and Dochinger, 1975), one has to know the sensitivity levels of plant to pollution. The indices for different plant types should be considered separately because of their different sensitivity levels to air pollutants. It is also expected that a plant species known to be sensitive or tolerant in one geographical region may not be the same in other geographical areas, which also depends on external ecological factors and internal physiological factors. In the present study certain physiological parameters like pH of the leaf extract, ascorbic acid contents, total chlorophyll and relative water content of leaves were used to determine APTI. The APTI is useful to some extent in selecting the species of plant for growing in urban, industrial areas where air pollution is predominant. The study gives the APTI for the plants growing at Hyderabad in pollution free environment. Table 19.1 also gives the information on total chlorophyll, ascorbic acid, relative water content and pH of the leaf extract used in compilation of APTI values for different species of trees (*Albizzia lebbeck, Annona squamosa, Azadirachta indica, Cassia fistula, Eugenia jambolana*), shrub (*Hibiscus rosans, Ixora coccina, Hemelia patens, Quisqualis indica, Asperagus racemoses*) and climbers (*Tinospora cordifolia, Pyrostegia grandusa, Adenocalymma ailiaceum, Thumbergia latifolia, Thumbergia alba*).

Table 19.1: Air Pollution Tolerance Index of Some Plants Growing at Hyderabad

Name of Plants	pH	Ascorbic Acid mg/g	Chlorophyll mg/g	Relative Water Content % (RWC)	Air Pollution Tolerance Index (APTI)
Trees					
Albizzia lebbeck	7.0	8.1	10.8	85.4	22.9
Annona squamosa	8.0	8.6	7.24	87.1	21.8
Azadirachta indica	6.8	4.8	12.3	38.8	13.0
Cassia fistula	8.5	3.5	11.6	45.6	11.5
Eugenia jambolana	7.0	2.7	8.3	72.8	11.4
Shrubs					
Hibiscus rosans	7.8	5.8	15.2	64.0	19.7
Ixora coccina	6.3	4.2	12.6	46.0	19.6
Hemelia patens	7.5	2.9	17.3	27.2	9.9
Quisqualis indica	6.5	2.4	13.8	37.3	8.6
Asperagus racemoses	7.7	3.3	8.6	25.9	7.9
Climbers					
Tinospora cordifolia	5.8	6.1	11.7	33.3	14.0
Pyrostegia grandusa	7.0	5.6	12.5	28.0	13.7
Adenocalymma ailiaceum	8.2	4.7	11.8	31.1	12.5
Thumbergia latifolia	5.5	4.5	11.9	41.1	11.9
Thumbergia alba	8.8	3.8	15.5	19.0	11.1

Conclusion

It is presumed from the index values that the plants whether they may be trees, shrubs, herbs or climbers, having higher index values *Albizzia lebbeck, Hibiscus rosans* and *Tinospora cordifolia* are highly tolerant to air pollution than the plants having lower index values *Eugenia jambolana, Asperagus racemoses* and *Thumbergia alba* are sensitive to air pollution.

References

Arnon, D.F. 1949. Copper enzymes in isolated chloroplasts, polyphenol oxidase in *Beta vulgaris. Pl. Physiol.*, 24: 1. In: *Proc. Symp. on Air pollution control* held at IIT, Delhi, November 1983, p. 176–180.

Rich, S. 1970. Effects of trees and forest in reducing air pollution. In: *Trees and Forests in an Urbanizing Environment*. Planning and Resource Developing Series No.17. Amberset, Holds Worth, Natural Resources Center, p. 29–33.

Smith, W.H. 1974. Air pollution effects on the structure and function of the temperate forest ecosystem. *Environ. Pollut.*, 6: 111–129.

Smith, W.H. and Dochinger, C.S. 1975. Capability of metropolitan trees to reduce atmospheric contaminants. In: *Better Trees for Metropolitan Landscapes*, (Eds.) Santamour, F.S. Jr., H.D. Grehold and S. Little. U.S. Forest Serv. Gen. Tech. Rep., NE 22: 49–50.

Singh, A. 1977. *Practical Plant Physiology*. Kalayani Publ., New Delhi, pp. 266.

Singh, S.K. and D.N. Rao. 1983. Evaluation of plants for their tolerance to air pollution. In: *Proc. Symp. On Air pollution control* held at IIT, Delhi. November, p. 218–224.

Chapter 20

Genetic Diversity Studies in Introgressed Lines of *Gossypium hirsutum* Cotton Using Cluster Analysis

J.S.V. Samba Murthy and N. Chamundeswari

Regional Agricultural Research Station,
Acharya N.G. Ranga Agricultural University, Lam, Guntur

ABSTRACT

Eighty two lines of *Gossypium hirsutum* along with two local checks developed out of introgression involving different wild species for tolerance to biotic and abiotic stresses and improvement in fibre quality at different Cotton Research Centres across the country were evaluated at Regional Agricultural Research Station, Lam during Kharif 2003-'04 to elucidate genetic divergence using a non-hierarchical Euclidean cluster analysis for yield and its components. The genotypes were grouped into ten clusters irrespective of geographic and genetic diversity. Cluster IX contained the largest number of sixteen genotypes and had the genotypes of heterogeneous origin, which showed that there was no parallelism between genetic and geographic diversity. The maximum genetic distance occurred between cluster III and VI. Cluster VI is monogenotypic and had high mean values for seed cotton yield, number of sympodia, boll weight, seed index, lint index, ginning out turn, 2.5 per cent span length and maturity coefficient. It is further suggested to go for a series of diallel analysis with the genotypes grouped in the above clusters for identifying superior heterotic combinations and isolating desirable recombinants in segregating generations. Genotypes were also identified which may serve as potent genetic donors for some metric and quality traits.

Introduction

Reduction in genetic variability of cultivated cotton makes it highly vulnerable to biotic as well as abiotic stresses. Besides above, changing national and international textile scenario, availability of quality cotton with better fibre qualities plays a major role. So genetic variability was created by introgression breeding utilizing wild sources of cotton. Mahalanobis D2 statistic designed by Rao (1952) has been utilized by a number of workers for estimating genetic divergence. Although D2 statistic has been widely used as a quantitative measure of genetic divergence, yet the clustering pattern of the genotypes is orbitrary (Spark, 1973). In the present study, eighty-two introgressed lines of *Gossypium hirsutum* cotton were subjected to non-hierarchical Euclidean cluster analysis to over come the limitations of D^2 statistics. Information on the nature and magnitude of genetic diversity present in superior parents for developing cotton varieties with multiple pest and disease tolerance besides arriving for fibre quality improvement.

Materials and Methods

The experimental material comprised of eighty-two variable introgressed lines along with two local checks was laid out in a randomized block design with two replications during Kharif 2003-04 at Regional Agricultural Research Station, Lamfarm, Guntur. Each entry was sown in two rows of six meters length. The spacing adopted between rows was 120cm while within the row it was 60cm. Normal agronomic practices recommended to the region were followed. Observations were recorded on five randomly selected plants in each entry for seventeen characters *viz.*, days to 50 per cent flowering, number of monopodia, number of sympodia, number of bolls/ plant, boll weight (g), number of seeds/boll, seed index (g), lint index (g), ginning out turn (per cent), 2.5 per cent span length (mm), maturity coefficient, uniformity ratio (per cent), micronaire (10–6 g/inch), bundle strength (g/tex), fibre elongation (per cent), fibre quality index, seed cotton yield/plant (g). The genetic divergence was carried out using non-hierarchal Euclidean cluster analysis as described by Beale (1969) and further elaborated by Spark (1973).

Results and Discussion

The analysis of variance revealed highly significant differences among the genotypes for all the seventeen characters under study. On the basis of Non-Hierarchical Euclidean cluster analysis, eighty-four genotypes were grouped into ten clusters (Table 20.1), which revealed existence of genetic diversity. The cluster IX had the highest sixteen genotypes followed by fourteen in cluster IV, thirteen in cluster I, seven in cluster V and VII, four in cluster VIII, three in cluster X and cluster VI was a single genotypic cluster. The clustering pattern of genotypes revealed no parallelism between genetic and geographic diversity. These findings are in agreement with Murthy *et al.* (1995b), Gururajan and Manickam (2002) and Altaher and Singh (2003).

Cluster means (Table 20.2) showed appreciable differences for all the characters, particularly in cluster VI which had high mean values for days to 50 per cent flowering, number of sympodia, boll weight, seed index, lint index, ginning out turn, 2.5 per cent span length, maturity coefficient and seed cotton yield per plant. Cluster I had high mean value for fibre elongation. Cluster II had the highest mean value for number of seeds per boll. Cluster V had the highest mean value for number of bolls per plant and bundle strength. Cluster VII had high mean value for fibre quality index. Highest mean value of uniformity ratio was observed in cluster VIII. Cluster X had high mean value for micronaire and number of monopodia per plant.

Inter and intra cluster distances are presented in Table 20.3. The magnitude of intra cluster distance measures the extent of genetic diversity between the genotypes of same cluster. Inter cluster distance is measuring of genetic distance between two clusters. The intra cluster distance was maximum in cluster IV (238.94) followed by cluster VIII (236.81) and minimum of zero in cluster VI. The maximum inter cluster distance of 1115.59 was observed between cluster III and VI followed by cluster IV and VI (961.84).

Table 20.1: Distribution of Genotypes in Different Clusters by Tocher's Method

Sl.No.	*Cluster No.*	*No. of Genotypes*	*Genotypes*
1.	I	11	IH 63, TCH 1649, IS 376/412/57, IS 376/411/31/27, GISV 25/1699XGCot 10, NA1325 X Palmeri, IS 376/411/20/25, IS 376/411/35-II, IS 376/411/34-I, IS376/411/37-I
2	II	14	GCot16XGISV 16, AKH 2053, AKH8828, IS376/411/13-II, GCot16XGISV61, TCH 1696, Abadita X Surat 5, RAI 9, MSH Sp91, AKH 2053-I, Hisutum X Barbadense X Wild-I, Hisutum X Barbadense X Wild-II, Abhadita X Surat 2, Hisutum X Barbadense X Wild-III
3	III	13	GISV 23XGCot16, 8-6PROG (Self) X AKH 2053, Abadita X TCH-1, Abadita X Surat 9, CWROK 165 X Palmeri, L 603 X Palmeri, TCH 1599, IS376/411/34-II, IS376/411/14, Abhadita X AKH 2053, Abadita X TCH-2
4	IV	8	GISV185 X Surat dwarf, GISV 61, MSM, 321-2PI (Self) X AKH2053, Hirusutm X Raimondii, Hirsutum X Anamolum, Rai 1, TCH 1654
5	V	7	Gcot 10 X GISV162, 8-6 PDx55 High fibre strength, IS376/411/33/21, GISV 154, TCH 1692, DSFH-I, IS376/411/1/87
6	VI	1	IH 35
7	VII	7	GCotl0X GISV79, TCH 1648, MCU 5 X Palmeri, GISV 2, Rai 7B-I, TCH 1650, TCH 1651
8	VIII	4	MAH 345, IS376/411/39, Abadita X Suart 2, TCH 1695
9	IX	16	GISV 197 X GISV61, IS376/411/7/64-II, TCH1652, GISV142, Abadita X Surat IV-3, TCH1691, MSH Sp53, IS376/411/68, Abadita X Surat IV-5, Abadita X TCH 6, Surabhi (Check 1), L 604 (Check 2), TCH 1653, TCH 1693, Abadita X Surat 5, GISV 201
10	X	3	Rai 7B-2, AKH 2031, GCot 10X GISV 61

Table 20.2: Intra and Inter Cluster Distances Values for 84 Genotypes

Cluster	*I*	*II*	*III*	*IV*	*V*	*VI*	*VII*	*VIII*	*IX*	*X*
I	193.16	245.38	236.30	336.66	431.63	780.59	404.63	520.88	325.47	378.38
II		121.72	212.83	258.20	516.05	816.49	298.16	317.54	230.81	359.32
III			141.95	267.47	593.42	1115.59	422.46	607.61	332.80	512.39
IV				239.94	574.16	961.84	371.09	552.55	384.16	658.50
V					188.06	422.00	535.90	552.93	362.52	492.01
VI						0.00	658.71	544.62	635.47	738.10
VII							178.01	338.50	314.74	612.99
VIII								236.81	329.03	433.16
IX									195.20	365.72
X										183.00

The clustering pattern of genotypes in the present study has not been distinctly influence by th source and origin. It should be logical to attempt the crosses between the genotypes of clusters separat by large estimated inter cluster distances to obtain high heterotic combinations and recombinants segregating generations. It is further suggested that the genotypes which fall under cluster III and may be subjected to a series of diallel analysis to identify superior parents for future breeding programmes based on their per se performance together with general and specific combining ability effects.

Table 20.3: Cluster Means for Eleven Characters of Cotton Based on D² Values

Sl.No.	Source	I	II	III	IV	V	VI	VII	VIII	IX	X
1.	Day to 50% flowering	56.73	57.50	55.54	56.50	57.71	60.00	56.71	58.50	56.56	56.83
2.	No. of monopodia	1.81	1.88	1.71	1.84	1.64	2.00	1.96	1.30	1.82	2.03
3.	No. of sympodia	15.58	14.19	16.60	12.83	16.63	18.10	15.49	13.98	15.13	14.93
4.	No. of bolls/plant	32.85	18.90	24.97	27.14	40.31	34.10	21.83	14.98	24.07	18.27
5.	Boll weight (g)	3.15	3.78	3.07	3.46	3.48	4.57	3.48	4.40	3.55	3.86
6.	No. of seeds/boll	27.56	28.45	27.09	26.84	26.99	28.20	27.33	27.45	27.06	25.97
7.	Seed index (g)	8.07	7.75	6.76	7.84	8.18	10.60	9.93	9.90	7.68	7.83
8.	Lint index (g)	4.63	5.18	4.46	4.46	6.03	6.41	5.62	6.09	6.08	5.85
9.	Ginning out turn (%)	34.79	35.82	35.67	35.33	34.64	37.60	33.41	36.10	37.10	36.00
10.	2.5% span length (%)	25.69	25.58	24.97	24.60	25.31	29.03	27.63	27.42	25.79	24.10
11.	Maturity coefficient	0.63	0.62	0.63	0.64	0.62	0.69	0.62	0.63	0.63	0.62
12.	Uniformity ratio	47.60	48.75	49.34	49.20	50.13	45.10	45.91	50.63	49.27	50.13
13.	Micronaire (10^{-6} g/inch)	4.11	3.88	3.65	3.18	3.69	4.02	3.08	3.82	3.64	4.75
14.	Bundle strength (g/tex)	23.49	24.06	23.95	23.95	26.36	18.20	22.80	23.93	23.62	21.83
15.	Fibre elongation (%)	6.57	6.47	6.48	6.45	6.50	6.30	6.45	6.50	6.48	6.43
16.	Fibre quality index	298.85	312.64	313.28	330.59	349.10	263.67	360.63	336.38	318.97	241.34
17.	Seed cotton yield/plant (g)	86.88	58.99	63.57	79.50	131.30	145.44	66.01	59.74	73.91	70.76

Acknowledgement

The authors are highly thankful to the authorities of Cotton Technology Mission and Acharya N.G Ranga Agricultural University for providing necessary financial help and infra structure facilities to undertake these studies.

References

Altaher, A.F. and Singh, R.P. 2003. Yield component analysis in upland cotton (*Gossypium-hirsutum* L. *J. Ind. Soc. Cotton Improv.*, 28(3): 151–157.

Beale, E.M.L. 1969. Euclidean cluster analysis. *Bull. Int. Stat. Inst.*, 43: 92–94.

Gururajan, K.N. and Manickam. 2002. Genetic divergence in Egyptian cotton (*Gossypium barbadense* L.). *J. Ind. Soc. Cotton Improv.*, 27(2): 77–83.

Murthy, J.S.V.S., Reddy, D.M. and Reddy, K.H.G. 1995. Genetic divergence for lint characters in upland cotton (*Gossypium hirsutum* L.). *Ann. Agrl. Res.*, 16(3): 357–359.

Spark, D.N. 1973. Euclidean cluster analysis algorithm. *Appl. Stat.*, 22: 126–130.

Rao, C.R. 1952. *Advance Statistical Methods in Biometrical Research.* John Wiley and Sons, Inc., New York.

Chapter 21

Seed Borne Fungi of *Celastrus paniculatus*, Willd. and their Control

R.K. Aher, L.S. Gund**, A.A. Kulkarni*** and R.R. Saswade****

**Department of Botany, New Arts, Commerce and Science College, Parner, Dist. Ahmednagar, M.S., India*

***Himachal Indage Limited, V.P.O. Nagwain, Dist. Mandi, H.P. India*

****Department of Botany, Ahmednagar College, Dist. Ahmednagar, M.S., India*

*****Department of Botany, Jijamata College of Arts, Commerce and Science, Bhenda, Tal. Newasa, Dist. Ahmednagar, M.S. India*

ABSTRACT

Celastrus paniculatus seed used as oil source which is important in ayurvedic as well as folk medicine. Moulded seeds were used to find out seed borne fungi and effect of different fungicides on seed borne fungi of freshly collected seeds. Total seven species of fungi were recorded in blotter as well as PDA, Czapek's and Waksmans medium. Three fungicides used for seed treatment Bavistin, Dithane M–45 and Bordeaux mixture were highly effective in controlling the seed borne fungi. It is recommended that the seeds of *Celastrus paniculatus* may be treated with above fungicides for controlling seed borne pathogens.

Keywords: *Celastrus paniculatus, Mycoflora, Fungicide control.*

Introduction

Celastrus paniculatus, Willd. (Jyotishmatii) is one of the important in siddha system of medicine i disease as well as widely used in Ayurveda system of medicine as a crude drug. Seeds used as o source which is important ayurvedic as well as folk medicine. In order to obtain healthy and qualit seeds, these should be collected from mother tree and extracted, processed and stored in scientif manner. Seeds may be damaged while on the plant, during collection, in transportation and storage b various injurious agency namely biotic, physical and pathogenic or microbial causes affecting the seed quality. To solve this problem seed testing becomes essential for determination of purity and freedom from harmful microorganisms according to rules laid dawn by ISTA (Anon, 1966). Gupta, *et al.* (1975) tabulated rules for germination tests. The present study was conducted to find out seed borne fungi and effect of different fungicides on seed borne fungi of freshly collected seeds. This information would be useful to plant breeders and researchers.

Materials and Methods

Seeds of *Celastrus paniculatus* were selected for present investigation. Seeds included collection from different localities of Pune area in different months by different methods of collection *i.e.*, hand plucking from plants, fallen fruits or seeds from ground. Infected seeds were used for mycoflora analysis. The fungi that were present on the moulded grains were isolated using agar plate and blotter technique. The agar media used were Potato Dextrose Agar (PDA), Czapeck's agar, Waksman's medium etc. by varying the pH in order to isolate the maximum number of fungi. Seed washings and plating of untreated seeds were done on agar media and on blotters at the rate of seeds per plate. The plates were incubated at 28 ± 2° C for 5–8 days and observations of fungal colonies and identified up to genera referring to standard references like Thomson and Raper (1945), Ellis (1976) etc.

To determine the effect of different fungicides in controlling seed borne fungi, seed samples were dressed separately with fungicides namely Bavistin, Dithane M–45 and Bordeaux mixture @ 3 g. per kg. seed weight. After 30–32 hours of treatment treated seeds were incubated. Untreated seeds used as a control.

Results and Discussion

A total seven species of fungi were isolated from seeds of tested species (Table 21.1). Among these the *Aspergillus flavus* dominated the flora. *A. flavus* had more number of colonies 16. Another dominant genus was *Penicillium* sp. Other dematious fungus was found on the grains were *Rhizopus stolonifer* which turned the seeds completely dark. The phytopathogenic *Drechslera euphorbiae*, *Curvularia lunata* and *Fusarium oxysporum* was also present in considerable number. Microorganisms associated with grains cause extensive damage to them in various ways. The major once is decrease in viability, production of toxins inside the seeds. Tripathi (1974) isolated six species of fungi from moulded seeds. Purohit *et al.* (1998) made a survey of the moulds occurring on some forest tree species and reported four genera of fungi. Screening of literature revealed that seed mycoflora of *Albizia lebbek* (Mittal and Sharma, 1976), *Butea monosperma* (Purohit and Jamaluddin, 1993) and *Acacia catechu* (Bharadwaj and Paul, 1995) have been studied.

Table 21.1: Ectophytic Mycoflora or Moulded Seeds or *Celastrus paniculatus*

Fungi	*No. of Colonies*
Aspergillus flavus	16
Aspergillus niger	10
Curvularia lunata	04
Drechslera euphorbiae	06
Fusarium oxysporum	09
Penicillium sp.	13
Rhizopus stolonifer	12

Mean value of three replicates.

Amongst the seed treatment with three fungicides, seed dressing with Bavistin, Dithane M–45 and Bordeaux mixture controlling most of the fungi on seeds of tested species where these were failed to control *A. niger* along with other fungi. Bavistin and Dithane M–45 were found superior to Bordeaux mixture in controlling the fungi on seeds of tested sp. (Table 21.2). Bharadwaj and Paul (1995) reported best control of large number of seed borne fungi of some leguminous tree species by using fungicides. Purohit, *et al.* (1998) also reported best control of seed borne fungi of some forest tree species by using fungicides Bavistin, Dithane M–45, Thyridine, Fytolan and Ceresan @ 3 g. per kg. seed.

Table 21.2: Effect of Fungicidal Seed Dressing on Occurrence of Fungi with Seed Surface of Freshly Collected Seeds of *Celastrus paniculatus*

Fungi	*Bavistin**	*Dithane M–45**	*Bordeaux Mixture**
Aspergillus flavus	–	–	–
Aspergillus niger	–	+	+
Curvularia lunata	–	–	–
Drechslera euphorbiae	–	–	–
Fusarium oxysporum	–	–	+
Penicillium sp.	–	–	–
Rhizopus stolonifer	–	–	–

Mean value of three replicates.

+: Fungal growth uncontrolled; –: Fungal growth controlled; *: @ 3 g./kg. seed weight.

Acknowledgement

The authors are thankful to Dr. B.A. Patil, New Arts, Commerce and Science College, Ahmednagar for encouragement. The authors also grateful to Dr. N.M. Kadam, Principal, Parner College, for providing research facilities.

References

Anon. 1966. International rules for seed testing. *Proc. Int. Seed Test. Assoc.*, 31(1): 1–132.

Bhardwaj, L.N. and V.S. Paul. 1995. Studies on seed mycoflora of some forest tree species and their control. *J. Trop. For.*, 11(2): 41–43.

Ellis, M.B. 1976. *More Dematiaceous Hyphomycetes*. Commonwealth Mycological Institute, Kew, England, pp. 607.

Gupta, B.N., P.G. Pattanath, Adarsh Kumar, R.C. Thapliyal and A.S. Raturi. 1975. Rules for germination test of tree seeds for certification. *Indian Forester*, 101(6): 320–327.

Mittal, R.K. and M.R. Sharma. 1976. Seed mycoflora of *Albizia lebbek*, Benth. *Indian J. For.*, 5(2): 156–157.

Purohit, Mamta and Jamuluddin. 1993. Seed mycoflora of Palas during storage. *Seed Res.*, 21(2): 126–127.

Purohit, Mamta, Jamuluddin and G.P. Mishra. 1998. Studies on germination and seed borne fungi of some forest tree species and their control. *Indian Forester,* 124(5): 315–320.

Thomson, C. and C.B. Raper. 1945. *A Manual of Aspergilli.* The Willi. and Wilk. Co.

Tripathi, R.K. 1974. Head fungi of sorghum. *Indian Phytopathol.*, 27: 499–501.

Chapter 22

Effects of Neem Leaf Extract on Growth Characteristics and Organic Constituents of *Trigonella foenum-graceum*, L.

S.L. Khapke, R.K. Aher, D.A. Bandal, G.V. Vyavahare, Y.B. Agre, D.B. Shelake and Miss R.K. Shaikh

Department of Botany, New Arts, Commerce and Science College, Parner, Dist. Ahmednagar, M.S., India

ABSTRACT

Trigonella foenum-graceum, L. plant has been valued as a medicinal in many parts of the world. The ayurvedic pharmacopoeia has ascribed many therapeutic uses to it. It has wide spread use as a therapeutic in chronic diseases. The leaves of the plant are commonly used as a vegetable in India. Morphological characteristics and biochemical investigation was conducted to study effects of neem leaf extract on *Trigonella*. Four per cent of leaf extract of neem was prepared as stock solution. Growth characteristics such as root length, shoot length and root nodules increased at 0.8 per cent concentration of leaf extract. There was significant increase in the organic constituents such as chlorophyll and proteins over control at 0.6 and 0.8 per cent concentrations.

Keywords: *Trigonella, Neem leaf extract, Morphological and biochemical changes.*

Introduction

Methi (*Trigonella foenum-graceum* L.) is an important food crop as well as a quick cash crop all over India especially in Maharashtra. The duration of crop varies from 25 to 40 days. It is grown for it's green leaves and young pods. The leaves are quite rich in protein (4.4 g), minerals (1.5 g), vitamin A (6450 IU) and vitamin C (54 mg) per 100 g of edible portion. Seeds are used in condiments. The whole plant is medicinally too much important.

The use of neem leaves (*Azadirachta indica* Juss) in pest control has been known since ancient time. But recently it has been studied scientifically and reported that it contains different chemicals which have insect repellant, insecticide antifeedant, nematicide, and antimicrobial properties. The chemical ingredients reported are azadirachtin, meliantrol and salanin.

Fujii *et al.* (2002) while taking the review of recent work in allelopathy explained that allelopathy now refers to any process involving secondary metabolites produced by plants, viruses, microorganisms and fungi that influences the growth and development of agricultural and biological systems. According to Bhat and Chauhan (2000) and Singh and Narsing Rao (2003) the chemicals present in the extract (aqueous or alcoholic) have stimulatory or inhibitory influence on seed germination, seedling growth, yield and physiology of subsequent crops.

Materials and Methods

Trigonella seeds were obtained from market. The earthen pots (20 × 30 × 40 cm) were used for growing the *Trigonella* seeds. The pots were filled with mixture containing garden soil and well decomposed compost (3 : 1). About 500 seeds of uniform size from cultivar were selected for the experiments. The seeds were surface sterilized with 0.1 per cent $HgCl_2$, washed thoroughly 3–4 times in sterilized distilled water and then soaked in distilled water for 12 hours. 25 well imbibed seeds were sown in each earthen pot. The seedlings were grown under natural and uniform conditions.

For the preparation of extract freshly mature green leaves of neem are used. About 4 g leaflets of neem were uniformly homogenized in 100 ml of 80 per cent ethanol and condensed on rotary evaporator to 10 ml at temperature below 60°C to avoid evaporation of volatile compound and denaturation of bioactive chemicals. This was filtered and the filtrate was made to 100 ml with sterile distilled water. This extract was diluted with distilled water to required concentrations and used for foliar spraying.

The experimental plants of *Trigonella* grown in pots were treated with freshly prepared alcoholic extract at the concentrations of 0.2 per cent, 0.4 per cent, 0.6 per cent, 0.8 per cent, 1.0 per cent. Spraying was usually done in the morning.

The controlled and treated seedlings were then used for further biochemical analysis. Chlorophylls were extracted and estimated by Arnon's (1949) method. The leaf proteins were estimated by using Lowry *et al.* (1951) method.

Results and Discussion

Results recorded in Table 22.1, indicate that the concentration from 0.6 per cent to 1.0 per cent of neem leaf extract has caused significant increase in root as well as shoot length in *Trigonella* seedlings over control. The maximum root length and shoot length were recorded in 0.8 per cent concentration (16.2±1, 25.1±1) over control.

Similar trend was observed in case of total number of leaves per plant. There is increase in total leaves found in 0.8 per cent and 1.0 per cent concentration. Highest number of leaves are found in 1.0 per cent concentrations pot (21 leaves).

Also the fresh weight and Dry weight was goes on increasing simultaneously in increasing concentration of extract over control. Maximum fresh weight and dry weight will be noted in 1.0 per cent concentrations seedlings (2.3 g, 0.4 mg).

As the nodulation is the part of productivity in legumes, the effect of neem leaf extract on nodulation is also studied. There is also a promotion in nodulation in treated plants over control. The maximum number of nodules are found in 0.8 per cent concentration plants.

Table 22.1: Effect of Neem Leaf Extract on Seedling Root Length, Shoot Length, Total Number of Leaves, Fresh Weight and Dry Weight and Root Nodules

Growth Parameters Per Plant ± SD	*Days from Spraying*	*Concentration of Neem Leaf Extract*					
		C	*0.2%*	*0.4%*	*0.6%*	*0.8%*	*1.0%*
Root length, cm	9	7.5±1	7.5±1	7.4±1	7.5±1	8±1	7.8±1
	16	10.5±1	14±1	12.2±1	10.9±1	13.9±1	14.1±1
	23	12.7±1	12.7±1	14.3±1	14.1±1	16.2±1	15.7±1
Shoot length, cm	9	10.5±1	10.4±1	10.4±1	10.7±1	10.8±1	10.7±1
	16	15.3±1	18.1±1	16.7±1	17.8±1	19.5±1	18.1±1
	23	20.1±1	21.3±1	21.5±1	21.6±1	25.1±1	22.2±1
Total no. of leaves	9	6	6	6	6	6	6
	16	12	12	12	15	15	15
	23	15	18	18	18	21	21
Fresh weight gm	9	0.7	0.7	0.8	0.7	0.9	0.8
	16	1.2	1.5	1.5	1.6	1.7	1.8
	23	1.6	1.8	2.1	2.3	2.3	2.4
Dry weight gm	9	0.08	0.08	0.09	0.08	0.1	0.1
	16	0.1	0.1	0.1	0.1	0.2	0.2
	23	0.2	0.3	0.3	0.3	0.4	0.4
Root nodule	9	8	9	10	9	11	10
	16	10	11	12	12	13	12
	23	11	13	13	14	15	14

Average of three determinations.

Sukul and Chaudhuri (1995 and 2001) reported increase in number of leaves per plant in rice due to treatment of leaf extracts of *Lantana camara*. Yamada *et al.* (2002) reported improved number of leaves per plant in various plant species due to alleochemicals present in *Lepidium sativum*.

Thakur *et al.* (1999) reported significant increase in plant height in bell pepper, when treated with bioextracts. Reddy and Rao (1995–96) also reported increased plant height in *Cajanus cajan* at lower concentrations, when treated with aqueous extracts of different plants. Sukul and Chaudhuri (1995 and 2001) noted increase plant height in rice when treated with the leaf extracts of *Lantana*.

The results of photosynthetic pigment indicated that the cultivar showed variation in their total chlorophyll contents as a effect of foliar spray. In general the increase in concentration showed the increase in total chlorophyll contents upto the 0.8 per cent concentration than control. But at 1.0 per cent concentration there is decrease in total chlorophyll content than 0.8 per cent concentration (Table 22.2). Bansal *et al.* (1976) reported enhanced chlorophyll contents in different plants like *Boerravia, Cenchrus, Convolvulus* and *Sesamum* due to allelopathic treatments. Rizvi *et al.* (1989) also recorded increased chlorophyll contents in maize due to treatment of nicotine.

Tripathi *et al.* (1999) reported that root extracts of *Tectona grandis* applied through soil stimulated the synthesis of chlorophyll a and chlorophyll b in soybean. Venkata Raju (1999) reported that low concentrations of aqueous extracts of root and shoot of *Plumbago zeylanica* stimulated the chlorophyll contents in *Amaranthus.* Bhalerao *et al.* (2003) observed same trend in strawberry with foliar application of *Tectaria* frond extract.

Table 22.2: Effect of Neem Leaf Extract on Increase in Total Chlorophyll Content and Total Protein Content in *Trigonella* in mg/g

Treatment	*Control*	*0.2%*	*0.4%*	*0.6%*	*0.8%*	*1.0%*
Chlorophyll	0.83	0.86	0.88	0.90	0.96	0.93
Protein	0.037	0.045	0.055	0.061	0.072	0.075

Average of three determinations.

The results regarding protein contents in *Trigonella* leaves as influenced by different concentrations of neem leaf extract recorded in Table 22.2. There is increase in the protein content over the control in different concentration and protein content goes on increasing simultaneously over control. The highest protein content should be found in 1.0 per cent concentration (0.075 mg/g).

Tripathi *et al.* (2000) studied the effects of leaf extracts of *Dalbergia sissoo* and *Eucalyptus* leaf litter or *Vigna radiata.* He noted decrease in protein content. However, Padhy *et al.* (2000) reported increase in protein content in 12 days old seedlings of finger millet, when treated with *Eucalyptus* leaf leachates. Similarly Bhalerao *et al.* (2001) noted increase in protein content in Mentha leaves when treated with *Pteridium* and *Aspidium* leaf leacaates. Bhalerao *et al.* (2003) also reported increase protein content in strawberry treated with *Tectaria* frond extract.

Sonali Bhargava and Ashok K. Bhargava (2004) observed increase in root nodules in *Cicer arietinum* treated with neem oil.

Acknowledgement

Authors are thankful to Principal of New Arts, Commerce and Science College Parner. Also authors are thankful to Department of Botany, New Arts, Commerce and Science College, Parner.

References

Bendre and Kumar. 1994–95. *Economic Botany.* p. 52, 53.

Bhalerao, E.B., S.L. Laware, R.R. Vaidya and K.N. Dhumal. 2001. Influence of leaf leachates of *Pteridium* and *Aspidium* in physiology of Mehtha. *J. Med. and Arom. Plant Sci.*, 22(4A) & 23(IA): 502–504.

Bhargava, Sonali and Ashok K. Bhargava. 2004. Effects of Neem oil on Growth characteristics of *Cicer arietinum* and control of pod borer. *Ad. Plant Sci.*, 17(11): 713–715.

Bhatt, B.P. and D.S. Chauhan. 2000. Allelopathic effects of *Quercus* spp. on crops of Garhwal Himalaya. *Allelopathy J.*, 7: 265–273.

Fujii, Y., S. Hirdate and H. Hrya. 2002. In: *Abstr. 3rd World. Cong on Allelopathy: Challenge for the New Millenium*, (Ed.) Tsukuba, Japan.

Kokate, C.K., A.P. Purohit and S.B. Ghokhale. 1999. *Pharmacognocy.* p. 250, 498.

Lowry, O.H., N.J. Rosebrrough, A.L. Farr and R.J. Randall. 1951. Protein measurement with Folin phenol reagent. *Biochemistry*, 15: 529–536.

Nandal, D.P.S., S.S. Bisla and S.S. Narwal. 1992. Allelopathic influence of Eucalyptus and *poplar* leaf extract on the germination and seedlings growth of winter vegetables. In: *Proc. First National Symposium on Allelopathy in Agroecosystems: Agriculture and Forestry*, (Eds.) P. Tauro and S.S. Narwal. Indian Society of Allelopathy, Hisar, India, p. 98–100.

Narswal, S.S., R. Pillaniraj, S.C. Sati, H.S. Kadian and D.S. Dahiya. 2003. Allelopathic plants: 8. *Parthenium hysterophours* L. *Allelopathy J.*, 11(2): 151–170.

Padhy, B., P.K. Patnaik and A.K. Tripathy. 2000. Allelopathic potential of Eucalyptus leaf litter leachates on germination and seedling growth of Finger millet. *Allelopathy J.*, 7(1): 69–78.

Panda, H. *Medicinal Plants: Cultivation and their Uses.* p. 201–213, 401–402.

Pandey, B.P. 1995. *Economic Botany*. p. 63, 306,307.

Singh, Deena and Y.B. Narsing Rao. 2003. Allelopathic evaluation of *Andrographis paniculata* aqueous leachates on rice (*Oryza sativa* L.). *Allelopathy J.*, 11(1): 71–76.

Sukul, S. and S. Chudhuri. 2001. Effect of foliar application of leaf extract containing plant growth promoting activity from *Lantana camara* on growth and yield of rice. *Indian J. of Plant Physiol.*, 6(2): 127–130.

Thakur, Anju, P.S. Thakur, Surekha Attri and S.P. Kanaujia. 1999. Effect of bioextracts on growth and yield of bell pepper under water stressed condition. *Indian J. Hort.*, 56(1): 46–51.

Tripathi, S.A., D.C. Tripathi, Kori and S. Paroha. 2000. Effect of *Dalbergia sissoo* extracts, rhizobium and nitrogen on germination, growth and yield of *Vigna radiata*. *Allelopathy J.*, 7(2): 255–264.

Venkata Raju, R.R. 1999. Allelopathic effect of *Plumbago-zeylanica* L. on it's associated plants. In: *Abstracts, 2nd World Congress on Allelopathy: Critical Analysis and Future Prospects.* (Ed.) A.U. Malik, pp. 181.

Yamada, K., K. Miyamoto, N. Goto, H. Kato-Naguchi, S. Kosemura, S. Yamamura and K. Hesegawa. 2002. Chemical and biological analysis of a hovel allelopathic substance, lepidimoide. In: *Abstr. 3rd World Congress on Allelopathy*, p. 237.

Chapter 23

Study of Important Medicinal Plants from Parner Tahasil, Dist. Ahmednagar, India

S.K. Aher, V.S. Shinde, R.B. Ghogare, A.R. Thube, Y.P. Date, S.R. Paimode and A.R. Kale

Department of Botany, New Arts Commerce and Science College Parner, Dist. Ahmednagar – 414 302, M.S.

ABSTRACT

The present study deals with exploration of medicinally important plants from Parner Tahasil of Maharashtra. Parner is bordering the west part of Ahmednagar District. The Tahasil has hilly as well as plain area with its distinct flora. 170 medicinally important plants belonging from 58 families have been recorded with their botanical name, vernacular name, family and plant part used in the medicines. Family Asteraceae, Cucurbitaceae, Papilionaeae were the richest in medicinal plants.

Keywords: *Medicinal plants, Parner Tahasil.*

Introduction

Plants have provided the means of curing diseases perhaps since the evolution of mankind on the earth. The methods of herbal treatment have achieved eminence especially in the Central and South Asia, China and countries surrounding the Mediterranean sea. In India the oldest record of the use of plants as medicine is mentioned in the Rigveda.

Maharashtra is the rich repository of medicinal wealth, which occupies on important place in vedic literature. Rich diversity of medicinal plants occurs in the Maharashtra growing naturally under diverse environmental conditions.

Parner Tahasil has an area of 1,77,676 hectares of land abundant in natural resources including immense potential for medicinal plants. The Tahasil is situated at 800 m above sea level and average rainfall is 577 m. The Tahasil has hilly as well as plain area. No scientific and systematic inventory of available resources of medicinal plants to their quantitative, qualitative parameters and their regeneration under natural conditions are available. Though a some works (Aher *et al.*, 2004) have been carried our relating to the study of few medicinal plants of the area, no exhaustive work in this valuable aspect have been so far studied.

The present study was undertaken to enlist, identify and quantity the important medicinal plants of the Parner Tahasil and gather information regarding the plant part used as medicine.

Materials and Methods

The study was carried out by adopting the methodology of Jain (1989). Ten field visits, each lasting for 2–3 days and spread over eight months (from June 2004 to January 2005) were undertaken. Information regarding medicinal plants, vernacular name and their part used were collected from local people also. Voucher specimens were collected, dried and mounted following conventional methods and preserved. Plants were identified taxonomically by using flora of Bombay Presidency (Cooke, 1901–1908) and Flora of Marathwada (Naik, 1998).

Results and Discussion

In the present study a total 170 species of medicinal plants were collected which belong to 58 different families of plant kingdom. Of these 170 species, 12 species belong to family Cucurbitaceae, 11 species to family Papilionaceae, 9 species to Asteraceae, 8 species to Apiaceae, 7 species to Caesalpiniaceae and remaining to other families. In 49 plant species leaves are used as medicines, in 41 species roots are used as medicines, in 15 species roots, in 14 species flowers where as in 9 species whole plant body is medicinally important (Table 23.1). Auti *et al.* (2004) recorded 50 medicinally important weeds from Shrirampur Tahasil, Dist. Ahmednagar. The uses, chemical constituents and the ayurvedic preparations made from many plants have also been given by Kaushik (1988) and Kaushi and Dhiman (2000).

Table 23.1

Sl.No.	*Botanical Name*	*Ver. Name*	*Family*	*Plant Part Used*
1.	*Annona squamosa*, L.	Sitaphal	Annonaceae	Seed
2.	*Annona reticulata*, L.	Ramphal	Annonaceae	Seed
3.	*Polyalthia longifollia*, L.	Ashok	Annonaceae	Leaf
4.	*Brassica compestris*, L.	Mohri	Cruciferae	Seed
5.	*Rhaphanus sativus*, L.	Mula	Cruciferae	Root
6.	*Datura strumarium*, L.	Dhotra	Solanaceae	Seed
7.	*Capsicum annum*, L.	Mirchi	Solanaceae	Seed
8.	*Nicotiana tobaccum*, L.	Tambakhu	Solanaceae	Leaf
9.	*Catharanthus roseus*, L.	Sadafuli	Apocynaceae	Flower
10.	*Nerium indicum*, L.	Kanher	Apocynaceae	Root
11.	*Eucalyptus globolus* L.	Nilgiri	Myrtaceae	Leaf

Contd...

Table 23.1–Contd...

Sl.No.	Botanical Name	Ver. Name	Family	Plant Part Used
12.	*Syzygium aromaticum*, L.	Lavang	Myrtaceae	Fruit
13.	*Cassia tora*, L.	Tarwad	Caesalpinaceae	Leaf
14.	*Syzygium cumini*, L.	Jambhul	Myrtaceae	Root, Seed
15.	*Tamarindus indica*, L.	Chinch	Caesalpinaceae	Fruit
16.	*Baulinia racemosa*, L.	Aapata	Caesalpinaceae	Leaf
17.	*Delonix resia*, L.	Gulmohar	Caesalpinaceae	Flower
18.	*Ocimum sanctum*, L.	Tulas	Labiatae	Whole plant
19.	*Ocimum bacilium*, L.	Subja	Labiatae	Whole plant
20.	*Mentha viridis*, L.	Pudina	Labiatae	Leaf
21.	*Accacia nilotica*, L.	Babhul	Mimoceae	Stem
22.	*Accacia catecheu*, L.	Kat	Mimoceae	Stem
23.	*Accacia concina*, L.	Shikakai	Mimoceae	Fruit
24.	*Citrus sinesis*, L.	Mosambi	Rutaceae	Fruit
25.	*Aegel marmalos*, Corr.	Bell	Rutaceae	Fruit
26.	*Citrus aurantifolia*, L.	Limbu	Rutaceae	Fruit
27.	*Citrus reticulata*, L.	Santri	Rutaceae	Fruit
28.	*Ferobia limoni*, L.	Kawath	Rutaceae	Fruit
29.	*Murraya konegii*, L.	Kadhipatta	Rutaceae	Leaf
30.	*Adhatoda vasica*, L.	Adulsa	Acanthaceae	Whole plant
31.	*Ricinus communis*, L.	Erand	Euphorbiaceae	Fruit
32.	*Phyllanthus emblica*, L.	Awala	Euphorbiaceae	Fruit
33.	*Barleria prionitis*, L.	Katekoranti	Acanthaceae	Fruit
34.	*Asparagus racemosa*, L.	Shatawari	Liliaceae	Leaf, Root
35.	*Allium sepa*, L.	Kanda	Liliaceae	Stem
36.	*Allium sativum*, L.	Lasun	Liliaceae	Stem
37.	*Aloe vera*, L.	Korpphad	Liliaceae	Leaf
38.	*Cocus nucifera*, L	Nariyal	Palmae	Fruit
39.	*Areca catecheu*, L.	Supari	Palmae	Fruit
40.	*Phoenix sylvestris*, L.	Sindhi	Palmae	Fruit
41.	*Sesbania sesban*, L.	Shivari	Papilionaceae	Leaf
42.	*Dalbergia sisoo*, Roxb.	Shisam	Leguminoceae	Leaf
43.	*Azadiracta indica*, L.	Kadunimb	Meliaceae	Whole plant
44.	*Lowsonia inermis*, L.	Mehandi	Lowsoniaceae	Leaf
45.	*Santalum album*, L.	Chandan	Santalaceae	Stem
46.	*Michelia champaka*, L.	Sonchafa	Magnoliaceae	Flower
47.	*Vitex nigunda*, L.	Nirgudi	Verbenaceae	Leaf
48.	*Eclipta alba*, L.	Maka	Asteraceae	Leaf

Contd...

Table 23.1–Contd...

Sl.No.	Botanical Name	Ver. Name	Family	Plant Part Used
49.	*Cynadon dactyolon*, L.	Durva	Poaceae	Whole plant
50.	*Calotropis procera*, R. Br.	Rui	Asclepidaceae	Leaf
51.	*Tridax prochumbers*, L.	Kurmudi	Asteraceae	Leaf
52.	*Lantana camara*, L.	Tantani	Verbenaceae	Leaf
53.	*Curcuma longa*, L.	Halad	Guttiferae	Stem
54.	*Arbus precatorius*, L.	Gunj	Papilionaceae	Leaf
55.	*Gyamopsis tetragonoloba*, L.	Gawar	Papilionaceae	Fruit
56.	*Trigonella foenum graeum*, L.	Methi	Papilionaceae	Seed
57.	*Ficus benegalensis*, L.	Wad	Urticaceae	Stem
58.	*Seasamum indicum*, L.	Til	Pedaliaceae	Seed
59.	*Coriandrum sativum*, L.	Kothimbir	Apiaceae	Seed
60.	*Helianthus anus*, L.	Sauryaphul	Asteraceae	Seed
61.	*Daucus carota*, L.	Gajar	Asteraceae	Root
62.	*Calocasia antiguorum*, L.	Alu	Araceae	Leaf
63.	*Melia azadirach*, L.	Bakan	Meliacesae	Leaf
64.	*Zingiber officinale*, L.	Aale	Meliacesae	Rhizome
65.	*Carus copticum*, L.	Owa	Apiaceae	Seed
66.	*Pongamia pinnate*, L.	Karan	Papilionaceae	Seed
67.	*Boerhavia repens*, L.	Punarnava	Nyctaginaceae	Seed
68.	*Prosopis julo flora*, L.	Wedi Babhul	Mimosae	Seed
69.	*Caesalpinia bondue*, L.	Sagargota	Leguminoceae	Fruit
70.	*Spirocia deraceae*, L.	Palak	Chenopodiaceae	Leaf
71.	*Pithecclobium duke*, L.	Vilayatichinch	Papavaraceae	Fruit
72.	*Terminalia catap*, L.	Badaam	Rosaceae	Seed
73.	*Curcuma aromaticum*, L.	Ranhalad	Scitaminae	Stem
74.	*Drimia indica*, Roxb.	Rankanda	Liliaceae	Tuber
75.	*Ficus religiosa*, L.	Pimpal	Uritaceae	Leaf
76.	*Ficus racemosa*, L.	Umbar	Uritaceae	Fruit
77.	*Guizotia abyssinica*, L.	Karle	Cucurbitaceae	Fruit
78.	*Coccinia grandis*, L.	Tondali	Apiaceae	Fruit
79.	*Bacoppa mannieri*, L.	Brahami	Umbelifevae	Leaf
80.	*Ocimum gratissimum*, L.	Ramtulas	Labiatae	Whole plant
81.	*Sapindus trifolious*, L.	Ritha	Sapindaceae	Seed
82.	*Foenicum vulgare*, L.	Badishep	Apiacae	Fruit
83.	*Panicum miliaceum*, L.	Rala	Apiacae	Seed
84.	*Hibiscus syriacus*, L.	Paidhari	Malvaceae	Whole plant

Contd...

Table 23.1–Contd...

Sl.No.	Botanical Name	Ver. Name	Family	Plant Part Used
85.	*Spilanthes paniculata*, L.	Akkalkara	Asteraceae	Flower
86.	*Terminalia belleria*, L.	Behda	Combretaceae	Leaf
87.	*Myristica beddomei*, L.	Jayphal	Myristicaceae	Fruit
88.	*Calotropis gigantea*, L.	Pandhari Rui	Asclepidaceae	Leaf
89.	*Benincasa hispida*, L.	Kohala	Cucurbitaceae	Fruit
90.	*Bryophyllum pinnatum*, L.	Panfuti	Crassulaceae	Leaf
91.	*Carica papaya*, L.	Papai	Caricaceae	Fruit
92.	*Cassia fistula*, L.	Bahava	Caesalpiniacae	Seed
93.	*Celosia argaentea*, L.	Kurdu	Amaranthaceae	Root
94.	*Clitoria ternatea*, L.	Gokarn	Papilionaceae	Leaf
95.	*Cleome gynandra*, L.	Tilwan Pandhari	Cleomaceae	Leaf
96.	*Jasminum auriculatum*, Vahl.	Jui	Oleaceae	Flower
97.	*Jasminum officinale*, Vahl.	Jai	Oleaceae	Flower
98.	*Mimosa pudica*, L.	Lajalu	Mimosae	Leaf
99.	*Nyctanthus arbortristis*, L.	Prajakta	Oleaceae	Flower
100.	*Passiflora foetida*, L.	Krishna kamal	Passifloraceae	Flower
101	*Portulacca oleracea*, L.	Ghol	Portulaccaceae	Leaf
102.	*Sida cordifolia*, L.	Chikana	Malvaceae	Leaf
103.	*Tejetus erecta*, L.	Zendu	Asteraceae	Leaf
104.	*Terminalia arjuna*, Bedd.	Arjun	Combrataceae	Stem
105.	*Zizyphus jujuba*, Lamk	Bor	Rhamnaceae	Seed
106.	*Tinospora cardifloia*, Miers	Gulvel	Menispermaceae	Leaf
107.	*Solanum melanogena*, L.	Wangi	Solanaceae	Fruit
108.	*Butea monosperma*, L.	Palas	Papilionceae	Flower
109.	*Solanum tubersum*, L.	Batata	Solanaceae	Tuber
110.	*Solanum virginianum*, L.	Bhuiringani	Asteraceae	Leaf
111.	*Regia repens, L.*	Ghatipitipapada	Acanthaceae	Leaf
112.	*Adiantum philippenes*, L.	Adiantum	Adiantaceae	Leaf
113.	*Achyranthese aspera*, L.	Aaghada	Amaranthaceae	Root
114.	*Amaranthus spinosus*, L.	Katemath	Amaranthaceae	Root
115.	*Mangifera indica*, L.	Amba	Anacardiaceae	Whole plant
116.	*Carum curvi*, L.	Shahajire	Apiaceae	Fruit
117.	*Cuminum cyminum*, L.	Jire	Apiaceae	Fruit
118.	*Tabernaemontana coronaria*, R.Br.	Tagar	Apocynaceae	Seed
119.	*Thevetia peruviana*, Pers.	Pivalakanher	Apocynaceae	Root
120.	*Borassus flabelliafer*, L.	Tad	Arecaceae	Fruit

Contd...

able 23.1–Contd...

Sl.No.	Botanical Name	Ver. Name	Family	Plant Part Used
121.	*Gymnema sylvestre*, R.Br.	Kawali	Asclepidaceae	Root
122.	*Cartnamus tinctorious*, L.	Karadi	Asteraceae	Fruit
123.	*Echinops echinatus*, Roxb.	Utkatar	Asteraceae	Root
124.	*Cordia dichotama*, F.	Bhokar	Caesalpinaceae	Leaf
125.	*Cassia sophera*, L.	Rantakalka	Caesalpinaceae	Bark
126.	*Chenopodium album*, L.	Chandanbatva	Chenopodiaceae	Leaf
127.	*Cleome viscosa*, L.	Ran-mohari	Cleomaceae	Seed
128.	*Quispualis densiflora*, Wall.	Lal-chemeli	Combrataceae	Leaf
129.	*Evolvululous hirsutus*, Lamk.	Shankpushpi	Convolvulaceae	Leaf
130.	*Benincasa cerifera*, Savi.	Khokali	Cucurbitaceae	Seed
131.	*Cucumis colocynthis*, L.	Kadu-wrundavan	Cucurbitaceae	Fruit
132.	*Cucumis melo*, L.	Kharbuj	Cucurbitaceae	Fruit
133.	*Cucumis sativus*, L.	Kakadi	Cucurbitaceae	Fruit
134.	*Cucurbita pepo*, L.	Dangar	Cucurbitaceae	Fruit
135.	*Piplocyclos palmatus*, Jeff.	Shivalingi	Cucurbitaceae	Leaf
136.	*Lagenaria siceraria*, L.	Kadubhopala	Cucurbitaceae	Root
137.	*Luffa cylindrica*, Roxb.	Dodaka	Cucurbitaceae	Leaf
138.	*Cyperus scariosus*, R.Br.	Lawhale	Cyperaceae	Tuber
139.	*Diospyros peregrina*, L.	Temburni	Ebenaceae	Bark
140.	*Arachis hypogea*, L.	Bhuimoong	Papilionaceae	Root
141.	*Cajanus cajan*, Linn.	Tur	Papilionaceae	Seed
142.	*Cicer aurientum*, L.	Harbara	Papilionaceae	Seed
143.	*Erythrina indica*, Lamk.	Pangara	Papilionaceae	Bark
144.	*Albemoschus esculentus*, L.	Bhendi	Malvaceae	Fruit
145.	*Abutilon indicum*, L.	Mudraa	Malvaceae	Stem
146.	*Gossypium arboreum*, L.	Kapasi	Malvaceae	Leaf, Seed
147.	*Cocculus cordifolius*, DC.	Amrutvel	Menispermaceae	Stem
148.	*Musa paradisiaca*, L.	Kel	Musaceae	Whole plant
149.	*Psidium guavajava*, L.	Mogara	Myrtaceae	Fruit
150.	*Jasminum sambac*, Ait.	Peru	Oleaceae	Flower
151.	*Cymbopogon citatus*, DC.	Gavati chaha	Poaceae	Whole plant
152.	*Sorghum bicolor*, L.	Lal-jowari	Poaceae	Seed
153.	*Clematis triloba*, Heyle ex. Roth	Ranjani	Rananculaceae	Root
154.	*Rosa damasana*, Mill.	Gulab	Rosaceae	Flower
155.	*Cyperus rotundus*, L.	Nagarmotha	Cyperaceae	Leaf
156.	*Tribulus terrestris*, L.	Sarata	Zygophyllaceae	Leaf

Contd...

Table 23.1–Contd...

Sl.No.	Botanical Name	Ver. Name	Family	Plant Part Used
157.	*Peucedanum gareolens*, L.	Balant Shepu	Apiaceae	Fruit
158.	*Hibiscus syriacus*, L.	Pandharijaswand	Malavaceae	Flower
159.	*Periploca aphylla*, L.	Ransher	Asclepidaceae	Stem
160.	*Citrulus colocynthi*, L.	Indrayan	Cucurbitaceae	Fruit
161.	*Vitiveria zizanioides*, L.	Wala	Poaceae	Flower
162.	*Boerrhavia diffusa*, L.	Getuli	Plantangenaceae	Root
163.	*Frerea indica*, L.	Makadi	Asclepidaceae	Stem
164.	*Solanum incanum*, L.	Karanda	Malvaceae	Leaf
165.	*Evolvulus alsinoides*, L.	Vishnukranta	Convolvulaceae	Leaf
166.	*Bacopa mannieri*, L.	Bramhi	Scrophulariaceae	Leaf
167.	*Albemoschus ficulmeus*, L.	Ranbhendi	Malvaceae	Fruit
168.	*Girardinia diversifolia*, L.	Agaya	Urticaceae	Root
169.	*Dalbergia latifolia*, L.	Beetee	Leguminoceae	Fruit
170.	*Tephrosia humiltonii*, L.	Unhali	Leguminoceae	Leaf

From the present study it indirectly becomes evident that:

1. Cucurbitaceae, Papilionaceae and Asteraceae families are rich in medicinal plants.
2. The density of those plants where roots or tubers are used as medicine is low.
3. Lack of scientific knowledge, has resulted in severe depletion of herbs and shrubs and severe biotic pressure of humans and grazing animals is detrimental to the regeneration of these plants.
4. Scientific studies may elaborate the prospects of growing more and more medicinal plants successively.

References

Aher, R.K., Auti, B.K., Deshmukh, R.N. and Borkar, G.B. 2004. Survey of medicinal plants from Parne (Ahmednagar District.). *Ad. Plant Sci.*, 17(1): 57–59.

Auti, B.K., Pingle, S.D., and Aher, R.K. 2004. Survey of weeds and their medicinal value from Shrirampur Tahasil (Ahmednagar District M.S.). *Ad. Plant Sci.*, 17(2): 395–401.

Cooke, T. 1901–1908. *Flora of Presidency of Bombay*, Vol. I–III. Adlard and Sons, Ltd., London.

Jain, S.K. (Ed.) 1989. *Methods and Approaches in Ethanobotany*. Society of Ethanobotanists, CDRI, Lucknow

Kaushik, P. 1988. *Indigenous Medicinal Plants Including Microbes and Fungi*. Today and Tomorrow's Printers and Publishers, New Delhi.

Kaushik, P. and Dhiman, A.K. 2000. *Medicinal Plants and Raw Drugs of India*. Bishan Singh Mahendra Pal Singh, Dehra Dun.

Naik, V.N. 1998. *Flora of Marathwada*, Vol. I–II. Amrut Prakashan, Aurangabad.

Chapter 24

Phytochemical Studies of Ethnobotanically Significant Plant, *Evolvulus alsinoides*, Linn. from the Tribal Region of Deoghar District of Santal Paraganas

B.N. Jha* and S.K. Chaudhury**

**Reader, P.G. Department of Botany, S.P. College, Dumka*
***Lecturer in Chemistry, S.S.P.Y College, Katoria, Banka*

ABSTRACT

In the present investigation phytochemical studies of *Evolvulus alsinoides* has been done. It is an important medicinal herb which is used for the treatment of diabetes in the tribal regions of Santal Paraganas. Besides diabetes, the plant is very much used by the tribal people for the treatment of syphilis, leucorrea, bronchitis, asthma and liver ailments. Perusal of the data of Table 24.1, it is clear that out of 24 amino acids only seven could be detected from the plant under study. They were lysine, L-valine, L-argenine, cystein, tyrosine and DL-tryptophan. In the above list it is clear that at least 4 essential amino acids are present. They were L-argentine, lysine, L-valine and DL-tryptophan. 10 essential amino acids are necessary for proper human growth and their metabolic functions in which 4 are present in this plant and rest three are non-essential. Non-essential amino acids have curative effect on a diseased persons (Alstead and Gird Wood, 1974). Perhaps due to presence of these non-essential amino acids this plant is ethnobotanically important.

Keywords: Amino acids, Diabetes, Syphilis, Leucorrea, Liver ailments etc.

Introduction

Santal Paraganas forests are rich in indigenous plants and they are still intact due to the dependence of tribal people on herbal medicines for the treatment of various diseases. For the present investigation, *Evolvulus alsinoides* Linn. was selected for study. The plant was collected from Inarabara forest of Deoghar district in the month of October–November. *Evolvulus alsinoides* is a member c convulvulaceae. The plant is rampantly used by the tribal people of Santal Paraganas for treatment c diabetes. Besides diabetes, the plants are very much used for the treatment of syphilis, leucorrhea bronchitis, asthma and liver ailments by the local Vaidyas.

Materials and Methods

The plants were collected from different places of Inarabaran forest in the month of October November. Simple chromatographic techniques were employed for detection of free and bound amin acids present in the plants following the methodology suggested by Ranjan and Loraya (1978). Th solvent used in the investigation was n-butanol, acetic acid and water in the proportion of 4 : 1 : 5. Th entire arrangement of apparatus was kept in close chamber for ascending fluid for 18 hours at roor temperature. The ascended region of solvent was marked with pencil and dried in open air. Th solution of 0.1 per cent (m/v) ninhydrin in n-butanol was sprayed on the chromatography paper an the paper was kept in oven at 80°C for 15 minutes. Amino acids of different colour were developec The Rf values were worked out on the basis of following formula:

$$Rf = \frac{\text{Distance travelled by solute}}{\text{Distance travelled by solvent}}$$

At a fixed temperature.

The various amino acids were identified by comparing coloration of the spots and also by R values of the known amino acids from the index. A set of 24 amino acids by 'Labo' of U.K. was used fo index purpose. On the basis of visual observation concentration of different amino acids were noted i the table as 4+, 3+, 2+, + and +tr(trace). To avoid any error in the analysis of three test plants (T_1, T_2 an T_3) were taken into consideration. These *Evolvulus alsinoides* were collected from three different place of the forest of Inarabaran in the month of October–November.

Observations

Evolvulus alsinoides is a plant of the family convulvulaceae. It is a branched, prostrate on th ground, 4 to 15" long with thinly or densely silky leaves. Leaves are broadly elliptical –0.25 to 0.5" Flowers are sub rotate up to 0.2" long. One tea spoon dried flowers and leaves is given to diabeti patients in the morning for lowering blood sugar level. One tribal patient of the Inarabaran localit told us that in one week only, this herbal medicine lowered my sugar level from 390 to 152 pp.

The data of amino acids analysis of *E. alsinoides* is given in Table 24.1. Perusal of the data of Tabl 24.1 it is clear that out of 24 amino acids only seven could be detected from the plants under study They were Lysine, L-argenine, Cystein, L-valine, Tryosine and DL-tryptophan among which cystein L-valine and DL-tryptophan could be detected in higher concentration *i.e.*, (3+) while lysine, L-argenine glycine and DL-tryptophan were present at (+) or (2+) level on visual observation. Besides this, L argenine, cystein, glycine and DL-tryptophan were present in both the forms *i.e.*, free form and bounc form while the rest three were restricted to free form only.

Discussion

India is blessed with incredible biological diversity. Out of 25 Global "hot spots" of biologica diversity, two–the Eastern Himalayas and Western ghat, are located in India (Murali Manohar, 1996)

The traditional knowledge based on Indian ethnics (Vedas, Satpathbrahman etc.) and communities traditions are perhaps richest in the world. Indian tribal and their folklore particularly the traditional medicine system are rich and unique. India's tribal community has long been the custodian of her biodiversity and traditional medicines system. Their notions, varied traditions and primitive state of economic backwardness forced tribal people to depend upon traditional medicinal flora for the treatment of various diseases. Some of their traditional medicines are so much effective that they have been accepted by allopathic doctors.

Table 24.1: Showing Data of Amino Acids Present in the Leaf Extract of *E. alsinoides* Collected from Inarabaran of Deoghar District

Amino Acids	*Rf Value*	*Concentration on Visual Observation*							
		Free Form			*Mean*	*Bound Form*			*Mean*
		T_1	T_2	T_3		T_1	T_2	T_3	
Lysine	0.04	+	+	+	+	–	–	–	–
L–histidine	0.06	–	–	–	–	–	–	–	–
Aspartic acid	0.09	–	–	–	–	–	–	–	–
L–argenine	0.12	+	+	+	+	+	+	+	+
Cysteine	0.16	3+	2+	3+	3+	+	–	–	–
DL–Aspargine	0.18	–	–	–	–	–	–	–	–
Proline	0.21	–	–	–	–	–	–	–	–
L–glutamine	0.23	–	–	–	–	–	–	–	–
Serine	0.24	–	–	–	–	–	–	–	–
Hydroxy–proline	0.26	–	–	–	–	–	–	–	–
DL–threonine	0.28	–	–	–	–	–	–	–	–
Glycine	0.31	2+	2+	2+	2+	+	+	–	+
L–glutamic acid	0.33	–	–	–	–	–	–	–	–
Ornithine	0.34	–	–	–	–	–	–	–	–
Dihydroxy–Phenylamine	0.37	–	–	–	–	–	–	–	–
DL–alanine	0.45	–	–	–	–	–	–	–	–
DL–methonine	0.49	–	–	–	–	–	–	–	–
Amino–n–butyric acid	0.56	–	–	–	–	–	–	–	–
L–Valine	0.61	2+	3+	3+	3+	–	–	–	–
Tryosine	0.63	2+	2+	+	2+	–	–	–	–
Phenylalanine	0.67	–	–	–	–	–	–	–	–
DL–tryptophan	0.71	2+	3+	3+	3+	+	2+	2+	2+
Iso–leucine	0.76	–	–	–	–	–	–	–	–
L–leucine	0.80	–	–	–	–	–	–	–	–

The traditions of tribals had received an excellent help from a Bengali press about 125 years ago and several books and articles published. In 1887 the famous book "Horkoren Mare Hapramka Reak Katha" was published. An English writer Bodding translated the book in 1942 and with great insight he divided the book into several chapters. One of the chapter was "The studies in Santhal medicines and connected folklore". This chapter mentions about the medicinal herbs used by tribals for the treatment of various diseases.

In the present investigation biochemical analysis of tribal medicinal herb used for the treatment of diabetes has been undertaken. *Evolvulus alsinoides* is a very important indigenous plant of tribal community used for the treatment of diabetes. Besides diabetes, the juice of the whole plant is also used for the treatment of syphilis, leucorrhea, bronchitis, asthma and liver ailments. This plant is also known as Shymkranta (Hindi), Kusha motha (Santhali), Vishnuklanthe (Marathi), Bichhamalia (Oriya), Shankhpuspi (Sanskrit), Vishnukirandi (Tamil) and Vishnukranthe in Telugu.

In the past many biochemists, cytogenetists and pharmacologists were undertaken phytochemical analysis of crop plants mainly for better quality and higher yield. Mention may be made in this regards to the works of Boge (1960), Joshi and Srivastava (1976), Power and Savagaon (1981), Venkataraman and Chari (1984), Jha (1988), Voila *et al.* (1991) and Smith (1997). In the present investigation, analysis of amino acids has been incorporated on the fact that amino acids represent a valuable reserve material which can be utilized for direct incorporation into essential proteins, as a source of energy, nitrogen and a mediator in many physiological reactions (Linskens and Pfahler, 1979). This is the first scientific study of this plant of tribal folklore. In the earlier, other plants of tribal folklore were studied for amino acids and protein contents. In this regard, mention may be made to the works of Riley and Ewart (1970), Anwar and Reddy (1974), Mohammad and Raghubanshi (1981), Gupta *et al.* (1984) and Patridge (1991).

Perusal of the data of Table 24.1, it is clear that in this plant out of 10 essential amino acids 4 types were present. They were argentine, lysine, valine and tryptophan. Ten essential amino acids most important for human growth and proper metabolic functions (Malhotra, 1985). Perhaps, due to the presence of these four essential acids in this plant, the plant is useful as medicine in various diseases. Rest three amino acids are non-essential amino acids present in this plant, these were cystein, tyrosine and glycine. Non-essential amino acids specially cystein, glycine, glutamic acid, DL metheonine and ornithine have profound curative effect on diseases persons (Alstead and Wood, 1994). Perhaps, due to the presence of four essential and two non-essential amino acids in *Evolvulus alsinoides*, this plant is ethnobotanically important. Cystein is a non-essential sulphur containing amino acid and its presence in the diet decreases the metheonine percentage in the body (Alstead and Wood, 1994). The metabolic role of this amino acid is responsible for the growth of thick hairs, curative effect in liver ailments, in pancreas ailments and in tanned skin (Malhotra, 1985). Besides this it is a rich source in human diet (Alstead and Wood, 1994). Excess of cystein in human diet can also cause "amino aciduria" (Martin *et al.*, 1981). Sinha (1988) reported that the plants of Shankhpuspi (*Evolvulus alsinoides*) are boiled in two glasses of water and after cooling, the extract is taken by those patients who feel burning sensation while urinating.

References

Alstead, S. and Girdwood, R.H. 1974. *Textbook of Medical Treatment* (English Language Book Society). Chirchil Living Stone Building, Great Britain.

Anwar, S.Y. and Reddy, G.M. 1979. Studies on free amino acid pattern in relation to polyploidy in *O. sativa*, L. *J. Cytol. Genet.*, 14: 95–98.

Boge, G. 1960. Amino acid composition herring and herring meal. Destruction of amino acids during processing. *J. Sc. Food. Agril.*, 11: 362–365.

Bodding, W.B. 1942. Some histological and ethnical aspects of Santhals. *J. Asiatic Society of Bengal*, 3: 27–37.

Gupta, P.P., Prakash, D. and Chaudhury, A.R. 1984. Effect of ploidy variation on amino acid composition of *Datura inoxia* (Solanaceae). *Plant Syst. Evol.*, 145(34): 165–168.

Josh, S.S. and Srivastava, R.K. 1979. Calorie and amino acid composition of Rice bran and rice. *Proc. Nat. Acad. Sci.*, 46A: 223–225.

Jha, B.N. 1988. Chemo-cytogenetical studies in cucurbitaceae. *Ph.D. Thesis*, Bhagalpur University, Bhagalpur.

Harborne, J.B. 1973. *Phytochemical Methods*. Singapore, Toppan Printing Co. Pty, Ltd.

Linskens, H.F. and Pfahler, P.L. 1977. Genotypic effects on amino acid relationship in maize (*Zea mays* L.) pollen and style. *Theo. Appl. Genet.*, 50: 173–177.

Malhotra, A. 1985. Assessment of amino acids in some ethnobotanical plants through chromatography. *J. Ind. Bot. Soc.*, 59: 69–74.

Martin, D.D., Mayer, P.A. and Rodwell, V.W. 1981. *Harper's view of Biochemistry*. Drawer Li Las Atlas, California, 94002, p. 261–273.

Mohammad, A. and Raghubanshi, S.S. 1981. Studies on free amino acid content of diploid and colchiploid *Citrus aurantifolia* cultivar Kagzi. *J. Plant Biochem.*, 8(1): 25–27.

Power, H.E., Savagaon, K.A., March, B.E. and Biely, J. 1969. Assortment of Nutrition content of Atlantic coast herring meals. *Feed Stuff*, 40: 77–80.

Patridge, S.M. 1991. *Nature*, 207: 271–275.

Rieley, R. and Ewart, J.A.D. 1970. The effect of individual rye chromosomes on the amino acids content of wheat grains. *Genet. Res.*, 15: 209–219.

Smith, F. 1997. *J. Amer. Chem. Soc.*, 110: 324–330.

Sinha, J.P. 1988. Ethnobotanical exploration of some cytologically significant Plants. *Biol. Bull. of India*, 9(2): 83–88.

Voila, S., Arieli, U., Rappaport, J. and Mokady, S. 1981. Experiment in nutrition of carp. Replacement of fish meal by soyabean meal. *Bamidgeh*, 33: 35–49.

Venkataraman, R. and Chari, S.T. 1957. Amino acid composition of some marine fishes. *Indian J. Med. Res.*, 45: 77–80.

Vedas and Brahman–Rigveda, 21.3.35, 4.83.9; Athar veda, 10.8.31; Yayurveda, 2.1.35; Sathpath Brahman, 2.2.45.

Chapter 25

Physicochemical Parameters and Elemental Analysis of the Soils of Sugarcane Fields with and without Red Rot Disease Incidence

S. Velmurugan*, R. Narayanaswamy** and S. Ravi**

**Department of Physics, Annamalai University, Annamalai Nagar – 608 002, Tamil Nadu, India*

***Engineering Physics Section, Faculty of Engineering and Technology, Annamalai University, Annamalai Nagar –608 002, Tamil Nadu, India*

ABSTRACT

In this study, an attempt has been made to find any possible variation in physicochemical parameters of the soils collected from various sugarcane fields with red rot disease prone and disease resistant areas. Also the concentrations of four important trace elements *viz.*, Fe, Mn, Cu and Zn are estimated using atomic absorption spectrometer. The results of the physicochemical parameters, the concentration of trace elements and their results are discussed.

***Keywords:** Sugarcane, Soil, Red rot disease, Physicochemical parameters and AAS.*

Introduction

The modern agriculture has considerably raised the output, but created problems like land degradation, atmospheric and water pollutions. Frequent and excess use of chemical fertilizers have created imbalance of nutrients in soil and adverse effect on soil physicochemical properties and ecology (Ramesh *et al.*, 2002).

The plant growth and yield primarily depends upon the nutrient status of soil (Zende, 1999). Now nutrients level is decreasing continuously in Indian soils due to extensive agricultural activities to meet the demand of increasing population (Aneg Singh *et al.*, 2001).

Sugarcane is one of the most important cash crops of India. It is along duration crop and intercroping is also recommended in the crop. The soil under sugarcane cultivation gets exhausted very soon. The nutrient uptake is also, very high in this crop. So this soil, under continuous sugarcane cultivation show deficiencies of different elements such as N, P, K and S (Aneg Singh *et al.*, 2001).

Diseases are one of major constraints in the profitable cultivation of sugarcane. Of these, red rot disease caused by "collectrotrichum falcutam went" is most destructive, because the fungal growth is encouraged by variations in temperature, pH, nutrients and environmental conditions (Raid and Lentini, 2002 and Sheokumar and Singh, 2003). Influence of environmental temperature has been studied by various researchers (Anilkumar *et al.*, 1998). But yet, only limited information is available on the role of infected soil in initiating red rot infection (Agnihotri *et al.*, 1979). Hence, the present study has been taken up to determine and to compare various physicochemical parameters and some major trace elements with respect to the disease prone and resistant soils from sugarcane fields.

Materials and Methods

The soil samples are collected from red rot disease prone and resistant sugarcane growing areas during the year 2003–2004. The sample areas are Bannari Amman Sugar Ltd., Sathiyamangalam, Erode district (Red rot disease resistant area) and EID Parry (I) Ltd., (red rot disease prone area) Nellikuppam, Cuddalore district, Tamil Nadu, India.

Representative composite soil samples are taken from the study area upto 15 cm depth as per standard procedure of IARI (Indian Agricultural Research Institute) New Delhi (Tripathy *et al.*, 1990). The location of sampling sites are randomized to avoid biasing in results. The collected samples after coning, quartering and sieving are used for various spectral investigations. The physical parameters of soil, like particle size, bulk density and pore space are determined by standard methods (Piper, 1966) and the chemical parameters of soil, like pH, EC, available N, available P, available K and organic carbon are determined by standard methods (Page, 1982). The soil samples are digested with diacid mixture using 10:4 nitric acid and perchloric acid. The resultant soil samples are subjected to chemical analysis for the estimating Fe, Mn, Cu and Zn through atomic absorption spectrometer (Linday and Norwell, 1978). This analysis has been carried out in soil testing laboratory, Cuddalore, Tamil Nadu, India.

Results and Discussion

Table 25.1 shows the physical parameters of red rot disease prone sugarcane growing soils. Table 25.2 shows the above parameters of red rot diseases resistant soils.

Soil texture is an important parameter. With the help of it, mineralogical and chemical composition are defined. The soil texture describes the size distribution of particles in soil. It indicates the relative percentages of sandy, silt and clay fraction of soil. In this study, the soil texture of all the samples fall under the type sandy clay loam. The total biomass of plants grown in sandy clay loam are always significantly greater, when compared with plants grown in gravel and clay (Paula Power, 1996).

All bulk density values are nearly equal both for red rot disease prone and resistant sugarcane growing areas soils. When bulk density values are more than 1.75 mg m^{-3} in sandy clay loam soil, the root growth of all the crops would be restricted (Richard, 2003).

Table 25.1: The Physical Parameters of Red Rot Disease Prone Sugarcane Growing Area Soils

*Sample No.**	*Sandy %*	*Clay %*	*Silt %*	*Textural Class*	*Bulk Density mg m⁻³*	*Pore Space %*
Sample 1	51.8	27.1	21.1	Scl	1.51	39.4
Sample 2	55.4	29.0	15.6	Scl	1.42	40.0
Sample 3	47.9	27.6	24.5	Scl	1.54	42.5
Sample 4	54.5	23.0	22.5	Scl	1.42	41.4
Sample 5	57.5	23.5	19.0	Scl	1.53	38.4

Note: *The samples collected from red rot disease prone sites.

Scl: Sandy clay loam.

Table 25.2: The Physical Parameters of Red Rot Disease Resistant Sugarcane Growing Area Soils

*Sample No.**	*Sandy %*	*Clay %*	*Silt %*	*Textural Class*	*Bulk Density mg m⁻³*	*Pore Space %*
Sample 6	66.4	29.5	4.1	Scl	1.42	48.5
Sample 7	63.4	28.2	7.9	Scl	1.44	47.8
Sample 8	64.3	21.9	13.8	Scl	1.46	45.5
Sample 9	73.2	22.0	4.8	Scl	1.53	43.0
Sample 10	72.3	29.8	2.9	Scl	1.56	42.1

Note: **The samples collected from red rot disease resistant sites.

Scl: Sandy clay loam.

The pore space of the soils in disease prone areas are very much lower when compared to red rot disease resistant area soils.

The chemical parameters of both red rot disease prone and resistant sugarcane growing soils are shown in Tables 25.3 and 25.4. In this study regarding the red rot disease resistant soils, it has been found that chemical parameters in electrical conductivity, available P, available K and organic carbon are high compared to red rot disease prone sugarcane growing soils, except available N. These results are in close conformity with the various observations of Jayapaual *et al.*, 2002 and Kumar *et al.*, 1999.

Table 25.3: The Chemical Parameters of Red Rot Disease Prone Sugarcane Growing Area Soils

*Sample No.**	*pH*	*EC (dsm⁻¹)*	*Available N (kg/ha)*	*Available P (kg/ha)*	*Available K (kg/ha)*	*Organic Carbon %*
Sample 1	8.25	0.40	261.85	53.70	376.75	1.34
Sample 2	7.60	0.30	275.50	48.15	444.70	1.80
Sample 3	8.25	0.30	252.65	69.10	338.65	0.99
Sample 4	7.70	0.15	272.95	87.40	290.30	1.51
Sample 5	8.30	0.25	243.35	84.55	297.70	0.65

Note: *The samples collected from red rot disease prone sites.

Table 25.4: The Chemical Parameters of Red Rot Disease Resistant Sugarcane Growing Area Soils

Sample No.*	pH	EC (dsm^{-1})	Available N (kg/ha)	Available P (kg/ha)	Available K (kg/ha)	Organic Carbon %
Sample 6	7.35	0.51	221.10	123.50	716.5	2.42
Sample 7	7.45	0.79	228.5	123.50	1238.4	2.23
Sample 8	7.50	0.78	161.6	123.50	1050.1	2.54
Sample 9	7.35	0.59	140.75	123.50	488.0	2.35
Sample 10	7.40	0.40	150.65	123.50	420.0	2.01

Note: **The samples collected from red rot disease resistant sites.

The pH of soils in red rot disease prone and resistant area soils are found to be primarily alkaline with some gradation (alkaline and weakly alkaline). It is recorded as 7.6 to 8.3 and 7.3 to 7.5 for red rot disease prone and resistant area soils respectively. The optimum range of pH values are 6.6 to 7.4, but sugarcanes will be grown in soils, with pH in ranges from 5 to 8.5 (Faostat, 2001).

Higher values of Fe (52.17 to 96.30 ppm), Mn (20.75 to 50.0 ppm), Cu (4.27 to 10.08 ppm) and Zn (2.1 to 5.28 ppm) have been found in red rot disease resistant sugarcane growing soils where as lower values of Fe (12.5 to 29.5 ppm), Mn (1.45 to 22.5 ppm), Cu (2.75 to 3.75 ppm) and Zn (1.5 to 3.2 ppm) are found in the red rot disease prone sugarcane growing soils as seen from Table 25.5 and Table 25.6. Similar results have been reported by Sheokumar *et al.*, 2003 and Rajakumar, 2002.

Table 25.5: The Trace Elements of Red Rot Disease Prone Sugarcane Growing Area Soils

Sample No.*	Iron (ppm)	Manganese (ppm)	Copper (ppm)	Zinc (ppm)
Sample 1	12.5	1.95	3.75	1.8
Sample 2	29.5	3.25	3.1	2.0
Sample 3	19.6	12.15	3.25	2.5
Sample 4	20.75	1.45	3.0	1.5
Sample 5	26.7	22.35	2.75	3.2

Note: *The samples collected from red rot disease sites.

Table 25.6: The Trace Elements of Red Rot Disease Resistant Sugarcane Growing Area Soils

Sample No.*	Iron (ppm)	Manganese (ppm)	Copper (ppm)	Zinc (ppm)
Sample 6	53.32	22.5	4.27	2.58
Sample 7	52.17	25.19	4.77	2.04
Sample 8	81.11	20.75	9.41	5.09
Sample 9	62.41	50.0	6.34	2.1
Sample 10	96.30	22.0	10.08	5.27

Note: *The samples collected from red rot disease sites.

From this study, it is seen that the observed concentrations of micronutrients (Fe, Mn, Cu, Zn) are higher in red rot disease resistant sugarcane growing soils. It leads to the fact that with proper management of micronutrient levels in soils *i.e.*, concentration of Fe, Mn, Cu and Zn are to be increased in order to control the disease. However the nitrogen content is found high in the disease prone areas when compared to diseased areas. So it is inferred that a higher value of nitrogen favours disease proliferation. Hence reduction of nitrogen may be one of the factors responsible in the reduction of the disease incidence.

References

Agnihotri, V.P., Budhayaraja, T.R and Krushana Singh. 1979. Role of diseased setts and soils in the annual recurrence of red rot in sugarcane. *Indian Sugar J.*, 263–265.

Aneg Singh, A.K., Gupta, R.N., Srivastava, K., Lal and Sing, S.B. 2001. Nutrient status of sugarcane growing soils in central Uttar Pradesh, India. *Sugar Tech.*, 3: 117–119.

Anil Kumar, Satyavir and Ram Niwas. 1998. Relationship between climatic factors and initiation of red rot infection in sugarcane. *Indian Phytopath.*, 51(4): 370–372.

Crop Water Management Sugarcane. 2001. FAOSTAT publication.

Jayapaul, G.S., Duraisingh, R., Senthilvel, T. and Joseph, M. 2000. Influence of population on stage of incorporation of intercrossed green manure (Daincha) and nitrogen levels on yield and quality of sugarcane. *Indian Sug.*, 60(1): 989–991.

Kumar, S., Singh, N.N.P., Dwivedi, N.B. and Kumar, S. 1999. Changes in chemical composition of cane juice due to pattotype red rot pathagne. *Bharathiya Sugar*, 24: 9–17.

Linday, N.L. and Norwell, W.A. 1978. Development of DTPA soil test for Zn, Fe, Mn and Cu. *Soil Sci. Soc. Am. Proc.*, 33: 62–68.

Page, A.L. 1982. *Methods of Soil Analysis*, Part 2. Soil Science Society of America, Madison, Wisconsion.

Paula Power. 1996. Growth of texas wildrice (*Zizania texana*) in three sediments from the san marcos River. *J. Agual. Plant Manage.*, 34: 21–23.

Piper, C.S. 1966. *Soil and Plant Analysis*. Hans Publication, Mumbai.

Raid, R.N. and Lentini, R.S. 2002. *Sugarcane Red Rot Disease*. UF/IFAS Publication.

Rajakumar, N. 2002. Spectral analysis of red rot disease incidence in sugarcane. *Ph.D. Thesis*, Department of Physics, Annamalai University.

Ramesh, T., Chinnusamy, C. and Sankaran N. 2002. Influence of integrated Nitrogen and phosphorus management on cane productivity and soil fertility status in sugarcane. *Cooperative Sugar*, 34: 7–38.

Richard, C. 2003. Cane planter (Soil compaction). *Sugar Journal*, p. 27–28.

Sheokumar, S. and Singh, R.R. 2003. Management of nutritional disorders in sugarcane. *Cooperative Sugar*, 34: 973–975.

Tripathy, D.P., Gurdeep, S. and D.C. 1990. Panigrahi proceeding of the 7th National symposium of environmental, India School of Mines, Dhansand, p. 204–205.

Zende, G.K. 1999. Recent trend in basic and applied research on soils and their utility in soils and crop management practices. *Bharathiya Sugar*, p. 9–29.

Chapter 26

Inhibitory Effect of *Hemidesmus indicus* R.Br. on Calcium Oxalate Crystal Growth *in vitro*

T. Vimala, S. Gopalakrishnan* and Y.M. Fazil Marickar***

**Department of Chemistry, Manonmaniam Sundaranar University, Abishekappatti, Tirunelveli – 627 012, Tamil Nadu, India*
***Professor of Surgery (Retired), Medical College Hospital, Trivandrum – 695 011, Kerala, India*

ABSTRACT

Hemidesmus indicus R.Br. (Fam. *Asclepiadaceae*) commonly known as "Nannari" in Tamil and "Ananta" in Sanskrit is used as a substitute for "Sarsaparilla". Roots are diaphoretic, demulcent, diuretic and blood purifier. They are also used for skin diseases, syphilis, rheumatism, scorpion-sting and snake-bite. The present work deals with the inhibitory studies of the ethanolic and aqueous extracts of the root of *H. indicus* on calcium oxalate crystal growth *in vitro*. Calcium oxalate crystals were grown in silica gel media in Hane's tubes by single diffusion method. The crystal growth after the addition of the ethanolic and aqueous extracts in 20 mg/5 ml and 10 mg/5 ml doses was studied. In both the extracts there were reductions in sizes of calcium oxalate crystal columns when compared with that of the control. Individual calcium oxalate crystals when viewed under microscope showed considerable reduction in size when compared with that of the control. Also the higher dose was found to be more effective when compared to the lower dose.

Keywords: *Inhibitory effect, Hemidesmus indicus, Calcium oxalate, in vitro.*

Introduction

Urinary stone disease (Urolithiasis) is common world wide. Most urinary stones probably begin as minute aggregates in the small collecting ducts of the kidney. They then migrate to the pelvicalyceal system, where they increase in size and may then move into the ureters and bladder. The general factors responsible for urinary stone formation are: environmental, genetic, the concentration of the relevant ionic species in the urine, the availability and chemical nature of the macromolecular components of the stone matrix and the concentration of urinary physiological inhibitors of crystallization (Werness *et al.*, 1981; Schrier *et al.*, 1978; Drach *et al.*, 1990). The types of urinary stones are calcium oxalate, calcium phosphate, magnesium–ammonium–phosphate, uric acid, sodium and ammonium urates, cystine, xanthine etc. (Prien, 1949). The higher incidence of urolithiasis in men than in women is due to their higher excretion rates of oxalate, calcium, uric acid and generally lower citrate excretion as well as the anatomical differences (Rajendran *et al.*, 1989). Occupations likely to cause either under hydration or increased vitamin D biosynthesis due to prolonged exposure to sunlight will predispose to urinary tract stones (Parry and Lister, 1975). Higher dietary intakes of dairy products, oxalate, vitamin C, refined carbohydrates, fats, purines, animal proteins and sodium chloride, together with low intakes of fibre and pyridoxine, have been suggested as factors that increase the incidence of stone formation (Clifford and Story, 1976; Robertson *et al.*, 1981).

Urinary stones can be either disrupted by extracorporeal shockwave lithotripsy or removed surgically (Chaussy *et al.*, 1982; Wickham and Miller, 1983). It has been reported that the possibility of stone recurrence in a person is 67–100 per cent (Coe *et al.*, 1977; Williams, 1963) and that mere stone removal cannot provide complete cure for the disease. The ultimate control of urolithiasis requires a proper application of both the approaches of stone removal and drug therapy. Drug therapy is mainly aimed at the inhibition of growth of the existing stone and the formation of new stone. It is also possible to study the effect of various inhibitors on crystal growth by performing *in vitro* experiments. Calcium oxalate is the major inorganic component of urinary calculi and in the present work an attempt is made to study the inhibitory effect of a medicinal plant, *viz.*, *Hemidesmus indicus* on calcium oxalate crystal growth *in vitro*. *H. indicus* is used as an alterative, demulcent, diaphoretic, diuretic and blood purifier. It is also employed in nutritional disorders, syphilis, chronic rheumatism, gravel and other urinary diseases and also in skin affections (Chopra *et al.*, 1956; Dymock *et al.*, 1891).

The root in combination with other drugs is prescribed in snake bite and scorpion-sting but the root is not an antidote to either snake-venom or scorpion-venom (Kirtikar *et al.*, 1993). An aqueous extract of *H. indicus* roots exhibited bacteriostatic activity in mice infected with *mycobacterium leprae* (Gupta, 1981). Hydrodistilled essential oil isolated from *H. indicus* was found to exhibit anti-bacterial activity against both gram-negative and gram-positive organisms even at the concentration of 0.2 per cent (Rajendra Prasad *et al.*, 1983). Two pregnane glycosides had been isolated from the dried stem (Prakash *et al.*, 1991) and three coumarinolignoids from the root of this plant (Mandal *et al.*, 1995). A new triterpene lactone was isolated from the hexane soluble portion of the ethanol extract of the stem of *H. indicus* (Gupta *et al.*, 1992). Chloroform and ethanol extracts of *H. indicus* root showed antifungal activity against *Aspergillus niger* (Hiremath *et al.*, 1997). Cell culture extract of *Hemidesmus indicus* prevented hypercholesterolemia in rats (Bopanna *et al.*, 1997). Oral treatment with the ethanol extract of the root significantly prevented rifampicin and isoniazid-induced hepatotoxicity in rats. The activity has been attributed to a free radical scavenging activity of the coumarinolignoids present in the extract (Prabakan *et al.*, 2000). Hexatriacontane, lupeol octacosanoate, β-amyrin acetate, lupeol acetate, α-amyrin, lupeol, α-Amyrin and sitosterol (Padhy *et al.*, 1973) were obtained from the petroleum ether extract of *H. indicus* root. 2-hydroxy-4-methoxy benzoic acid isolated from the methanolic extract of *H. indicus* root significantly neutralized the viper venom-induced lethality and hemorrhagic activity in albino rats and mouse (Alam and Gomes, 1998).

Materials and Methods

The roots of *H. indicus* were collected from Thenmalai, Tirunelveli District, Tamil Nadu, India during the month of September. The botanical identity was confirmed by comparing the sample with the Herbarium specimens preserved in the Dept. of Botany, by Dr. V. Chelladurai, Government Siddha Medical College, Palayamkottai, Tirunelveli District, Tamil Nadu, India. A voucher specimen of the plant had been deposited at the Herbarium, Department of Chemistry, Manonmaniam Sundaranar University, Tirunelveli, Tamil Nadu, India (Voucher specimen number MSU 052). The plant material was thoroughly washed with water, cut into small pieces, dried under shade and powdered.

The powdered root was successively extracted with petroleum ether (40°–60°C), benzene, chloroform, ethanol and water and they were condensed and evaporated to dryness under vacuum. Ethanolic and aqueous extracts alone had been subjected to the inhibitory studies.

Calcium oxalate crystals were grown *in vitro* in silica gel media in Hane's tubes by single diffusion method. In this method, gel was prepared by treating sodium metasilicate solution of density 1.03 g/ml with 3M acetic acid and the pH was adjusted to 6.3. One of the reactants, calcium chloride (1 M) was incorporated inside the gel. After gellation, 1 M oxalic acid was slowly added over the gel along with alcohol/water as the supernatant solution for control group and alcoholic/aqueous extracts of the test plant for other groups. The crystals appeared as cloudy precipitate and in due course, the depth of the column was found to increase. The thickness (length) of crystal column was measured in centimetre on days 1, 3, 7, 10, 15, 20, 25 and 30. The sizes of the crystals formed were noted in microns (μm) on days 1, 3, 7, 14, 21 and 28 using an optical microscope and the values thus obtained were compared with each other. The whole experiment was carried out at room temperature (27°±3°C). The inhibitory effect of alcoholic and aqueous extracts were studied at 20 mg/5 ml and 10 mg/5 ml doses. The experiments were repeated six times and the average value was taken. All the results were expressed as means±SE. The test of significance was statistically analyzed using Student's t-test (Fisher, 1950). After the crystal growth, the crystals were cleared off the gel by repeated washing with distilled water, filtered and air-dried. The purity of the crystals thus obtained was assessed by infra-red spectroscopy using FT-IR-410 JASCO-Japan instrument (KBr pellet method). Scanning electron microscopy was used for studying the morphology of the crystals. For SEM studies a JEOL JSM 35-C scanning electron microscope was used.

Results and Discussion

The inhibitory effect of the alcoholic and aqueous extracts of the roots of *H. indicus* on *in vitro* oxalate crystal growth have been studied at 20 mg/5 ml and 10 mg/5 ml doses and the results are presented in Figures 26.1 and 26.2 and Tables 26.1 and 26.2. Concentration dependent inhibitory effect was observed in both alcoholic and aqueous extracts. In the case of alcoholic extracts both these concentrations did not produce any significant change on crystal growth upto the 3rd day. On the third day the growth was almost equal in control and drug added tubes. From the 5th day onwards significant decrease in growth rate was observed in tubes containing alcoholic extracts. Aqueous extracts did not produce significant reductions in crystal columns. The mean lengths of crystal columns in the control set up were 3.48 cm and 3.97 cm respectively on the 30th day when alcohol and distilled water were used. The mean lengths of crystal columns in the extract set up were 3.00 cm (20 mg/5 ml) and 3.13 cm (10 mg/5 ml) for alcoholic extracts and 3.87 cm (20 mg/5 ml) and 3.90 cm (10 mg/5 ml) for aqueous extracts. The degree of inhibition was more with 20 mg/5 ml concentration in both the extracts.

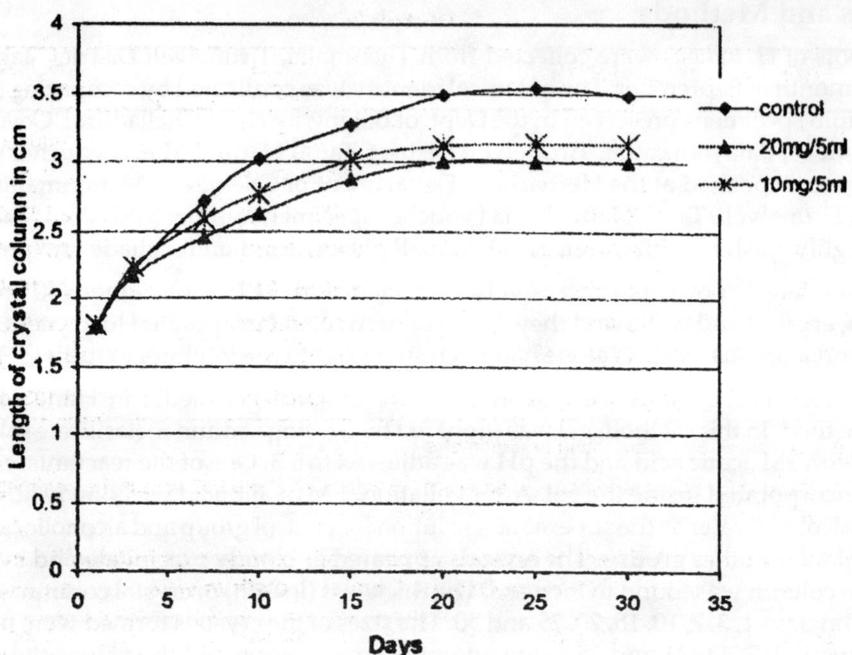

Figure 26.1: Effect of Alcoholic Extract of *H. indicus* on Crystal Growth *in vitro*

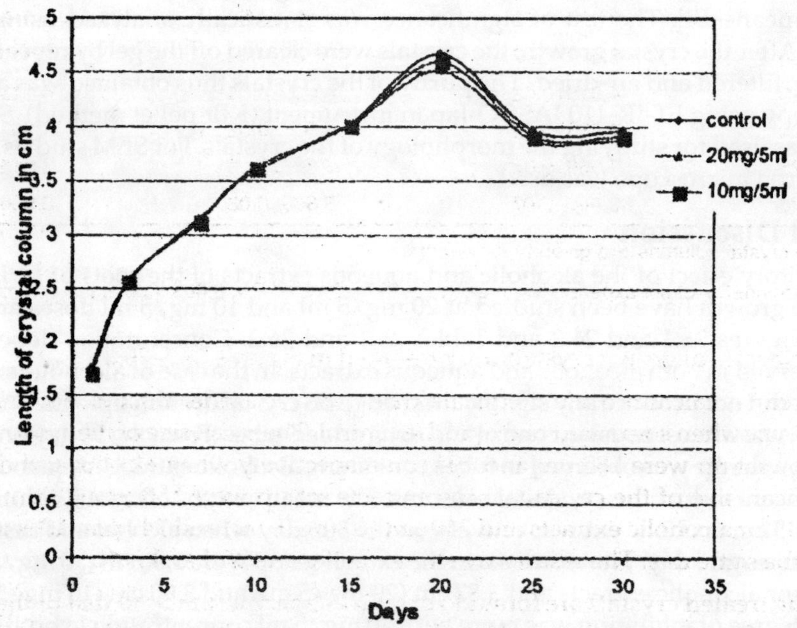

Figure 26.2: Effect of Aqueous Extract of *H. indicus* on Crystal Growth *in vitro*

Table 26.1: Effect of Alcoholic Extract of *Hemidesmus indicus* Root on Calcium Oxalate Crystal Growth *in vitro*

Days	*Control*	*H. indicus* 20 mg/5 ml	*H. indicus* 10 mg/5 ml
1	1.783±0.08	1.783±0.01	1.800±0.00
3	2.200±0.05	2.166±0.03	2.233±0.01
7	2.716±0.01	2.450±0.02**	2.583±0.08
10	3.016±0.03	2.616±0.01**	2.766±0.08
15	3.266±0.03	2.866±0.01**	3.016±0.08
20	3.483±0.04	3.000±0.00**	3.116±0.08*
25	3.533±0.04	3.000±0.00**	3.133±0.08*
30	3.483±0.04	3.000±0.00**	3.133±0.08*

Lengths of the crystal columns are given in centimetre.

All values are mean±SE of six experiments in each set * $P < 0.05$; ** $P < 0.005$ versus the control group.

Table 26.2: Effect of Aqueous Extract of *Hemidesmus indicus* Root on Calcium Oxalate Crystal Growth *in vitro*

Days	*Control*	*H. indicus* 20 mg/5 ml	*H. indicus* 10 mg/5 ml
1	1.733±0.02	1.666±0.02	1.700±0.06
3	2.566±0.02	2.550±0.03	2.566±0.04
7	3.133±0.02	3.150±0.03	3.116±0.04
10	3.616±0.02	3.583±0.04	3.616±0.04
15	4.016±0.02	4.000±0.03	4.016±0.04
20	4.683±0.04	4.516±0.02	4.600±0.06
25	3.966±0.07	3.866±0.03	3.916±0.04
30	3.966±0.07	3.866±0.03	3.900±0.05

Lengths of the crystal columns are given in centimetre.

All values are mean±SE of six experiments in each set * $P < 0.05$; ** $P < 0.005$ versus the control group.

The IR Spectra of grown crystals are presented in Figures 26.5–26.8.

Drug treated calcium oxalate crystals when viewed under microscope showed considerable reduction in size when compared to that of the control. The mean size of the grown crystals on 28th day in the control set up were 180 μm and 294 μm respectively when alcohol and distilled water were used. The mean size of the crystals in the extracts set up were 150 μm (20 mg/5 ml) and 162 μm (10 mg/5 ml) for alcoholic extracts and 240 μm (20 mg/5 ml) and 240 μm (10 mg/5 ml) for aqueous extracts on the same day. The results are shown in Figures 26.3 and 26.4.

The drug treated crystals are found to exhibit IR spectra similar to that of the control, indicating no possibility of any type of chemical bonding between the drug and the oxalate crystals. Hence the crystals were subjected to SEM analysis for understanding the morphological changes if any by the plant extracts. Figures 26.9–26.12 show the SEM of *in vitro* grown calcium oxalate crystals in silica gel media.

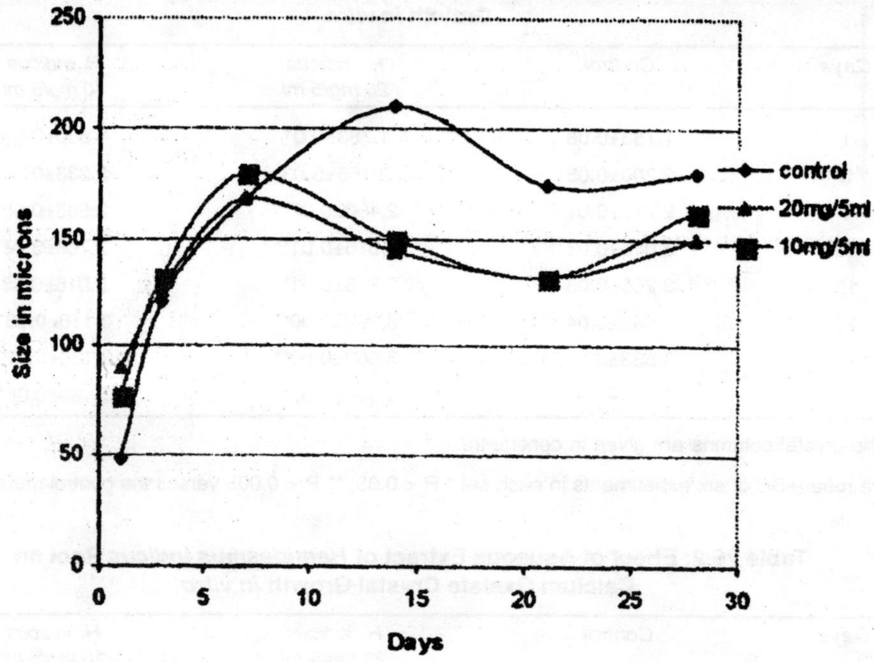

Figure 26.3: Effect of Alcoholic Extract of *H. indicus* on the Size of the Crystals

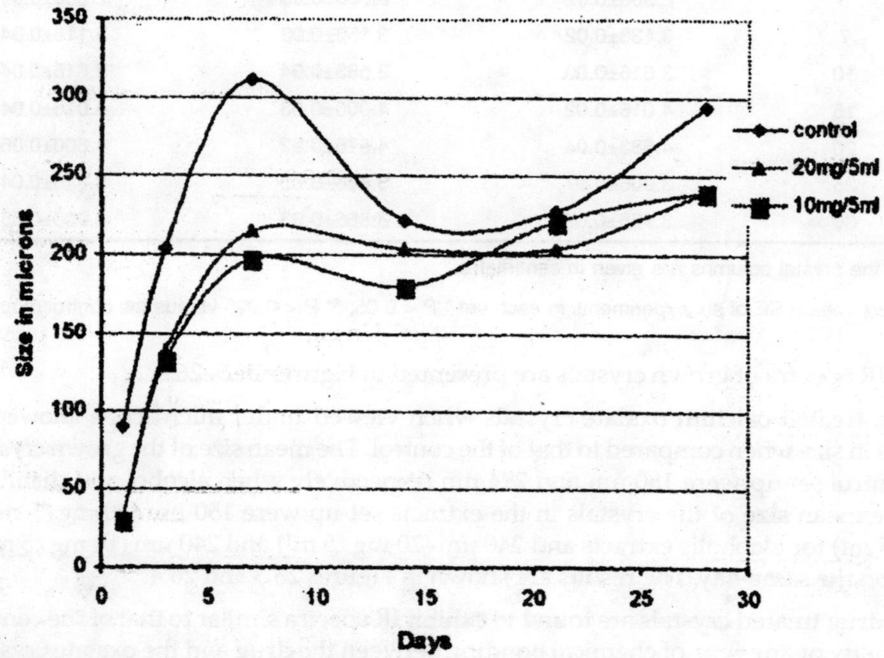

Figure 26.4: Effect of Aqueous Extract of *H.indicus* on the Size of the Crystals

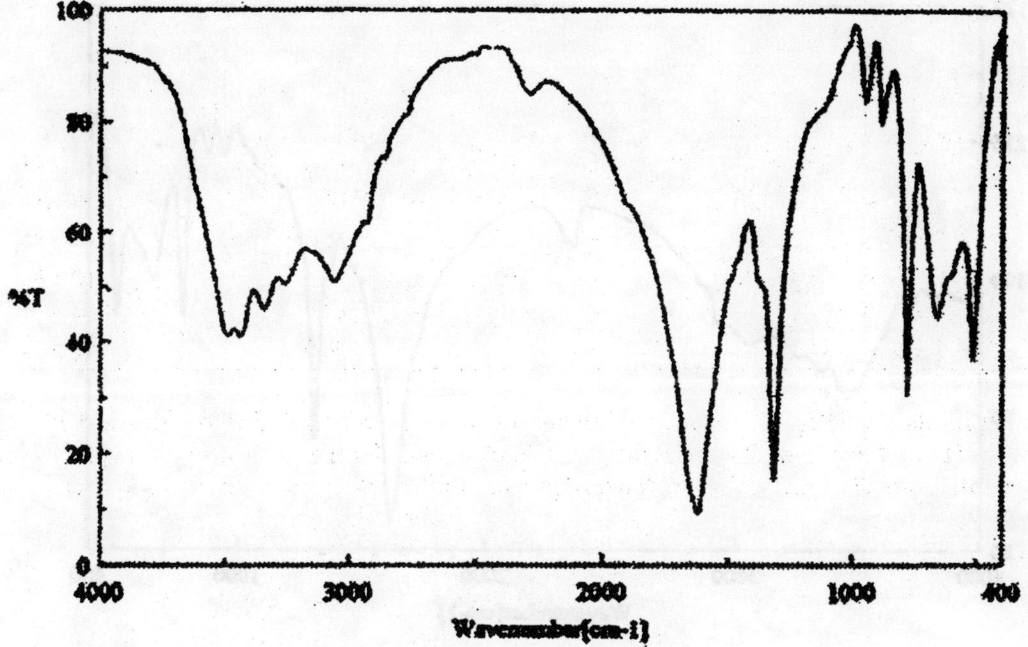

Figure 26.5: Control (Âlcohol)

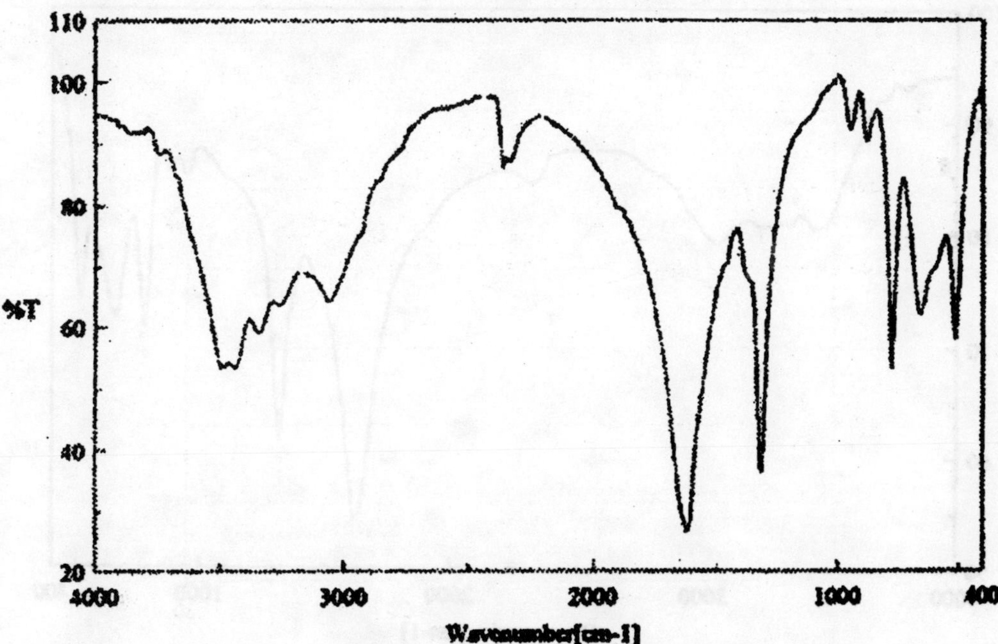

Figure 26.6: Control (Water)

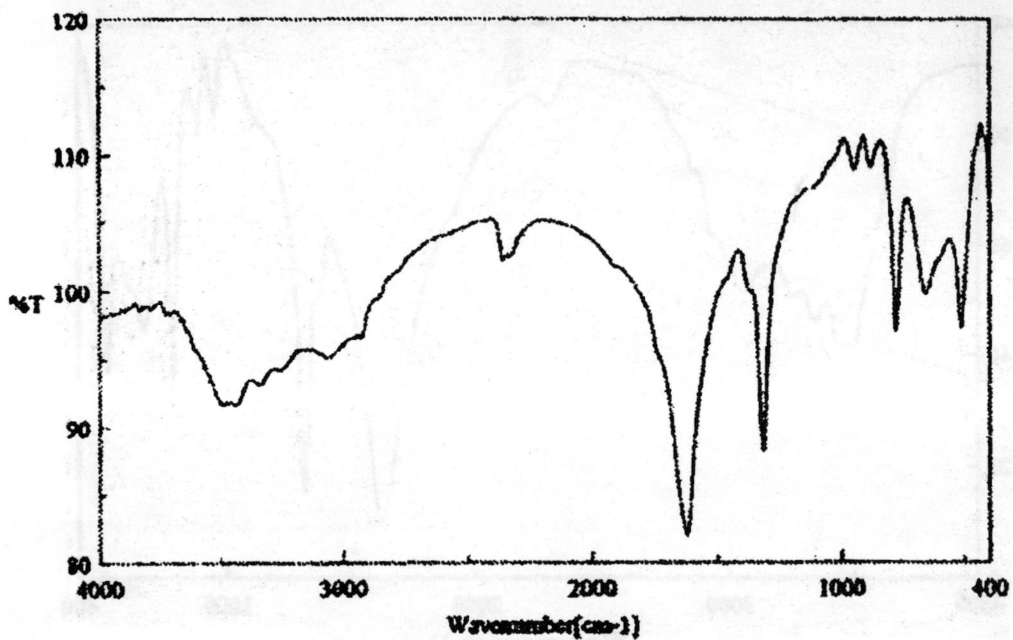

Figure 26.7: *H. indicus* Alcoholic Extract

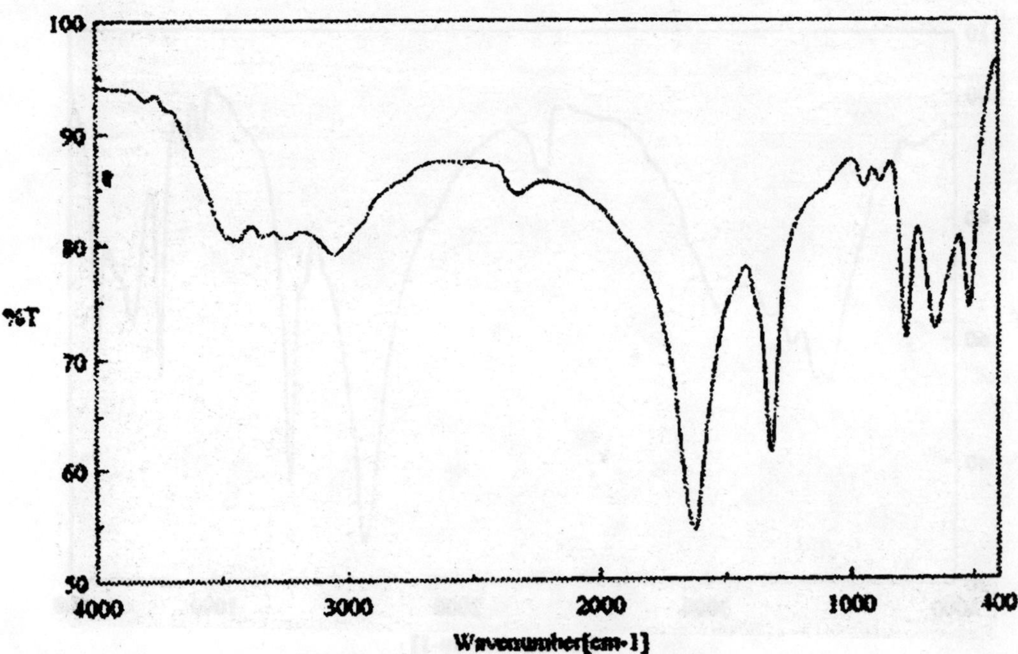

Figure 26.8: *H. indicus* Aqueous Extract

Figure 26.9(a): Calcium Oxalate Crystals on Adding Alcohol

Figure 26.9(b): Calcium Oxalate Crystals on Adding Alcohol Showing Sharp Edges

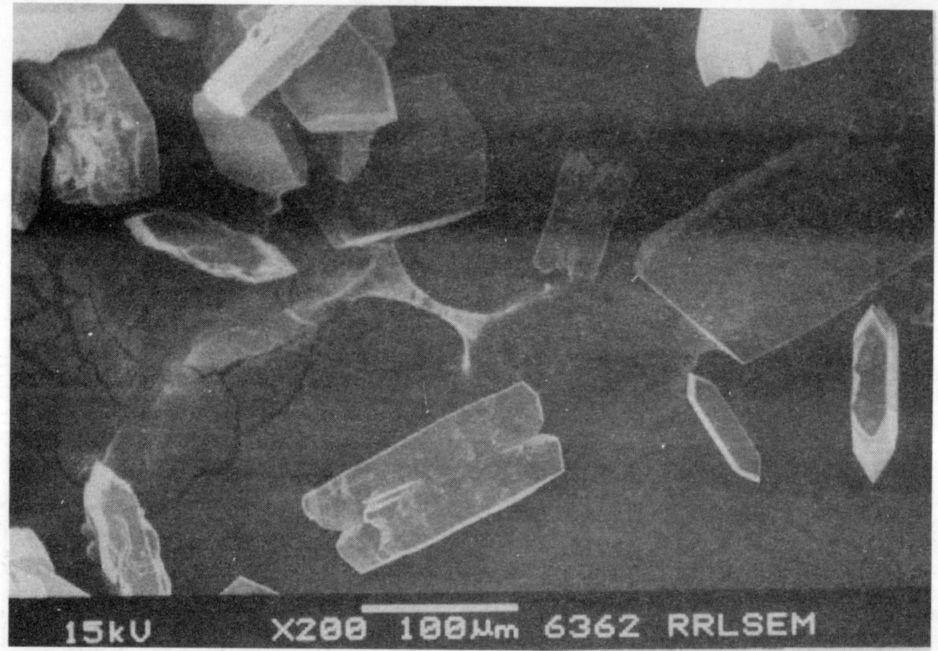

Figure 26.10(a): Calcium Oxalate Crystals on Adding Alcoholic Extract

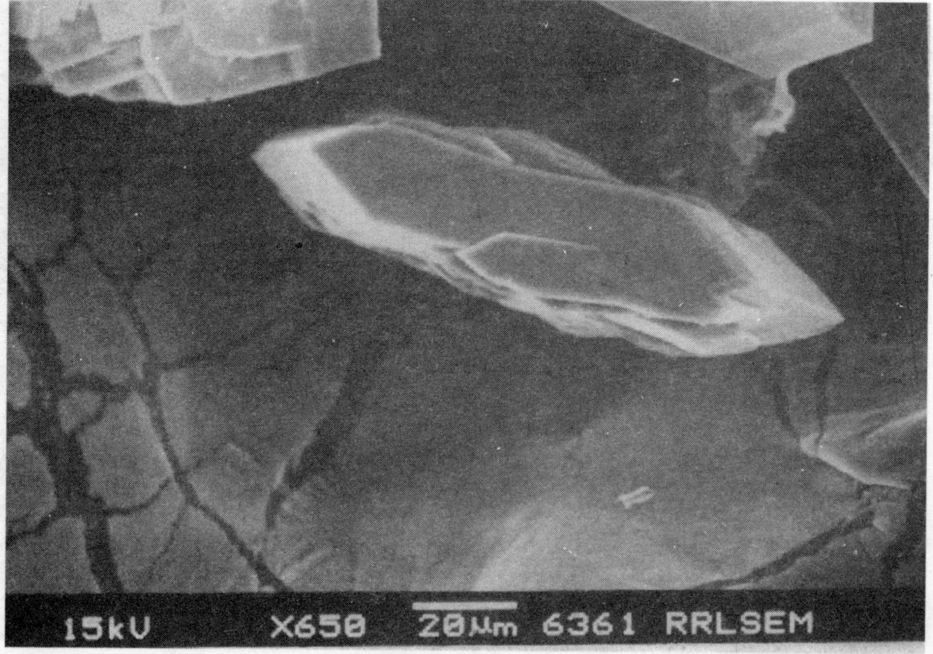

Figure 26.10(b): Calcium Oxalate Crystals on Adding Alcoholic Extract Showing Less Defined Edges

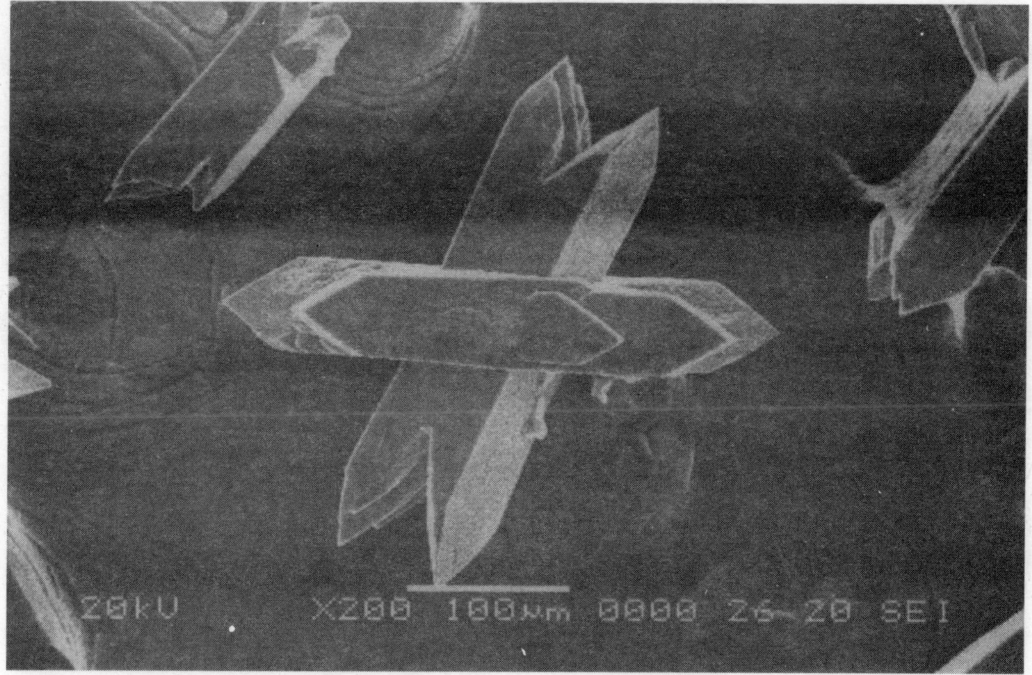

Figure 26.11(a): Calcium Oxalate Crystals on Adding Water Showing Twinning

Figure 26.11(b): Calcium Oxalate Crystals on Adding Water Showing Aggregation

Figure 26.12(a): Calcium Oxalate Crystals on Adding Aqueous Extract Showing Broken Edges

Figure 26.12(b): Calcium Oxalate Crystals on Adding Aqueous Extract Showing Crystals Without Aggregation

The crystals from control experiments contained individual prismatic crystals, twinned crystals and some aggregates. Crystals grown with alcoholic extract of *H. indicus* showed less defined edges, cracks, crevices and dots of corrosion which could mean a tendency to reduce the size by crushing the stones into small bits.

Addition of aqueous extract of *H. indicus* resulted in an increase in the number of single crystals and decreased the tendency of agglomerate formation. The agglomeration of calcium oxalate crystals plays a major role in urinary stone formation and has the potential to produce very large crystals in a short period of time with no reduction of supersaturation (Hounslow *et al.*, 1988).

The SEM study reveals that the drug treated oxalate crystals are smaller in size and show a tendency for chipping away of the parts as compared to that of the controls. Thus the results of *in vitro* experimental models, IR and SEM together attribute an inhibitory capacity with respect to calcium oxalate crystals for *H. indicus* extracts.

Acknowledgement

One of the authors (T.V.) wishes to thank the University Grants Commission, New Delhi and the Principal and Management of St. Jude's College, Thoothoor, Kanyakumari District for selecting her under FIP Programme. The authors also thank Dr. Peter Koshy, RRL, Trivandrum for SEM analysis.

References

Alam, M.I. and Gomes, A. 1998. Viper venom–induced inflammation and inhibition of free radical formation by pure compound isolated and purified from anantamul root extract. *Toxicon.*, 36: 207–215.

Bopanna, K.N., Bhagyalakshmi, W., Rathod, S.P., Balaraman, R. and Kannan, J. 1997. Cell culture derived *Hemidesmus indicus* in the prevention of hypercholesterolemia in normal and hyperlipidemic rats. *Indian Journal of Pharmacology*, 29: 105–109.

Chaussy, C., Schmiedt, E., Jocham, D., Brendel, W., Forssmann, B. and Walther, V. 1982. First clinical experience with extracorporeally induced destruction of Kidney stones by shock waves. *J. Urol.*, 127: 417–420.

Chopra, R.N., Nayar, S.L. and Chopra, I.C. 1956. *Glossary of Indian Medicinal Plants*. Council of Scientific and Industrial Research, New Delhi.

Clifford, A.J. and Story, D.L. 1976. Levels of purines in foods and their metabolic effects in rats. *J. Nutr.*, 106: 435–442.

Coe, F.L., Keck, J. and Norton, E.R. 1977. The natural history of calcium urolithiasis. *J.A.M.A.*, 238: 1519–1523.

Drach, G.W., Thorson, S. and Randolph, A. 1990. Effects of urinary macromolecules on crystallisation of calcium oxalate. Enhancement of nucleation. *J. Urol.*, 143: 123–127.

Dymock, W., Warden, C.J.H. and David Hooper, 1891. *Pharmacographia Indica: A History of the Principal Drugs of Vegetable Origin*. Thacker, Spink and Co., Calcutta.

Fisher, R.A. 1950. *Statistical Methods for Research Workers*. Oliver and Boyd, Edinburgh.

Gupta, M.M., Verma, R.K. and Misra, L.N. 1992. Terpenoids from *Hemidesmus indicus*. *Phytochemistry*, 31: 4036–4037.

Gupta, P.N. 1981. Antileprotic action of an extract from 'Anantamul' (*H. indicus* R. Br.). *Lepr. India*, 53: 354–359.

Hiremath, S.P., Rudresh, K. and Badami, S. 1997: Antimicrobial activity of various extracts of *Striga sulphurea* and *Hemidesmus indicus. Indian Journal of Pharmaceutical Sciences*, 59: 145–147.

Hounslow, M.J., Ryall, R.L. and Marshall, V.R. 1988. A discretized population balance for nucleation, growth and aggregation. *AICHE J.*, 34: 1821–1832.

Kirtikar, K.R., Basu, B.D. and An, I.C.S. 1993. *Indian Medicinal Plants.* Bishen Singh Mahendra Pal Singh, Dehra Dun.

Mandal, S., Das, P.C., Joshi, P.C. and Chatterjee, A. 1995. Chemistry of coumarino lignoids, a rare class of plant products having anticancer and antihepatotoxic activities. In: *Seminar on Research in Ayurveda and Siddha*, CCRAS, New Delhi, p. 20–22.

Padhy, S.N., Mahato, S.B. and Dutta, N.L. 1973. Triterpenoids from the roots of *Hemidesmus indicus. Phytochemistry*, 12: 217–218.

Parry, B.S. and Lister, I.S. 1975. Sunlight and Hypercalciuria. *Lancet.*, 1: 1063.

Prabakan, M., Anandan, R. and Devaki, T. 2000. Protective effect of *Hemidesmus indicus* against rifampicin and isoniazid-induced hepatotoxicity in rats. *Fitoterapia*, 71: 55–59.

Prakash, K., Sethi,A., Deepak, D., Khare, A. and Khare, M.P. 1991. Two pregnane glycosides from *Hemidesmus indicus. Phytochemistry*, 30: 297–299.

Prien, E.L. 1949. Studies in Urolithiasis: II Relationships between pathogenesis, structure and composition of calculi. *J. Urol.*, 61: 821–836.

Rajendran, R., Sachidev, K., Thomas, J. and Marickar, Y.M.F. 1989. Changing trend in Paediatric Urolithiasis in Kerala, India. In: *Urolithiasis*, (Eds.) Walker, V.R., Sutton, R.A.L., Cameron, E.C.B., Pak, C.Y.C. and Robertson, W.G. Plenum, New York, p. 779–780.

Rajendra Prasad, Y., Alan Kararao, G.S.J.G. and Baby, P. 1983. Antimicrobial studies on essential oil of *Hemidesmus indicus* R. Br. *Indian Perfum.*, 27: 197–199.

Robertson, W.G., Peacock, M., Heyburn, P.J. and Bambach, C.P. 1981. Risk factors in calcium stone disease. In: *Urinary Calculus*, (Eds.) Brockis, J.G. and Finlayson, B. PSG, Littleton Mass, p. 265–273.

Schrier, E.E., Lee, K.E., Rubin, J.L., Werness, P.G. and Smith, L.H. 1978. Macromolecular inhibitors of calcium oxalate crystal growth and aggregation in the urine. In: *Oxalate in Human Biochemistry and Clinical Pathology*, (Eds.) Rose, G.A., Robertson, W.G. and Watts, R.W.E. Proc. Int. Symp. Welcome Foundation, London, p. 22–27.

Werness, P.G., Jaw, H., Bergert, S., Karen, E. and Lie, K.E. 1981. Urinary crystal growth effect of inhibitor mixtures. *Clin. Sci.*, 61: 487–491.

Wickham, J.E.A. and Miller, R. 1983. *Percutaneous Surgery of Renal Calculi.* Churchill Living Stone, London.

Williams, R.E. 1963. Long-term survey of 538 patients with upper urinary tract stone. *Br. J. Urol.*, 35: 416–437.

Chapter 27

Effect of Saline Irrigation on Chlorophyll and Proline Contents in Selected Tree Species

B.R. Gurumurthy*, S.P. Nataraja*, Naryan S. Mavarkar* and M.B. Shivanna**

**College of Agriculture, Shimoga – 577 204*
***Department of Applied Botany, Kuvempu University, Jnana Sahyadri – 577 451, Karnataka, India*

ABSTRACT

Seedlings of six-selected tree species grown in polybags were irrigated with saline solutions ranging from EC 4 to 20. Increased levels of salinities significantly reduced chlorophyll contents in all the tree species while the proline contents were substantially enhanced. The jack, tamarind and eucalyptus recorded low chlorophyll content on 360 DAT but other species acacia, neem and pongamia had increased relative chlorophyll contents at this time. By 60 DAT all the species recorded higher proline contents, which decreased by 360 DAT. Among the six tree species acacia followed, by neem and pongamia had high relative chlorophyll and proline contents by 360 DAT. Hence, they could be considered as relatively tolerant under the present experimental conditions.

***Keywords:** Relative tolerance, Proline, Chlorophyll.*

Introduction

Irrigated land account for 15 per cent of the cultivated land but produces 36 per cent of world's food (Anon, 1989). In India 62 m ha of land is provided with irrigation facilities. The excessive irrigation has resulted in the conversion of cultivated land into marsh and finally salinisation of soil. Plants differ in their ability to grow and yield under saline situations. The crops are generally relatively tolerant during germination but become sensitive during emergence and early seedling stage (Mass, 1990). Information on the effect of water salinity on growth and yield in perennial tree species is very scant. The present study is aimed to enrich our knowledge in this direction, through the study of biochemical changes that might occur in seedlings of selected tree species grown under induced salinity.

Materials and Methods

Greenhouse Experiments were conducted during June 2000–May 2001 and repeated in June 2001–May 2002 at Agricultural College, Shimoga. Six tree species–neem, jack, acacia, eucalyptus, pongamia and tamarind grown in agro/social forestry were chosen for the study. Two seeds each of jack, tamarind, neem or pongamia and ten seeds each of eucalyptus or acacia were sown in small poly bags (12 × 5 cm) containing potting mixture, placed in the greenhouse and watered regularly to raise uniform and healthy seedlings. Thinning was done to retain only one seedling, as soon as the seedlings attained a height of 5 cm. Further, large poly bags (40 × 15 cm) filled with potting mixture were watered and after a little compaction of soil, a hole was made in the center of surface of soil into which 90-day-old seedling raised in small poly bag was transplanted and grown up to the age of 450 days. During this period, they were supplied with distilled water. Saline solutions were artificially prepared by adding NaCl and $Cacl_2$ (2 : 1) to the distilled water (Todd and Reed, 1998) to obtain EC of 4, 8, 12, 16 and 20 dsm^{-1}. Four hundred and fifty-day-old plants were irrigated once with saline water of EC 4, 8, 12, 16 or 20 dsm^{-1} or distilled water (control) depending upon the treatment requirement and the field capacity of soil was determined. Later, once in three days, all seedlings were irrigated with a measured amount of distilled water to bring the soil to field capacity. Care was taken to avoid leaching due to excess watering. Plants were labeled in each treatment. The experiment was laid out in a complete randomized design. The chlorophyll and proline contents plants were estimated at 5 stages (0, 60, 120, 240 and 360 days) by the spectrophotometric (Arnon, 1949) and colorimetric method (Bates *et al.*, 1973) respectively. The data of two years was subjected to pooled statistical analysis and significance was tested at P 0.01 as per the procedure outlined by Gomez and Gomez (1984).

Results and Discussion

Effect of Salinity on Chlorophyll (mg g^{-1} fresh weight) Content

The chlorophyll contents (Table 27.1) were affected in plants when treated with different strengths of salinity at all the 5 different stages (0, 60, 120, 240 and 360 DAT). In case of neem, chlorophyll content in control plants increased from 0.73 mg on zero day to 1.83 mg on 360 DAT. Following salinity treatments, except at EC 4, chlorophyll content was significantly affected. At salinity > EC 12, the chlorophyll content was considerably reduced as early as 60 days but it increased gradually by 360 days but it was less than control. In case of jack, upon treatment with saline water of > EC 8, chlorophyll content started declining as early as 60 days and continued to decrease until 120 days and later increased gradually. Only in case of treatment EC 4, there was no decreasing trend. When acacia was irrigated with saline solution of EC 4 to 12 the total chlorophyll content was reduced to a greater extent at 120 DAT, while at EC 16 and 20, the content started decreasing as early as 60 DAT. The chlorophyll content of plants treated with EC 20 was 3 times lower than that of the control plants when observations were recorded on 360 DAT but slightly more than the content at zero day. In eucalyptus, the chlorophyll content increased from 0.72 to 2.91 mg in control plants on 360th day. And, it started decreasing by 120 DAT when treated with saline of EC 8 and 12 however, it was decreased to a great extent by 60 DAT itself when treated with saline solution of EC 20. Only plants grown in salinity upto EC 8 could overcome the salinity stress and increase the chlorophyll content to a high level by 360 DAT. However, treatments EC 12, 16 and 20 caused decrease in the content of chlorophyll in plants at 360 DAT as compared to zero day. With respect of pongamia, although saline treatments EC 4 and EC 8 did not reduce the chlorophyll content to a great extent, the reduction was significant (P 0.01) in comparison to control Treatment at EC 12, 16 and 20 caused a considerable decline in the chlorophyll content by 60 DAT but soon the chlorophyll content increased significantly by 360 DAT. In case of EC 20, the chlorophyll was almost equal at zero day and at 360 DAT. The saline irrigation had detrimental effect on the chlorophyll content of tamarind seedlings at 60 DAT and the recovery from salinity stress was poor even at 360 DAT. Only in case of EC 4, there was a slight increase in chlorophyll content by 360 DAT.

Table 27.1: Effect of Saline Irrigation on Chlorophyll Content (mg g^{-1}) in Selected Tree Species

	Neem					*Jack*					*Acacia*				
DAT	0	60	120	240	360	0	60	120	240	360	0	60	120	240	360
C	0.73	0.91	1.37	1.75	1.83	0.83	0.96	1.27	1.63	1.93	0.54	0.73	1.23	1.73	2.13
EC4	0.74	0.90	0.92	1.30	1.57	0.83	0.89	1.14	1.35	1.45	0.55	0.65	0.60	0.80	1.06
EC8	0.75	0.81	0.81	1.29	1.38	0.81	0.65	0.29	0.75	0.94	0.53	0.50	0.42	0.62	0.92
EC12	0.73	0.58	0.44	0.85	1.10	0.84	0.52	0.20	0.64	0.80	0.55	0.45	0.35	0.76	1.10
EC16	0.73	0.26	0.43	0.83	1.13	0.82	0.32	0.37	0.61	0.84	0.55	0.20	0.22	0.52	0.75
EC20	0.75	0.17	0.40	0.72	1.03	0.83	0.24	0.16	0.17	0.22	0.51	0.16	0.27	0.46	0.71
Mean	0.74	0.61	0.73	1.12	1.34	0.83	0.60	0.57	0.86	1.01	0.54	0.45	0.52	0.82	1.11
P 0.01[4]	NS[5]	0.16[6]	0.02	0.02	0.02	NS	0.02	0.02	0.02	0.02	NS	0.02	0.02	0.02	0.02
CV (%)	1.36	1.65	1.37	0.89	0.75	1.21	1.68	2.63	2.15	1.48	1.86	2.23	1.94	1.23	0.90
	Eucalyptus					*Pongamia*					*Tamarind*				
DAT	0	60	120	240	360	0	60	120	240	360	0	60	120	240	360
C	0.72	0.94	1.53	2.30	2.91	1.27	1.33	2.47	3.13	3.53	0.63	0.63	0.87	1.03	1.13
EC 4	0.71	1.13	1.23	2.00	2.73	1.33	1.17	2.17	2.60	3.03	0.60	0.51	0.87	0.76	0.82
EC8	0.71	0.77	0.53	1.13	1.50	1.23	1.03	1.67	2.10	2.40	0.63	0.41	0.25	0.48	0.59
EC12	0.73	0.91	0.23	0.29	0.53	1.23	0.97	1.43	2.07	2.13	0.63	0.23	0.26	0.32	0.38
EC16	0.73	0.73	0.70	0.32	0.62	1.23	0.87	1.17	1.83	2.03	0.67	0.18	0.22	0.27	0.40
EC20	0.72	0.20	0.50	0.41	0.61	1.20	0.87	0.87	1.17	1.43	0.57	0.17	0.22	0.33	0.36
Mean	0.72	0.79	0.791	1.08	1.48	1.26	1.04	1.63	2.18	2.43	0.61	0.36	0.42	0.53	0.61
P 0.01	NS	0.02	0.02	0.09	0.02	NS	0.05	0.02	0.01	0.06	NS	0.05	0.01	0.02	0.02
CV (%)	2.29	1.93	1.89	0.12	0.67	0.80	0.96	0.61	0.43	1.72	1.68	2.17	2.41	1.87	1.63

All the saline irrigation treatments affected the chlorophyll contents in different tree species at the seedling stage. There was a linear response to the increased levels of salinities and a single saline irrigation substantially affected plant pigments. The effects were prolonged in case of tree species–jack, eucalyptus and tamarind. However, in neem, acacia and pongamia the effect of salinity stress was for a short period and by 360 DAT these species had increased chlorophyll content suggesting the recovery from salinity stress over a period of time. Such observations were also made by Alamgir *et al.* (1989) in rice cultivars where in decreased leaf chlorophyll content and its stability index was noticed following salinity stress.

Effect of Salinity on Proline Accumulation (μg g^{-1} fresh weight)

Under saline stress, water potential of soil solution is reduced compared to the root system and this might result in exosmosis unless osmotic adjustments are done by the plant system. In field crops important osmolytes synthesized include glycine betaine, proline, aspargine, valine and others. Detailed studies have been mode on osmotic adjustments of field crops but such studies are scant in tree species. In the present investigation proline accumulation was assessed to study the osmotic adjustments and the differences among the tree species.

The proline contents (Table 27.2) were significantly affected by saline irrigations in all species tested. The data indicated that proline content reached its highest level by 60th day which further declined slightly by 120 DAT and later considerably by 240 and 360 DAT irrespective of saline treatments. The control plants of all the species did not show any variation in proline content through out the experimental period. In case of jack, proline content at 360 DAT did not vary much as compared to zero day. However, neem and acacia had proline, which was nearly double at 360 DAT as compared to zero day. In case of other tree species the proline content increased at 360 DAT over zero day in the saline concentration of EC 12 and above. Neem, acacia and pongamia recorded higher accumulation of proline (500 μg and above) on 60 DAT when soil was irrigated with EC 20.

Table 27.2: Effect of Saline Irrigation on Proline Content (μg g^{-1}) in Selected Tree Species

	Neem					*Jack*					*Acacia*				
DAT	0	60	120	240	360	0	60	120	240	360	0	60	120	240	360
C	24	23	25	22	23	26	27	28	28	28	25	24	26	23	22
EC4	23	223	210	62	53	27	72	63	43	28	25	184	155	68	65
EC8	25	227	219	65	63	28	146	129	73	29	26	218	193	79	78
EC12	26	320	300	68	65	29	152	130	75	32	24	410	363	83	78
EC16	26	336	321	75	70	29	170	145	83	37	29	427	385	90	81
EC20	23	520	480	80	75	30	173	143	60	39	24	560	490	92	88
Mean	24	275	259	62	58	28	123	106	60	32	25	304	270	73	69
P 0.01	NS	3.0	7.3	8.5	6.6	NS	5.8	3.0	4.4	9.5	NS	14.5	6.7	8.2	7.7
CV (%)	4.1	0.76	1.9	16.1	7.8	5.9	3.7	2.0	5.0	18.0	7.2	3.3	1.7	7.8	7.9

	Eucalyptus					*Pongamia*					*Tamarind*				
DAT	0	60	120	240	360	0	60	120	240	360	0	60	120	240	360
C	40	41	38	42	40	33	34	32	35	30	35	34	30	34	35
EC4	39	201	180	75	42	34	146	140	80	38	33	81	70	45	35
EC8	40	121	200	72	48	35	290	280	75	39	34	162	139	53	45
EC12	38	227	210	83	50	34	430	415	90	42	32	176	153	55	47
EC16	38	245	220	91	58	32	490	480	100	51	33	188	162	60	50
EC20	39	282	232	100	60	29	500	475	110	65	34	200	170	75	52
Mean	39	200	180	77	50	33	316	304	82	44	34	140	I21	54	44
P 0.01	NS	15.2	13.4	9.8	10.2	NS	6.7	8.3	7.2	6.6	NS	7.7	4.2	8.9	4.9
CV (%)	4.8	6.9	7.8	8.2	9.5	5.2	1.5	3.9	6.2	7.3	8.3	4.3	7.8	3.3	6.2

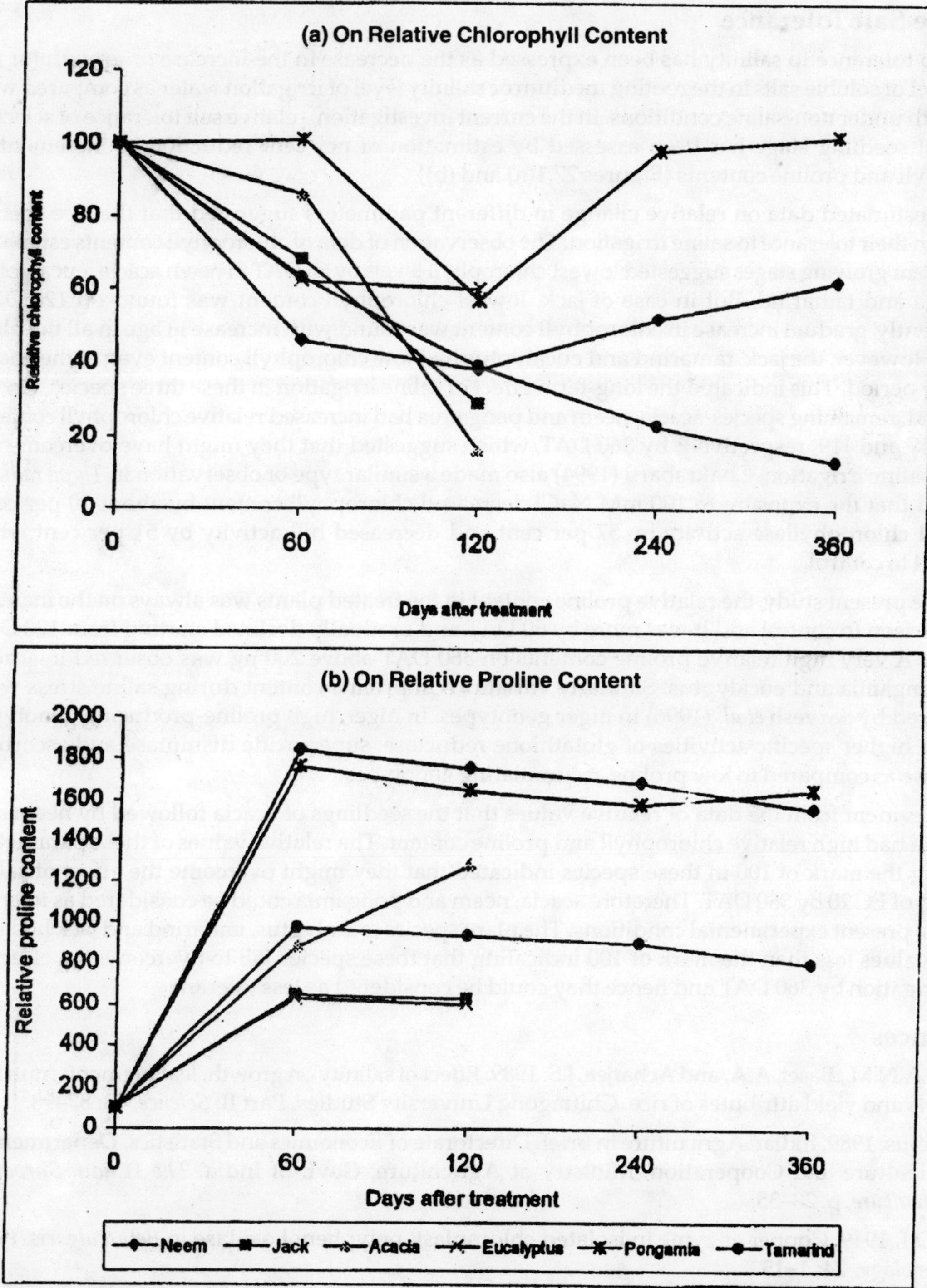

Figure 27.1: Effect of Induced Salinity by Saline Irrigation of EC 20 in Different Tree Species

Relative Salt Tolerance

Crop tolerance to salinity has been expressed as the decrease in the increase or growth for the given level of soluble salts in the rooting medium or salinity level of irrigation water as compared with the growth under non-saline conditions. In the current investigation, relative salt tolerance of selected species at seedling stage has been assessed by estimation of per cent reduction or increment in chlorophyll and proline contents (Figures 27.1(a) and (b)).

The estimated data on relative change in different parameters suggested that the tree species differed in their tolerance to saline irrigation. The observation of data of chlorophyll contents estimated at 5 different growing stages suggested lowest chlorophyll levels by 60 DAT in neem acacia, eucalyptus, pongamia and tamarind. But in case of jack, lowest chlorophyll content was found on 120 DAT. Subsequently, gradual increase in chlorophyll content was found with increase in age in all the plant species. However, the jack, tamarind and eucalyptus had low chlorophyll content even at the end of the study period. This indicated the long-term effect of saline irrigation in these three species. On the other hand, remaining species-acacia, neem and pongamia had increased relative chlorophyll contents of 139, 137 and 119, respectively, by 360 DAT, which suggested that they might have overcome the effect of saline irrigation. Chakrabarti (1994) also made a similar type of observation in *Tigna radiata*. He found that the exposure to 100 mM NaCl decreased chlorophyll content by about 50 per cent, increased chlorophyllase activity by 57 per cent and decreased hill activity by 51 per cent when compared to control.

In the present study, the relative proline content in the treated plants was always on the increase in comparison to control and it was more by 60 DAT and gradually declined starting from 120 DAT onwards. A very high relative proline contents on 360 DAT above 200 µg was observed in acacia, neem, pongamia and eucalyptus. Similarly variations in proline content during saline stress were also noticed by Sarvesh *et al.* (1996) in niger genotypes. In niger, high proline-producing genotypes exhibited higher specific activities of glutathione reductase, super-oxide dismutase and ascorbate peroxidase as compared to low proline accumulating genotypes.

It is evident from the data of relative values that the seedlings of acacia followed by neem and pongamia had high relative chlorophyll and proline content. The relative values of these parameters exceeding the mark of 100 in these species indicated that they might overcome the effect of saline irrigation of EC 20 by 360 DAT. Therefore acacia, neem and pongamia could be considered as tolerant under the present experimental conditions. The plant species–eucalyptus, tamarind and jack had low relative values less than the mark of 100 indicating that these species fail to overcome the effect of saline irrigation by 360 DAT and hence they could be considered as less tolerant.

References

Alamgir, A.N.M., Baset, A.A. and Acharjee, J.S. 1989. Effect of salinity on growth, leaf pigments, nutrient levels and yield attributes of rice. Chittagong University Studies. Part II: *Science*, 13: 87–98.

Anonymous. 1989. Indian Agriculture in Brief. Directorate of Economics and Statistics, Department of Agriculture and Cooperation, Ministry of Agriculture, Govt. of India. *The Hindu: Survey of Agriculture*, p. 28–35.

Arnon, D.I. 1949. Copper enzyme in isolated chloroplast, polyphenol oxidase in *Beta vulgaris*. *Plant Physiology*, 24: 1–15.

Bates, L.S., Waldren, R.P. and Teare, I.D. 1973. Rapid determination of free proline in water stress studies. *Plant and Soil*, 38: 205.

Chakrabarti, N., Subhendu, M. and Mukherji, S. 1994. Effect of IAA, Gibberlic acid and kinetin foliar spray in C3 glycophyte *Vigna radiata* under salt stress. *Plant Physiology and Biochemistry*, 21: 59–65.

Gomez, A.K. and Gomez, A.A. 1984. *Statistical Procedure for Agricultural Research II*. John Wiley and Sons, Inc., New York, p. 464–470.

Gupta, S.K. and Gupta, I.C. 1997. *Management of Saline Soils and Waters*. Scientific Publishers, Jodhpur, p. 98–103.

Mass, E.V. 1990. Crop salt tolerance. In: *Agricultural Salinity Assessment and Management Manual*, (Ed.) Tanji NSCE, New York, p. 262–304.

Sarvesh, A., Avkadha, M., Pullaiah, T., K. Reddy, T.P. and Kishore, P.B.K. 1996. Salt stress and antioxidant response in high and low proline producing cultivars of niger. *Indian Journal of Experimental Biology*, 34: 252–256.

Todd, N.M. and Reed, D.W. 1998. Characterizing salinity limits of New Guinea impatiens in recirculating sub irrigation. *Journal of American Society of Horticultural Science*, 123: 156–160.

Chapter 28

Interaction Studies on Bioinoculant and Inorganic Fertilizers on Growth Performance of Cole Crops

N.S. Kulkarni*, J.V. Jaiswal* and M.G. Bodhankar

**Department of Microbiology, Rajasthan Aryan College, Washim – 444 505, M.S., India*
***Department of Microbiology, Bharati Vidyapeeth's M.B.S.K. Kanya Mahavidyalaya, Kadegaon, Sangali – 416 416, M.S., India*

ABSTRACT

As bioinoculant have number of advantages over chemical fertilizers and significantly contribute to improve crop productivity. An attempt has been made to study the effect of bioinoculant singly and in combination with different levels of fertilizers on cauliflower. The study revealed that maximum germination count was observed with bioinoculant treatment. Seed germination percentage was more with co-inoculation as compared to single inoculation. Co-inoculation with 3/4th dose of fertilizer was found to be equally effective to full dose of fertilizer over control.

Keywords: *Co-bioinoculant, Chemical fertilizer, Cole crops.*

Introduction

The main practice of obtaining maximum yield in agriculture production is judicious use of chemical fertilizers, which at present is proving a set back in increasing the yield. The extensive use of chemical fertilizer is wasting thousands of hectares of land in India secondly due to high cost and short supply use of chemical fertilizer is restricted to rich farmers only, (Motsara, 1995). Therefore more attention should be given to harness the potential benefits of bioinoculant as plant nutrient supply and for improvement of agriculture productivity.

Hence, use of bioinoculant formulation had recently started to achieve better productivity. However, frequently bioinoculant were practiced for cereals, pulses, grains etc. Among all the vegetable cole crops, especially cauliflower and cabbage removes large quantities of major nutrients from soil. An average crop of cauliflower produces 16 t/ha of marketable curd, removes 46 kg of nitrogen, 16 kg of phosphorous and 55 kg of potassium. (Nath, 1976). On other hand owing to lack of initial soil nutrient the plant remain stunted and put on poor leaf growth and ultimately fails to form curd, (Purewal, 1957). Insufficient manure and fertilizer supply also results in arrest of plant growth.

Review of literature shows that much of the attention has been given on the use of biofertilizer for cereals and grain legumes. However, very few citations on use of biofertilizer for vegetable crops could be traced out, so more attention is needed to use biofertilizer for vegetable crops in India. Thus, an attempt has been made to study the effect of bioinoculant on growth parameters of cauliflower using *Azotobacter, Azospirillum,* and *Pseudomonas* individually and in combination with different levels of fertilizer. Similarly, interaction study between level of fertilizers and bioinoculant treatment were carried out.

Materials and Methods

Pure cultures of *Azotobacter chorococum, Azospirillum brasilense* and *Pseudomonas strita* were obtained rom Dr. Panjabrao Deshmukh Krishi Vidyapeeth, Akola. The seeds of variety (cauliflower Pusa Snow all K-1) were purchased from National Seeds Corporation Limited. The experiments were carried out private farm at Patur, Akola.

Germination studies were carried out in triplicate on seed beds of size 1 m × 1 m. 100 seeds were treated with carrier-based bioinoculant *viz., Azotobacter, Azospirillum,* and *Pseudomonas* singly and in combination (Table 28.1). Seeds were dried and immediately sown on respective seeds beds and covered to a thickness of 0.5 cm with fine shifted soil. Simultaneously, seed mixing recommended dose, 3/4th dose of fertilizer count and average number of leaves in raised seedlings were recorded on 14th DAS.

Table 28.1: Effect of Different Treatments on Germination Count and Number of Leaves in Seed Bed on 14 DAS

Sl.No.	Particulars	Germination (%) (Mean of Triplicate)	Average No. of Leaves
1.	Seed Treatment with Azotobacter (S1)	79	7
2.	Seed Treatment with Azospirillum (S2)	65	4
3.	Seed Treatment with Pseudomonas (S3)	72	6
4.	Seed Treatment with Azotobacter + Azospirillum (S4)	72	6
5.	Seed Treatment with Azotobacter + Pseudomonas (S5)	82	7
6.	Seed Treatment with Azospirillum + Pseudomonas (S6)	68	5
7.	Seed Treatment with Azotobacter + Azospirillum + Pseudomonas (S7)	72	6
8.	No seed treatment (Control) (S8)	66	4
9.	Full Dose of fertilizer (NPK) (T1)	69	5
10.	3/4th dose of fertilizer (T2)	67	4
11.	No dose of fertilizer (T3)	65	4

Young seedlings were carefully excavated, treated with bioinoculant and transplanted to new beds. Similarly, the young seedlings from the fertilizer-enriched beds were transplanted to respective beds.

Results and Discussions

The results obtained by the effect of seed treatment and fertilizer treatments on seed germinatior are presented in Table 28.1. The data reveals that seed treatment with biofertilizer resulted in increasec germination. Maximum germination was obtained in *Azotobacter* + *Pseudomonas* (S5) treatment followec by *Azotobacter* (S1). Maximum number of leaves on 14th DAS was obtained in S5 and S1 treatment respectively. It was observed that due to increased dose of fertilizer germination percentage als increased but the increase was not superior over seed treatment with *Azotobacter* + *Pseudomonas* (S5) Combined inoculation of *Azotobacter* + *Pseudomonas* showed satisfactory improvement in germinatio count and number of leaves among all other treatments. Chhonkar *et al.* (1964) reported that applicatio of gibberellic acid is effective in increasing growth along with improvement in quality and yield o cauliflower. *Azotobacter* and *Pseudomonas* are non-symbiotic free living aerobic bacteria having goo capacity to fix nitrogen and phosphate solubilization respectively, as well as have special advantag of secreting growth promoting hormones like thiamin, riboflavin, pyridoxine, pantothenic acid, indo acetic acid and gibberellins or gibberllin like substances and this may be the reason for beneficial effec of seed colonization with *Azotobacter* + *Pseudomonas* or *Azotobacter* alone on improved germination an number of leaves in cauliflower. Hence, our findings can be correlated with the findings of abov workers.

Effect of different treatments on growth parameters was studied and mean effect of randoml selected 10 plants in a row are presented in Table 28.2. Seedling root treatments with bacterial inoculant showed improvement in growth parameters over uninoculated control (Herandez, 1994). Treatmen with *Azotobacter* (Nitrogen fixer) in combination with *Pseudomonas strita* gave maximum number c leaves on maturity followed by *Pseudomonas* and *Azotobacter* respectively but it was not superior ove full dose of fertilizer. *Pseudomonas strita* and 3/4th dose of fertilizer gave same number of leaves o maturity among all the plants tested.

Full dose of fertilizer (T1) treatment showed faster growth rate in curd initiation than bioinocular treatments. It was observed that treatment with *Azotobacter, Pseudomonas* and their combination ar similarly effective as compare to treatment with 3/4th dose of fertilizer in increasing growth rate fc curd initiation. The curd diameter and fresh weight of curd was found to be improved over uninoculate control. Treatment with *Azotobacter* + *Pseudomonas* gave maximum curd diameter and fresh weight c curd among all bacterial treatments. But the results obtained by bioinoculant treatments were nc superior than fertilizer treatment alone (T1). Application of fertilizer had significant effect on cur diameter and fresh weight of curd. Curd diameter and weight on curd was maximum in T1 treatmer (full dose of fertilizer) followed by T2 (3/4th dose of fertilizer) and was minimum in the treatmer where no fertilizer was applied. Even though bioinoculant has not showed superior results over fu dose of fertilizer the results are considerable as compared to them and are encouraging in most of th parameter studied with combine application of *Azotobacter* + *Pseudomonas* followed by othe combinations of biofertilizer over single culture inoculations similar report on co-inoculation ha been given by Yadav *et al.* (1994); however, their work was with Cicer arietinum. The results of interactic between levels of fertilizer and bioinoculant treatments were found to be significant for the grow parameter. The details are presented in Table 28.3. It could be seen that full dose of fertilizer + *Azotobact* + *Pseudomonas* (T1, S5) showed maximum improvement in all growth parameters as compared to a

Table 28.2: Flow Sheet for Field Trials

No. of Rows	No. of Plant/Row	Treatment	Particulars		
1	20	S1	S1 = Azotobacter chroococum		
2	20	S2	S2 = Azospirillum brasilense		
3	20	S3	S3 = Pseudomonas strita		
4	20	S4	S4 = Azotobacter + Azospirillum		
5	20	S5	S5 = Azotobacter + Pseudomonas		
6	20	S6	S6 = Azospirillum + Pseudomonas		
7	20	S7	S7 = Azotobacter + Azospirillum Pseudomonas		
8	20	S8	S8 = Control (I) No treatment		
9	20	TI	Fertilizer treatment		
10	20	T2	T1 = Full dose of Fertilizers		
11	20	T3	N75 : P60 : K60		
12	20	T1S1	T2 = 3/4th dose of fertilizers		
13	20	T1S2	N56.25 : P 54 : K45		
14	20	T1S3	T3 = Control (II) No Treatment		
15	20	TIS4			
16	20	TIS5	Meteorological Data 1998–99		
17	20	T1S6	(A) Temperature °C		
18	20	T1S7		Min	Max
19	20	T1S8	July	29.78	23.86
20	20	T2S1	Aug	30.4	22.9
21	20	T2S2	Sept	31.8	26.4
22	20	T2S3			
23	20	T2S4	(B) Rainfall in mm (Average)		
24	20	T2S5	July–53.0 mm		
25	20	T2S6	Aug–50.75 mm		
26	20	T2S7	Sept–26.4 mm		
27	20	T2S8			
28	20	T3S1	(C) Soil pH means at 1% Level		
29	20	T3S2	July–6.8		
30	20	T3S3	Aug–5.9		
31	20	T3S4	Sept–6.3		
32	20	T3S5			
33	20	T3S6	Design: Factorial Randomized		
34	20	T3S7	Block Design		
35	20	T3S8	Replication: 3		

other combinations. The results of T1 S5 treatment are surprising and are superior over T1 (full dose alone) as well as T1S1, T1S3 combination was also found to be more significant for growth parameters

Table 28.3: Effect of Different Treatment on Growth Parameters of Cauliflower

Sl.No.	*Treatment*	*Mean Effect of 10 Plant in the Row*			
		No. of Days of Curd Initiation from Sowing	*No. of Leaves on Maturity*	*Curd Diameter (cms)*	*Fresh wt of Curd (gms)*
1.	S1	66	21	11.8	600
2.	S2	70	16	7.8	400
3.	S3	68	22	11.8	575
4.	S4	74	18	11.4	580
5.	S5	65	24	13.5	625
6.	S6	72	17	9.4	490
7.	S7	70	20	11.8	595
8	S8	75	16	8.6	445
9.	T1	60	28	15.6	800
10.	T2	66	22	13.2	650
11.	T3	79	18	9.7	498
12.	Mean	69.54	20.18	11.32	568.90
13.	SE ±	1.85	0.81	0.46	15.12
14.	CD at 5%	5.27	2.31	1.29	43.11

Table 28.4: Effect of Interactions Between Levels of Fertilizers and Bioinoculant Treatment

Treatments	*No. of Leaves (In No.) (A)*			*Curd Initiation (Days) (B)*			*Curd Diameter (cms) (C)*			*Fresh wt. of Curd (cms) (D)*		
	T1	*T2*	*T3*	*T1*	*T2*	*T3*	*T1*	*T2*	*T3*	*T1*	*T2*	*T3*
S1	29	21	19	60	68	28	15.8	12.5	10.5	800	625	536
S2	28	19	16	69	65	70	15.2	10.5	8.5	799	512	436
S3	27	22	19	61	64	69	15.7	12.5	10.5	810	615	524
S4	28	20	17	60	69	68	15.8	12.3	10.3	800	628	526
S5	29	27	21	58	60	69	15.8	12.6	11.3	816	815	548
S6	27	21	17	63	69	73	15.5	11.3	9.3	805	637	491
S7	26	22	18	62	70	72	15.5	12.5	10.5	806	622	533
S8	26	19	18	69	68	75	15.2	10.9	8.9	802	547	469

Interaction Studies for comparing	SE±	CD at 5%
1. Leaves	1.02	2.55
2. Curd initiation	3.19	8.48
3. Curd diameter	0.55	1.37
4. Fresh Weight of Curd	26.0	74.07

than T1 full dose alone and is considerable for crop productivity improvement. Treatment with ¾ per cent dose of fertilizer (T2) along with bioinoculant showed compensating results for plant NPK requirement as compared to full dose of fertilizer. The data reveal that combined action of T2 + *Azotobacter* + *Pseudomonas* on all growth parameters studied are superior over full dose of fertilizer (T3) indicating that large amount of fertilizer could be saved if biofertilizer are used in combination with fertilizer. Also they help to improve crop growth, moreover the activity of bioinoculant was found to be unchanged at even full dose of fertilizer. Tandon (1995) has suggested the same findings on comparative study of fertilizer, organic manure and bioinoculant for plants. Even though our results are encouraging however, for proper economic and commercial implementation of the concept extensive research have to be carried out.

Chapter 29

Sol-gel Synthesis and Photoelectrochemical Behaviour of Si-doped α-Fe_2O_3

Anshul Agrawal*, Yatendra S. Chaudhary*, Vibha R. Satsangi**, Sahab Dass* and Rohit Shrivastav*#

**Department of Chemistry, **Department of Physics and Computer Science Dayalbagh Educational Institute, Dayalbagh. Agra – 282 005, India*

ABSTRACT

This study deals with the synthesis and characterization of Si-doped α-Fe_2O_3 by a new sol-gel route and its use as photocatalyst in photoelectrochemical splitting of water. The preparatory method involves the use of hexamethylene tetraamine–urea as key reagent. Resulting samples of sintered pellets were found photosensitive in the visible range of spectrum and the photocurrent values increased with the increase in electrolyte–pH.

Keywords: *Photosensitive materials, Sol-Gel synthesis, Hematite, Photoelectrochemical cell.*

Introduction

Doped oxides of transition metals find extensive use in gas sensors. dielectric ceramics, photovoltaic solar cells, pigment and also as photocatalysts in initiating and regulating chemical processes of industrial and environmental significance (Zhang, *et al.*, 2001; Song, *et al.*, 2001; Licht, 2001; Turner, 1999). Hence, many physical, chemical and physicochemical methods for their synthesis/ preparation have been developed (Finkela, 1988; Nizhusis *et al.*, 2001; Shinar and Kennedy, 1982). In

Corresponding Author, E-mail. rohitshrivastav_del@yahoo.com.

the conventional dry powder method of preparation, sintering behaviour and quality of doped oxides are greatly influenced by the quality of starting powders. Fine unagglomerated powders with narrow size distribution, possessing the requisite chemical and phase purity are essential for obtaining improved microstructure required in advanced photocatalysts. In contrast solution based routes closely meet these requirements, but most of them are tedious, and involve several intermediate stages of preparation, which raises the cost of material prepared. Besides many solution based methods are material specific and cannot be treated as general method of preparation (Bjorksten *et al.*, 1994 and Cherapy *et al.*, 1998).

The physical and chemical characteristics of doped/mixed metal oxides can be controlled by their particle size, morphology and crystalline phase. The wet-chemical synthesis, particularly the sol-gel approach, allows compositional and microstructural tailoring through controlling the precursor chemistry and processing conditions (Brinker and Schrere, 1990 and Livage *et al.*, 1988). Sol-gel processing provides for excellent chemical homogeneity and even the possibility of deriving unique metastable structures at low reaction temperatures. In essence, it involves the formation of a metal-oxo-polymer network from molecular precursors such as metal alkoxides or metal salts (Sanchez and Ribot, 1994). The relative rates of hydrolysis and precondensation strongly influence the structure and properties of the resulting metal oxides. Factors affecting the sol-gel process include the reactivity of metal salts, pH of the reaction medium, reaction temperature, nature of solvent and additives. Hence, by varying these processing parameters, material with different microstructure and surface chemistry can be obtained. Typically, sol-gel derived precipitates are amorphous in nature, requiring further heat treatment to induce crystallization, which may also lead to phase transformation.

This report deals with a new sol-gel route, involving the use of hexamethylene tetramine (HMT A), as key reagent, for the synthesis of doped metal oxides. The method reported here is simple, fast and has been employed, successfully, to synthesize Si-doped (1 and 2 per cent at.) Fe_2O_3 (hematite). The material, so prepared, has been characterized and its possible use as photocatalyst in photoelectrochemical (PEC) splitting of water has been explored.

Experimental

Stock solution ($\approx$ 650 cm^3) of ferric nitrate (2.18M) was prepared by dissolving $Fe(NO_3)_3 9H_2O$ crystals (99.99+%, Sigma Aldrich) in 1M nitric acid. Similarly nitrate solution of Si (0.58M) was prepared from tetra-ethoxy silane (Fluka, Chemie). Exact concentrations of above precursor solutions were determined gravimetrically. To synthesize doped Fe_2O_3, mixed nitrate solution of Fe and Si, with total volume $\approx$ 20 cm^3, was gradually gelled at room temperature by the slow addition of 3M HMT A-urea solution (prepared by dissolving 210 g HMTA and 90 g urea in water with solution volume 500 cm^3), with continuous stirring, The change in pH was recorded continuously employing a digital pH meter (Systronics, India). At pH $\approx$ 4.1, the ferric ions formed the hydrous oxide gel, with dopant ions entrapped within the gel matrix. The gel, obtained, was firstly soaked in ammonia solution to cause homogeneous precipitation of dopant ions at pH > 7, and then washed with dilute ammonia solution to remove water soluble compounds without causing any loss of metal hydroxides by peptisation. The washed gel when dried in air at 400°C resulted in precursor powder, which was subsequently calcined at a higher temperature ($\approx$ 800°C) to get doped Fe_2O_3 powder (A) free from carbonaceous matter.

In order to use as photocatalyst in PEC cell, 8–10 pellets were prepared, from obtained doped Fe_2O_3 powder (A), using a 15 mm die, by applying a load of $\approx$7 tons cm^{-2}. The pellets were sintered in air, in a SiC furnace, with temperature rising slowly to the maximum of 1200°C, at which these were kept for one hour. It was essential to avoid heat shock. Afterwards, the pellets were further heated at 1350°C in air, in a molybdenum disilicide furnace and then cooled quickly in air. Few pellets of each type were heated and cooled in partial vacuum of 5×10^5 millibar with a soak time of one hour at the maximum temperature of 1350°C. Sintered pellets were polished both sides using alumina powder and characterized for the nature of majority charge carriers (by observing Seebeck effect), and electrical resistance. Ohmic electrical contact, through a copper wire, was established from the surface of each pellet by employing conductive silver paste after rubbing the pellet's surface gently with gallium-indium eutectic. Excluding the central area of ≈ 1.0 cm² on the pellets surface, which is opposite to the contact surface, the pellets were covered by a non-conducting opaque epoxy resin, hysol (Dexter, Singapore). The pellets, converted into working electrodes, as above, were used in conjunction with platinum counter and saturated calomel reference electrodes. The PEC cell was filled with 40 cm³ of electrolytic solution ($NaOH/H_2SO_4$). The cell was also fitted with a quartz window for illuminating the photocatalyst with a 100W white light source (Bentham). With this set-up, along with a 5 cm thick water jacket to prevent the passage of infrared radiations, the illumination at the photocatalyst surface was found ≈ 350 lumens. PEC studies were conducted maintaining electrolyte pHs 4, 11 and 13 at 27±2°C. Using a potentiostat (ECDA 001, Conserv Enterprises), each set of experiment was repeated 3–5 times and the results were found reproducible within±4.5 per cent deviation. Only A.R. grade reagents (Unless otherwise specified) and double distilled deionized water were used throughout 7 the study.

Results and Discussion

With the adopted route of synthesis, the yield of doped Fe_2O_3 was > 96 per cent of the theoretically computed yield of 3.488 g. Table 29.1 depicts some physico-chemical properties of doped Fe_2O_3 pellets prepared, which were dark-brown coloured, shiny, smooth, had high resistance and exhibited n-type conductivity. Compared to air-sintered pellets, resistances of pellets sintered in partial vacuum were found less nearly by an order. The resistances of doped samples were significantly less compared to undoped samples, but the decrease in resistance could not be equated to the amount of dopant present. XRD analysis of all the synthesized sintered Fe_2O_3 samples, employing Philips X-ray diffractrometer and CuK-alpha 1 wavelength revealed a similar pattern (Figure 29.1), showing strong peaks from planes (104), (110), (116), (024), (012), (214), (300) and (113), with relative intensities approximately 100, 65, 47, 35, 27, 31, 28 and 23 per cent, respectively. This confirmed the formation of polycrystalline α-Fe_2O_3 (hematite) in chemically homogenized form. Average particle size in doped Fe_2O_3 samples, determined by light scattering technique using a particle size analyzer, ranged from 200 to 300 nm, which indicated a fine distribution of particles in the samples.

Table 29.1: Some Physico-chemical Properties of Doped Fe_2O_3 Pellets Sintered in Air

Dopant Concentration (% at.)	Sintered Pellet Size			Density		2-probe Resistance Across the Diameter
	Diameter (mm)	Thickness (mm)	Weight (g)	(g cm⁻³)	% TD*	
1	13.0	1.09	0.4242	3.14	60 %	10.2 Ω (995 Ω)
2	12.0	119	0.5290	3.93	76%	2.9 K Ω (410 Ω)
Undoped	13.5	0.67	0.3001	3.13	70%	12.1 K Ω (1.3 K Ω)

Values, in parenthesis, are for vacuum-sintered samples

* Percentage theoretical density (based on the theoretical density of 5.2 g cm⁻³ for α-Fe_2O_3).

Probable use of doped Fe_2O_3 pallets for PEC splitting of water has been investigated by observing current (I) – voltage (V) characteristics of the PEC cell, with applied bias voltage ranging from +2 V to –2 V. By computing the difference in current values recorded under illumination (*i.e.* when the semiconductor working electrode was illuminated with the light source) and under darkness (*i.e.*, when the PEC cell was enclosed in a dark chamber), the net photocurrent generation was determined (Figure 29.2). PEC measurements also included the determination of open circuit photopotential (η, defined as the difference of potentials recorded, between semiconductor working electrode and saturated calomel reference electrode, under illumination and in darkness) and onset voltage (V_{os}, where V_{os} is the minimum anodic bias voltage needed at which significant current *i.e.*, $\geq 10\ \mu A\ cm^{-2}$ was noticed to flow through the PEC cell. Results obtained at three different electrolyte pHs have been presented in Tables 29.2 and 29.3.

Table 29.2: Observed Negative Values of Photopotential (η), Recorded with Air-sintered Samples (Unless otherwise stated)

Dopant Concentration (% at.)	*Observed Photopotential (η)* (mV)*		
	pH 4	*pH 11*	*pH 13*
1	24 (21)	192 (205)	268 (266)
2	29 (32)	203 (210)	274 (284)
Undoped	11 (15)	161 (153)	183 (198)

Values, in parenthesis, are for vacuum-sintered samples.

Table 29.3: Observed Values of V_{onset}

Dopant Concentration (% at.)	*Electrolyte pH*	*V_{onset} (mV)*	
		Under Darkness	*Under Illumination*
Air sintered samples			
1 :	4	1601	1499
	11	1302	1200
	13	603	304
2 :	4	1303	898
	11	1000	597
	13	1102	503
Undoped:	4	1750	1580
	11	1315	1100
	13	925	610
Vacuum sintered samples			
1 :	4	1200	1003
	11	902	603
	13	503	205
2 :	4	1197	998
	11	902	603
	13	603	201
Undoped :	4	1765	1615
	11	1340	1135
	13	915	555

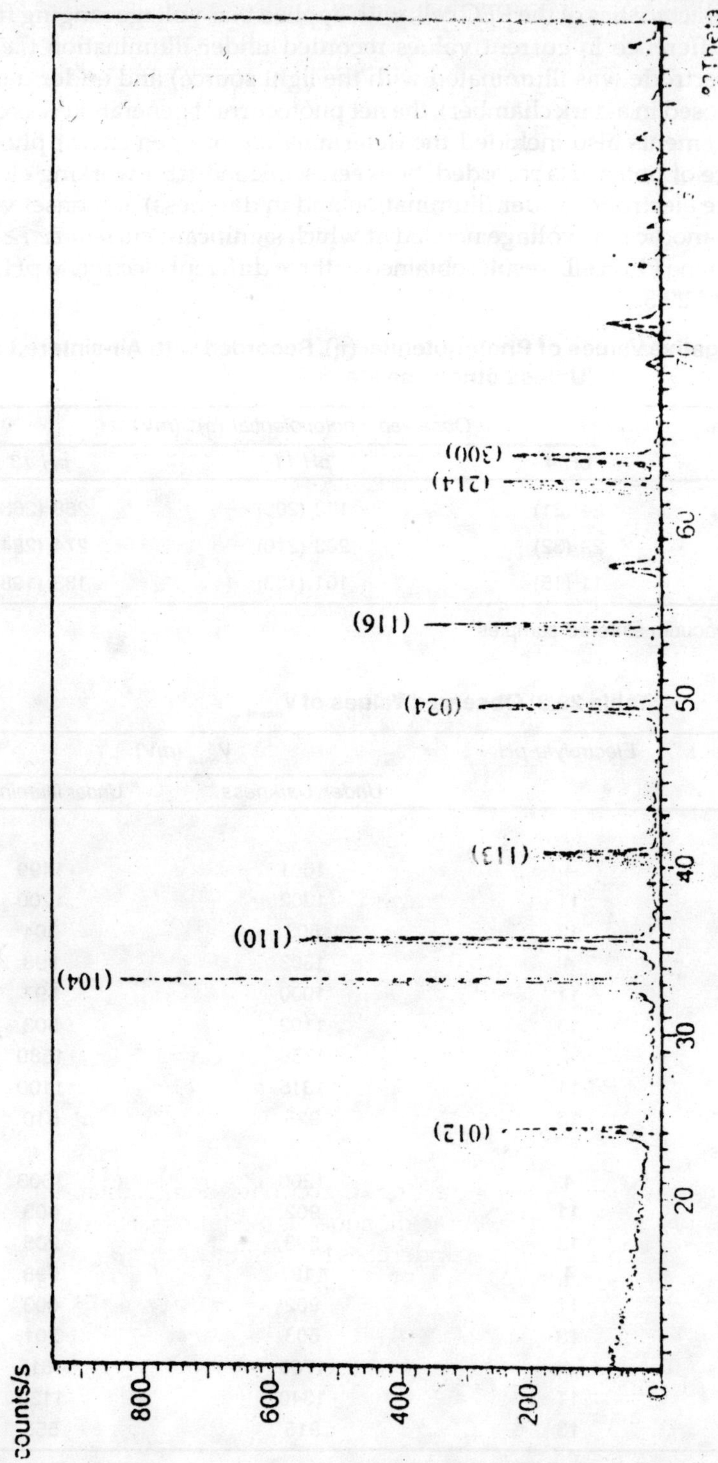

Figure 29.1: XRD Pattern Observed with the Air-sintered Samples of 2% at. Si Doped Fe_2O_3

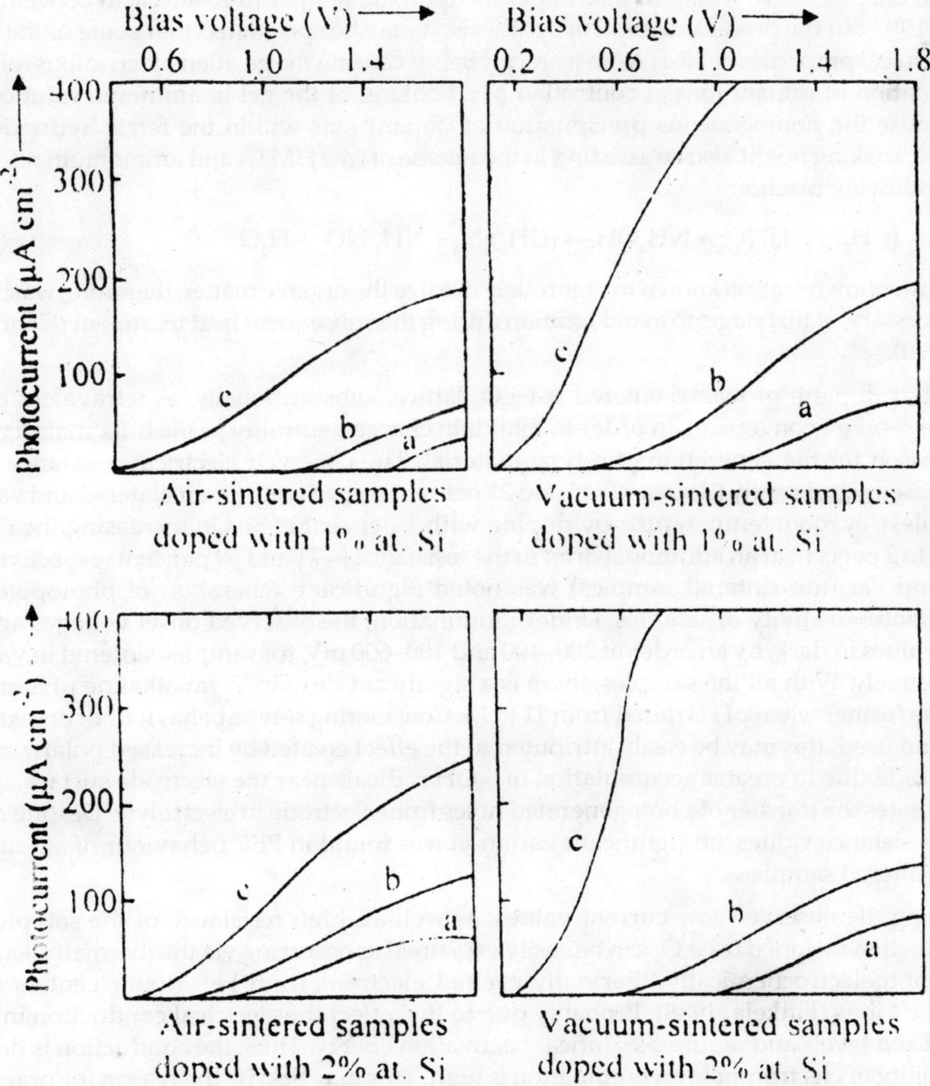

Figure 29.2: The Observed Net Photocurrent Values Under Varying Anodic Bias Electrolyte pH: 4(a), 11(b), 13(c)

The pH of HMTA solution ranges 7–9 depending on its concentration (Kamat *et al.*, 1996). In aqueous medium HMTA is quickly protonated, but in acidic aqueous medium, particularly in presence of complexing agent, *viz.*, urea, protonated HMTA undergoes hydrolytic decomposition at pH ≈ 3.8 (Collins *et al.*, 1987).

$$(CH_2)_6N_4 + H^+ \rightarrow (CH_2)_6N_4H^+ \quad (1)$$

$$(CH_2)_6N_4H^+ + 3H^+ + 6H_2O \rightarrow 4NH_4^+ + 6HCHO \quad (2)$$

The occurrence of above reactions also depends upon the temperature and acid concentration (Collins *et al.*, 1987). In the present method of synthesis, it may be speculated that some of the above mentioned chemical properties of HMTA-urea are, probably, causing the gelation of ferric ions followed by the precipitation of dopant ions at controlled pH. Soaking of the gel in ammonia solution was intended to cause the homogeneous precipitation of dopant ions within the ferric hydroxide gel matrix. Besides, soaking might also be assisting in the release of free HMTA and ammonium nitrate, as given by the following reaction.

$$(CH_2)_6N_4\,HNO_3 + NH_4OH \rightarrow (CH_2)_6N_4 + NH_4\,NO_3 + H_2O \qquad (3)$$

Since, ammonium nitrate is known to vigorously oxidize the organic matter, therefore, washing of the gel was necessary, at this stage, to avoid ignition during the subsequent heat treatment (Cotton and Wilkinson, 1990).

Si, added as dopant, probably, entered α-Fe_2O_3 lattice, substitutionally, as tetravalent cation, resulting in $Fe^{+3} \rightarrow Fe^{+2}$ conversion, in order to maintain charge neutrality in the bulk material. This may be the reason for the generation of n-type material. The observed electrical resistance of the material decreased significantly (decrease ≈ 16 and 23 per cent respectively for air sintered and vacuum sintered samples), at room temperature, by doping with 1 per cent at Si. On increasing the dopant concentration to 2 per cent at an additional drop in the resistance (≈ 71 and 59 per cent respectively, for air sintered and vacuum sintered samples) was noted Significant generation of photopotentials suggests the photosensitivity of samples. Under illumination, the observed onset voltages are less, compared to values in dark, by an order of 200–400 and 100–600 mV, for samples sintered in vacuum and air, respectively. With all the samples, there is a significant drop in V_{os} at alkaline pHs and the values decrease further when pH is raised from 11 to 13. Considering n-type behaviour of doped Fe_2O_3 synthesized and used, this may be easily attributed to the effect created by increased polarization of working electrode due to greater accumulation of – OH radicals near the electrode surface, which, probably, facilitates the transfer of photogenerated holes from electrode to electrolyte. Despite a large difference in resistance values, no significant variation was found in PEC behaviour of air-sintered and vacuum-sintered samples.

Considering the observed low current values, as well as, high resistance of the samples, the electrical conduction in doped α-Fe_2O_3 can be easily explained as occurring via the thermally activated hopping of photoelectrochemically/thermally excited electrons from Fe^{2+} donor centers to the surrounding Fe^{3+} ions (Finkela, 1988). Probably due to this effect the electrical conduction in Fe_2O_3 occurs in localized levels and requires significant activation energy. Thus, the conduction is delayed and the possibility of electron-hole recombination is high. This may also be the reason for practically no photocurrent observed at low bias voltages. However, at bias voltage > V_{os}, the emergence of photocurrent is clearly evident by a significant gain in current recorded under illumination compared to darkness. In addition to the above stated hopping mechanism, the observed low conductivity of synthesized samples may also be attributed to the interferences created by impurities, grain boundaries and crystal defects. Although, this is a typical problem associated with most submicro-nanostructured semiconductors, particularly prepared by sol-gel route, but their role in suppressing current values looks more severe with α-Fe_2O_3.

The continuous use of photocatalysts, synthesized, in PEC cell, did not show any measurable corrosion/dissolution of the material even after 30–60 days. This was confirmed by microscopic examination, which did not indicate any deterioration in the electrode surface exposed to electrolyte.

Besides, weights of working electrodes were also observed to remain constant (within±0.08 per cent error) during this period, which again confirmed the high stability of the material synthesized.

Finally, it may be concluded that through the route of synthesis adopted in this study homogenized samples having fine particle size distribution are generated and, hence, the method may be treated as a general purpose method for the low-cost synthesis of doped metal oxides, which have many applications as photocatalyst The synthesized photocatalyst powder may be used in the form of pellets of appropriate dimensions, as employed here, or alternatively may also be spin coated on suitable substrates to have thin films of photocatalyst, It is also hoped that by regulating the experimental conditions of reactions (1)–(3), probably the nanoparticles of metal oxides can be generated quite economically. Work on this direction is in progress and results will be communicated soon.

Acknowledgements

Financial assistance from All India Council of Technical Education, Govt. of India, New Delhi (F.No. 8020/RID/R&D–118/2001–02) is gratefully acknowledged. Authors are also grateful to Dr. V.N. Vaidya, Dr. R.V. Kamat and Dr. K.T. Pillai, Fuel Chemistry Division, Bhabha Atomic Research Center, Bombay, for the initial help in the work and to Dr. Umesh Tiwari of Solid State Physics Laboratory, Delhi, for XRD analysis of samples.

References

Bjorksten, U., Moser, J. and Gratzel, M. 1994. *Chem Mater*, 6: 858.

Brinker, C.J. and Scherer, G.W. 1990. *Sol-Gel Science.* Academic Press, San Diego.

Cherepy, N.J., Liston, D.B., Lovejoy, J.A., Deng, H. and Zhang, J.Z. 1998. *J. Phys. Chem.*, 102: 770.

Collins, J.L., Lloyd, M.H. and Fellows, R.L. 1987. *Radiochim Acta*, 42: 121.

Cotton, F.A. and Wilkinson, G. 1990. *Advanced Inorganic Chemistry.* Wiley Eastern Limited, New Delhi, India, pp. 349.

Finkela, H.O. 1988. *Semiconductor Electrodes: Studies in Physical and Theoretical Chemistry.* Elsevier, Amsterdam, p. 147–202.

Kamat, R.V., Pillai, K.T., Vaidya, V.N. and Sood, D.D. 1996. *Mat. Chem. Phys.*, 46: 67.

Licht, S. 2001. *J. Phys. Chem.*, 105: 6281.

Livage, J., Henry, M. and Sanchez, C. 1988. *Prog Solid State Chem.*, 18: 259.

Nizhusis, T.A., Beers, A.E.W., Vergunst, T., Hoek, I., Kapteijn, F. and Moulijn, J.A. 2001. *Catal Rev.*, 43: 345.

Sanchez, C. and Ribot, F. 1994. *New J. Chem.*, 18: 1007.

Shinar, R. and Kennedy, J.H. 1982. *Solar Energy Mat.*, 6: 323.

Song, K.Y., Park, M.K., Kwon, Y.T., Lee, K.W., Chung, W.J. and Lee, W.I. 2001. *Chem., Mater.*, 13: 2349.

Turner, J.A. 1999. *Science*, 285: 687.

Zhang, L., Kanki, T., Sano, N. and Toyoda, A. 2001. *Solar Energy*, 70: 331.

Chapter 30

Assessment of Relative Efficacy and Economics of Chemical and Bioinsecticides for Management of Fruit Borer of Okra

Rabindra Prasad, Muneshwar Prasad, Rajesh Kumar and Devendra Prasad

Department of Entomology, Birsa Agricultural University, Kanke, Ranchi – 834 006

ABSTRACT

In order to search out multiple choice of alternative insecticides, to be used during each and every spray belonging to the varied group within the purview of IPM, bio-efficacy of thirteen insecticides of diverse categories was evaluated alongwith their economics in the field for the two consecutive years against the highly injurious pest, *Earias vitella* Fab. infesting okra. The results revealed that all the insecticides proved to be effective in minimising the fruit damage and maximising the yield of marketable fruits of okra and net economic return. In this respect, the performance of different categories of insecticides were found in order of: synthetic phrethroids (alphamethrin > deltamethrin > fluvalinate > cypermethrin etc.) > microbial product (Btk) > OP insecticides (quinalphos > chlorpyriphos) > carbamate insecticide (thiodicarb) > IGR (diflubenzuron) > botanicals (Neem and Karanj oil). Hence, it would be wiseful to choose insecticides of varied group by altering them during each spray on the basis of eco-toxicological and economical grounds in order to avoid the chance of development of insecticidal resistance in the pest species with a view to apply the insecticide(s) as need based, judicious and supplementary part of IPM for the sustainable crop production.

***Keywords:** Okra, Earias vitella, Alternative insecticides, Economics, IPM.*

Introduction

Fruit borer (*Earias vitella* Fab.) is the most important, constantly occurring and dreaded pest of okra. The larvae of the insect damage any developing stage of okra fruits, resulting them to become unfit for human consumption. The pest infestation are responsible for the very high loss in the fruit yield amounting to 35 per cent (Krishnaiah, 1980). However, earlier the yield loss was recorded even upto 40 to 50 per cent (Sriniwasan and Gowder, 1959). With the objective to minimise the avoidable yield loss, the present experiment was conducted to combat the pest menace by applying only need based foliar sprayings of insecticides. For the purpose, comparative efficacy and economics of some chemical as well as bio-pesticides for the control of *E. vitella* was estimated in the field conditions for two years.

Materials and Methods

The experiments were conducted with okra, variety 'Prabhani Kranti' in the two consecutive years, 2003 and 2004 during spring-summer season (February to June) in the farmers fields, as operational research programme, nearby main campus of Birsa Agricultural University, Ranchi. Total of fourteen treatments, including one untreated control were accommodated in a randomized block design with three replications, details of which are given in Table 30.1. Among the thirteen insecticides used, five belong to synthetic pyrethroids, two belong to organophosphorous (O.P.) compound, three fall under botanical pesticides, one each belongs to chitin inhibitor (IGR), carbamate and microbial group. The foliar spray-treatments were applied three times at three weeks intervals, commencing first spray 40 days after sowing of okra. Okra was sown in the second week of February, during 2003 and 2004. The plot size was kept 4.0×2.5 m^2.

Recommended packages of practices were adopted for raising the crop. After 40 days of sowing, okra plants were sprayed with knap sach sprayer with respective insecticidal solution in the required concentrations as per the planned treatments set right in the layout plan.

Plucking of okra fruits was done thrice in a week for recording fruit-yield in terms of kg per plot basis and finally calculated into q/ha. Total number of fruits and number of damaged fruits were recorded treatment and replication wise to calculate per cent fruit damage at the time of each picking/plucking of fruits. Data of all observations were ultimately added together treatment and replication wise for the respective parameters, after tabulating, to workout the pooled mean of the two years. Accordingly, percent figure were transformed into angular values for the statistical analysis. The data obtained in the compiled forms, treatment and replication wise, for the respective years and that of pooled values (*i.e.* means) of the two years were subjected to the statistical analysis. The pooled mean(s) of the two year's data on yield of healthy marketable fruits and percentage of damaged fruits were used to estimate the comparative economics (net return) and benefit cost ratio (BCR) of each of the insecticidal application.

Results and Discussion

The results data pertaining to incidence of fruit borer (percentage of fruit damage) recorded during 2003 and 2004 are presented in Table 30.1; while the estimate of comparative economics of respective insecticides are shown in Table 30.2.

Fruit Borer Infestation

During 2003, all the insecticidal treatments were significantly effective in reducing the pest incidence (<16.30 per cent fruit infestation in case of Karanj oil) as compared to the untreated control

(24.83 per cent). Deltamethrin proved to be significantly the best in minimising the fruit damage (1.43 per cent). Other synthetic pyrethroids *viz.*, alphamethrin, fenvalerate, cypermethrin and fluvalinate remained at par in reducing the fruit damage which ranged from 2.13 to 4.32 per cent. It is worthwhile to note that almost all the synthetic pyrethroids proved superior over other insecticides, excepting the microbial pesticide *e.g. Bacillus thuringiensis* var. *kurtaski* (Btk) applied @ 0.75 kg/ha in the form of Delfin (4.03 per cent fruit infestation). Next to all these synthetic pyrethroids and Btk, a carbamate insecticide *e.g.* thiodicarb @ 0.75 per cent provided highly significant protection to okra against fruit borer, resulting to as lower as 7.85 per cent fruit damage. The rest of the insecticides including the chitin inhibitor (IGR) *e.g.* diflubenzuron remained at par, excepting the botanical pesticides. Neemark applied at 0.06 per cent afforded the significant protection to okra against *E. vitella* and Neem oil @ 4 per cent remained on par. Karanj oil applied at 4 per cent appeared to be the least effective giving rise to 16.30 per cent fruit infestation, but significantly superior to untreated plants which were found to be suffering the highest fruit damage (24.83 per cent).

Table 30.1: Effect of Various Insecticides on the Incidence of Fruit Borer (*Earias vitella* Fab.) and Yield of Okra, Recorded During 2003 and 2004

Treatment(s)		*Fruit Damage (%) Caused by E. vitella***			*Yield of Marketable Okra Fruits** (q/ha)*		
Insecticide	*Dose (% a.i./ha)*	*2003*	*2004*	*Pooled Mean*	*2003*	*2004*	*Pooled Mean*
Karanj oil	(4.0%)	16.30 (23.82)	18.62 (25.58)	17.48 (24.70)	59.88	64.70	62.29
Neem oil	(4.0%)	12.88 (20.98)	8.96 (17.37)	10.92 (19.17)	64.21	69.30	66.75
Chlorpyriphos	(0.06%)	9.06 (17.48)	7.08 (15.46)	8.07 (16.47)	71.36	77.60	74.48
Alphamethrin	(0.006%)	1.89 (7.74)	1.05 (5.86)	1.47 (6.80)	105.50	108.76	107.13
Deltamethrin	(0.004%)	1.43 (6.82)	1.03 (5.75)	1.23 (6.28)	103.75	106.50	105.12
Fenvalerate	(0.006%)	2.13 (8.34)	1.38 (6.56)	1.75 (7.45)	96.30	99.37	97.83
Cypermethrin	(0.006%)	4.32 (11.98)	2.35 (8.74)	3.33 (10.36)	86.97	90.20	88.58
Btk (Delfin)*	(0.750 kg/ha)	4.03 (11.56)	2.06 (8.15)	3.04 (9.85)	82.66	87.38	85.02
Fluvalinate	(0.006%)	3.88 (11.42)	1.97 (8.02)	2.92 (9.72)	91.23	93.80	92.51
Thiodicarb	(0.75%)	7.85 (16.26)	5.73 (13.84)	6.79 (15.05)	75.66	78.80	77.23
Quinalphos	(0.06%)	10.36 (18.85)	8.38 (16.79)	9.25 (17.82)	77.60	79.75	78.67
Neemark Azadira-chtin (0.03%)	(0.05%)	11.62 (19.97)	8.77 (17.18)	10.19 (18.57)	65.30	68.70	67.00
Diflubenzuron	(0.03%)	8.96 (17.28)	6.74 (15.05)	7.85 (16.16)	77.35	80.82	79.08
Untreated check (control)		24.83 (29.88)	22.03 (27.98)	23.43 (28.93)	42.85	48.57	45.71
SEm (±)		(1.76)	(1.96)	(1.02)	3.06	3.28	3.02
C.D. (0.05)		(5.32)	(5.71)	(3.08)	9.62	10.16	9.08
C.V. (%)		(6.81)	(7.04)	(8.72)	6.23	8.54	9.75

Figures under the parentheses are angular transformed values.

*Dose of formulated product as Delfin.

**Means of 25 observations in terms of number of plucking of fruits.

Table 30.2: Comparative Economics of Various Insecticidal Treatments in Protecting Okra Against Fruit Borer, Based on the Pooled Results of 2003 and 2004

Insecticidal Treatments		Fruit Damage Caused by Fruit Borer (%)	Reduction in Fruit Damage Due to Fruit Borer, Control (%)	Yield of Marketable Okra Fruits (q/ha.) Over	Money Value of Okra Fruits @ Rs. 600/q i.e. Gross Income (Rs./ha.)	Yield Increase Over Untreated Crop		Value of Increased Produce @ Rs. 600 per q. (Rs./ha.)	Cost of Insecticidal Control of Fruit Borer, Comprosing Cost of Insecticides and Labour (Rs./ha.)	Net Profit Over Control (Rs./ha)	Benefit Cost Ration (BCR)
Insecticides	Dose (% a.i./ha.)					q/ha.	Increase %				
Karanj oil	4.0%	17.48	5.95	62.29	37374.00	16.58	36.27	9948.00	3080.00	6868.00	2.22
Neem oil	4.0%	10.92	12.51	66.75	40050.00	21.04	46.03	12624.00	4450.00	8174.00	1.84
Chlorpyriphos	0.06%	8.07	15.36	74.48	44688.00	28.77	62.94	17262.00	3640.00	13622.00	3.74
Alphamethrin	0.006%	1.47	21.96	107.13	64278.00	61.42	134.36	36852.00	2970.00	33882.00	11.41
Deltamethrin	0.004%	1.23	22.20	105.12	63072.00	59.41	129.92	35646.00	2500.00	33146.00	13.25
Fenvalerate	0.006%	1.75	21.68	97.83	58698.00	52.12	114.02	31272.00	2680.00	28592.00	10.67
Cypermethrin	0.006%	3.33	20.10	88.58	53148.00	42.87	93.78	25722.00	2870.00	2845.00	7.95
Btk (Delfin)	0.750 (kg/ha)*	3.04	20.39	85.02	51012.00	39.31	85.99	23586.00	4045.00	19541.00	4.83
Fluvalinate	0.006%	2.92	20.51	92.51	55506.00	46.80	102.38	28080.00	3490.00	24590.00	7.04
Thiodicarb	0.75%	6.79	16.64	77.23	46338.00	31.52	68.96	18912.00	3960.00	14952.00	3.78
Quinalphos	0.06%	9.25	14.18	78.67	47202.oo	32.96	72.11	19776.00	3465.00	16311.00	4.71
Neemark (Azadirachtin 0.03%)	0.50%	10.19	13.24	67.00	40200.00	21.29	46.57	12774.00	4090.00	8684.00	2.12
Diflubenzuron	0.03%	7.85	15.58	79.08	47448.00	33.37	73.73	20022.00	5075.00	14947.00	2.94
Untreated check (control)		23.43	–	45.71	27426.00	–	–	–	–	–	–

* Dose of the formulated commercial brand of Btk in the name of Delfin.

During 2004, results in terms of fruit damage caused by *E. vitella* followed almost the similar trend as to that of results obtained during 2003.

So far as the pooled results of 2003 and 2004 in terms of fruit damage is concerned, it was observed that all the treatments were found to be significantly efficacious against the incidence of fruit borer in almost the similar trends as that of recorded during both of the years, 2003 and 2004 (Table 30.1).

Earlier, Mishra (1989) reported that promising control of fruit borer (*E. vitella*) could be obtained by foliar spray of fenvalerate and deltamethrin. Recently, Sreelatha and Divakar (1998) recorded significant reduction in fruit damage through foliar application of alphamethrin. More recently, Prasad and Prasad (2000) obtained significant management of *E. vitella* by spraying of Btk (Delfin) @ 0.50 and 0.75 kg/ha. Sardana and Tiwari (1987) reported that difluebenzuron proved effective against *E. vitella.* Dhamdhere *et al.* (1988) noticed that thiodicarb at 0.15 per cent was found to be highly effective in reducing fruit borer infestation. Raja *et al.* (1997–1998) obtained promising control of *E. vitella* of okra with Neem oil (4 per cent).

Yield of Healthy and Marketable Fruits of Okra

The perusal of yield data revealed that maximum yield of 105.50 and 108. 76 q/ha were recorded during 2003 and 2004 respectively by treating the plants with alphamethrin, while deltamethrin and fenvalerate remained at par followed by fluvalinate. Btk (Delfin) applied @ 0.75 kg/ha was also responsible for the significantly higher yields of 82.66 and 87.38 q/ha, obtained during 2003 and 2004. All other chemical insecticides remained almost at par but superior over botanicals and inferior to synthetic pyrethroids in terms of production of healthy marketable fruits (Table 30.1). Earlier, Mishra (1989) obtained higher yield of okra by spray of fenvalerate and deltamethrin as compared to the conventional insecticides. Sreelatha and Divakar (1998) registered rewarding yield of okra through application of alphamethrin.

Comparative Economics of Insecticides

The results on the comparative economics of the application of the respective insecticide(s) computed on the basis of the pooled data of 2003 and 2004 are presented in Table 30.2.

A perusal of Table 30.2 reveals that significant reduction in fruit damage, ranging from 5.95 to 22.20 per cent over untreated check were registered, that in turn resulted to 36.27 to 134.36 per cent enhancement in yield of marketable fruits of okra. The maximum net returns of Rs. 33,882.00 ha. was obtained from the plots treated with alphamethrin, which was closely followed by Deltamethrin. (Rs. 33,146/ha), fluvalinate (Rs. 24,590/ha) and fenvalerate (Rs. 28,592/ha), cypermethrin (Rs. 22,845/ha). All the synthetic pyrethroids resulted to higher net return (Rs. 22,845 to 33,882/ha) with higher benefit cost ratios (7.04 to 13.25) in comparison to other treatments.

The highest benefit cost ratio of 13.25 was achieved in respect of deltamethrin which was closely followed by alphamethrin (11.41), fenvalerate (10.67), cypermethrin (7.95) and fluvalinate (7.04). The other promising treatment were Btk giving rising to net return of Rs.19,541/ha with BCR of 4.83, followed by quinalphos, thiodicarb and chlorpyriphos.

As far as both the efficacy and the economics are concerned, the synthetic pyrethroids as a whole took the lead, followed by quinalphos, thiodicarb and chlorpyriphos. However, diflubenzuran and botanical pesticides also can be taken into account as the considerably ecofriendly pesticides, against *E. vitella,* giving rise to net return of Rs. 6,868/ha to Rs. 14,494/ha along with BCR of 1.84 to 2.94. Some workers obtained somewhat or almost similar results, with the different chemicals as in the case of present results. Mishra (1989) reported that significant control of *E. vitella* by the foliar application of either fenvalerate or deltamethrin was noticed with the higher yields of okra fruits. Sreelatha and Divakar (1998) obtained the promising control of the fruit borer by treating the crop plants with alphamethrin that in turn gave rise to higher yield and net return. Dhamdhere *et al.* (1998) found that thiodicarb at 0.15 per cent, proved to be highly effective against the pest species that resulted to significantly higher yield of okra alongwith the higher net profit. More recently, Prasad and Prasad (2000) registered good results with Btk (Delfin) applied @ 0.500 and 0.750 kg/ha that proved to be significantly effective against the *E. vitella* for realising higher yield of okra. In the present study, all the thirteen treatments proved to be significantly effective against the fruit borer (*E. vitella*) infesting okra.

Thus, the overall results widen the scope, choice and option for the selection of alternative and substitute of varied insecticides belonging to different groups, without making repeatation of the same category during spray of the one after the another, if multiple or frequent spraying would be required to suit the time and space to protect the crop against the pest species for obtaining higher yields and profits.

Acknowledgement

The authors are grateful to the Vice-Chancellor, Dean (Agric.) and Director of Research, B.A.U., Kanke, Ranchi for providing the facilities for conducting this experiment.

References

Dhamdhere, S.V., Bahadur, J. and Mishra, U.J. 1988. Effect of carbamate insecticides against okra fruit borer, *Earias vitella* Fabricius. *Indian J. Agric. Res., India*, 22(1): 19–22.

Krishnaiah, K. 1980. Methodology for assessing crop loss due to pest of vegetables. In: *Assessment of Crop Losses Due to Pest and Diseases,* (Eds.) Govindu, H.C. *et al.* Tech. Series, U.S.A., 33: 259–267.

Mishra, P.N. 1989. Bio-efficacy of some newer insecticides against the pest complex of okra in India. *Bangladesh Hort.*, Bangladesh, V17(1): 1–4.

Prasad, Rabindra and Prasad, Devendra. 2000. Eco-friendly management of fruit borer (*Earias vitella* Fab.) through application of bio-pesticides for sustainable production of okra. In: *Abstracted Research papers, International Meet Entomocongress–2000: Perspective for the New Millennium,* held at University of Kerala, Thiruvanthapuram, India, pp. 52.

Raja, J., Rajendra, B., Pappiah, C.M., Reddy, P.P., Kumari, N.K.K. and Verghese, A. 1997–98. Advances in IPM for horticultural crops. In: *Proceedings of the 1st National Symposium on Pest Management in Horticultural Crops, Environmental Implications and Thrusts,* held during 15–17 October, at IIHR, Bangalore, India, p. 118–120.

Sardana, H.R. and Tewari, G.C. 1987. The effects of diflubenzuron on the pupa of okra shoot and fruit borer, (*Earias vitella* Fab.). *International J. Tropical Agric., India*, V5(2): 150–153.

Sreelatha and Divakar, B.J. 1989. Impact of pesticides on okra fruit borer, *Earias vitella* Fab. (Lepidoptera: Noctuidae). *Insect Environment*, 4(2): 40–41.

Sriniwasan, P.M. and Gowder, R.B. 1959. A preliminary notes on the control of bhendi shoot and fruit borer, *Earias fabia* and *E. insulana. Indian J. Agric. Sci.*, 30(1): 55–57.

Chapter 31

Fertilizer and Spacing Affect Cabbage Quality

Devesh Tripathi and Rakesh Kumar Upadhyay

C/o Dr. J.S. Tripathi, Department of Agronomy, N.D.U.A.T., Kumarganj, Faizabad, U.P.

ABSTRACT

The experiment was conducted during the year 2003–2004 at Research Farm of Allahabad Agriculture Institute, Deemed University, Allahabad. It was found that percent acidity, ascorbic acid, cooking quality was significantly higher on 200 : 120 : 80 Kg NPK/ha application. Where as ascorbic acid content was significantly higher at 40 × 60 cm spacing as compared to 45 × 30 cm spacing but cooking quality was at par with respect to spacing.

Introduction

Cabbage is the most popular vegetable around the world in respect of area, production and availability almost round the year. In India, its rank is second next to cauliflower in area and production. Cabbage is commonly used fresh as salad, boiled vegetable, cooked in curries. Cabbage is known to possess medicinal value. It was used against ailments like gout, diarrhoea, stomach and coeliac troubles. Cabbage is rich in minerals and vitamins. Therefore the quality of cabbage under varying fertilizer levels and spacing was studied. Srivastava *et al.* (1970) reported that application of 300 kg N/ha was effective in increasing ascorbic acid content in cabbage which improve quality of cabbage.

Materials and Methods

Experiment was laid out in the field adopting randomized block design in factorial concept with three replication. There were four fertility level *viz.*, F = 0 : 0 : 0 kg N.P.K./ha, F_1 = 150 : 80 : 60 kg N.P.K./ha, F_2 = 175 : 100 : 70 kg N.P.K./ha and F_3 = 200 : 120 : 80 kg N.P.K/ha, and three plant spacing *viz.*, S_1 = 45 × 30 cm, S_2 = 45 × 45 cm. and S_3 = 45 × 60 cm. The total treatment combination was 12 and total experimental unit was 36.

Results and Discussion

Effect of Fertilizer on Quality of Cabbage

Table 31.1 revealed that significantly higher per cent acidity was found in 200 : 120 : 80 kg NPK/ ha where as cooking quality was at par in F_1, F_2 and F_3 but these are significantly higher as compared to when apply 0 : 0 : 0 kg NPK/ha and same trend was also observed in case of ascorbic acid content in cabbage.

Table 31.1: Effect of Fertilizer on Quality of Cabbage

Treatment	*Acidity (%)*	*Cooking Quality (Minutes)*	*Ascorbic Acid (mg/100 g pulp)*
F_0	0.54	23.82	18.30
F_1	0.78	25.55	22.18
F_2	0.89	25.41	22.60
F_3	1.17	25.22	22.58
SEm±	0.024	0.313	0.234
C.D. at (5%)	0.04	0.92	0.714

Effect of Spacing on Quality of Cabbage

Table 31.2 revealed that per cent acidity was at par in each spacing and same result was observed, in case of cooking quality. The ascorbic acid content was significantly higher at spacing of 45 × 60 cm as compared to 45 × 30 cm and 45 × 45 cm.

Table 31.2: Effect of Spacing on Quality of Cabbage

Treatment	*Acidity (%)*	*Cooking Quality (Minutes)*	*Ascorbic Acid (mg/100 g pulp)*
S_1	0.83	24.5	21.25
S_2	0.83	25.22	21.08
S_3	0.86	25.28	21.91
SEm±	0.02	0.27	0.21
C.D. at (5%)	NS	NS	0.61

Reference

Srivastava, D.C., Verma, A.N., Mishra, H.R. and Sharma, R.K. 1970. Growth, yield, protein and ascorbic acid content of cabbage (*Brassica oleracea* L. Var. capitula) as affected by different levels of nitrogen. *Pl. Sci.*, 2: 132–134.

Chapter 32

Effect of Fertilizer and Spacing on Days Taken to Head Formation and Yield of Cabbage

Devesh Tripathi and Rakesh Kumar Upadhyay

C/o Dr. J.S. Tripathi, Department of Agronomy, N.D.U.A.T., Kumarganj, Faizabad, U.P.

ABSTRACT

The experiment was conducted during the year 2003–2004 at Research Farm of Allahabad Agriculture Institute, Deemed University, Allahabad on well level field with proper drainage facility. Application of 200 kg N, 120 kg P_2O_5 and 80 kg K_2O/ha significantly gave higher cabbage yield as compared to lower doses, but higher cabbage yield was obtained with narrow spacing *i.e.*, 45 × 30 cm. Number of days taken to head formation increased with increasing dose of fertilizer application, where as earlier head formation occur at narrow spacing.

Materials and Methods

Experiment was laid out in the field adopting randomized block design in factorial concept with three replications. There were four fertilizer level *viz.*, F_0 = 0 : 0 : 0 kg NPK/ha, F_1 = 150 : 80 : 60 kg NPK/ha, F_2 = 175 : 100 :70 kg NPK/ha and F_3 = 200 : 120 : 80 kg NPK/ha and three plant spacing *viz.*, S_1 = 45 × 30 cm, S_2 = 45 × 45 cm and S_3 = 45 × 60 cm. The total treatment combination is 12 and total experimental unit is 36.

Results and Discussion

Effect of Fertilizer on Days Taken to Head Formation and Yield of Cabbage

With increasing dose of fertilizer application the number of days taken increased. The significant number of days taken to head formation at 200 kg N, 120 kg K_2O and 80 kg P_0O_5/ha application as compared to lower doses the same trend was observed in yield of cabbage. The significant highest yield of cabbage was found in 200 : 120 : 80 kg NPK fertilization (Table 32.1). This is supported by Singh and Naik (1988).

Table 32.1: Effect of Fertilizers on Days Taken to Head Formation and Yield of Cabbage

Sl.No.	Treatment	Days Taken to Head Formation	Yield of Cabbage (q/ha)
1.	F_0	42.37	199.61
2.	F_1	45.08	320.90
3.	F_2	47.16	362.59
4.	F_3	50.10	420.94
	SEM±	0.88	7.98
	CD at (5%)	2.59	23.41

Effect of Spacing on Days Taken to Head Formation and Yield of Cabbage

Data presented in Table 32.2 revealed that there was significantly less number days taken to head formation at narrow spacing as compared to wider spacing. When spacing increased then cabbage yield reduced. Significantly highest cabbage yield was obtained at narrow spacing as compared to wider spacing. Singh *et al.* (1988) observed that closer spacing gave maximum cabbage yield as compared to wider spacing.

Table 32.2: Effect of Spacing on Head Formation and Yield of Cabbage

Sl.No.	Treatment	Days Taken to Head Formation	Yield of Cabbage (q/ha)
1.	S_1	44.80	340.31
2.	S_2	46.07	324.83
3.	S_3	47.675	312.89
	SEM±	0.76	6.91
	CD at (5%)	2.24	20.28

References

Singh, R.V. and Naik, L.B. 1988. Response of cabbage to plant spacing, nitrogen and phosphorus levels. *Indian J. Hort.*, 45(3–4): 325–328.

Srinivas, K. 1984. Note on response of cabbage to plant density and fertilizer. *Indian J. Hort.*, 41(3/4): 227–235.

Chapter 33

Yield and Quality of *Gossypium arboreum* Hybrid (AKDH-7) as Influenced by Plant Density and Nitrogen Levels

P.V. Mahatale, T.K. Tiwari and A.V. Patil

Department of Agronomy, Dr. Panjabrao Deshmukh Krishi Vidyapeeth, Akola – 444 104, India

ABSTRACT

Yield contributing characters like number of bolls picked per plant, seed cotton yield per plant and seed cotton per boll was higher under wider spacing (60 × 75 cm) than closer spacing (60 × 30 cm). The values of these parameters were maximum at higher level of nitrogen (75 kg ha^{-1}). The various levels of spacing as well as nitrogen did not affect quality parameters significantly. The maximum seed cotton yield, lint yield and biological yield were recorded under 60 × 45 cm spacing with the application of 50 kg N ha^{-1} (18.65 q ha^{-1}).

Keywords: *Gossypium arboreum, Nitrogen levels, Spacing, Yield, Quality.*

Introduction

The most appropriate spacing is one that enables plant to take best advantage of growth condition, as it is ultimately connected with the root development shoot growth and fructification. Plant may show better growth or development and give higher yield per plant may not give maximum yield per unit area because of inadequate plant population. Thus, for increasing economic yield the optimum spacing is essential.

Nitrogen is an important nutrient required to achieve the good yield of cotton crop. It is an important constituent of organic compounds in the plant, which occur in the protoplasm, nucleus and other tissues. It plays vital role in energy transfer.

It is closely related with growth, maturity and yield of cotton plant. Thus the present investigation was carried out to find out the effect of different levels of plant density and nitrogen on yield and quality attributes of deshi cotton hybrid (AKDH-7).

Materials and Methods

The experiment was conducted under rainfed condition during *kharif* season of 2001–2002 in Factorial Randomized Block Design with three replications at Cotton Research Unit, Central Research Station, Dr. Panjabrao Deshmukh Krishi Yidyapeeth, Akola. The treatment consist of combination of four spacing levels *viz.*, 60 × 30, 60 × 45, 60 × 60 and 60 × 75 cm and four levels of nitrogen namely 0, 25, 50 and 75 kg N ha^{-1} (N_0, N_1, N_2 and N_3 respectively).

The cotton was sown on 25th June 2001. The deshi cotton hybrid AKDH-7 was used for this study. The soil of experimental plot was clayey. The crop was grown as per recommended agronomic package of practices. The observations regarding yield attributes like number of bolls picked plant^{-1}, seed cotton yield plant^{-1}, seed weight boll^{-1} were recorded treatment wise on the basis of five observation plants, selected randomly from net plot of each replication. Whereas the observation regarding quality parameters like lint index, ginning percentage were recorded by using following formula.

$$\text{Lint Index Plot}^{-1} = \frac{\text{Seed cotton yield plot}^{-1} \times \text{Ginning percentage}}{100}$$

$$\text{Ginning Percentage (\%)} = \frac{\text{Lint weight}}{\text{Seed cotton yield}} \times 100$$

While, earliness index was calculated by formula.

$$\text{Earliness Index} = \frac{(P1) + (P1 + P2) + \text{.........}(P1 + P2 + \text{....................} + Pn)}{n\,(P1 + P2 + \text{....................} + Pn)}$$

where,

P1, P2, Pn are the weight of seed cotton collected in 1st, 2nd ....nth picking and n is total number of pickings taken.

Results and Discussion

Effect of Plant Spacing

The number of boll picked plant^{-1}, yield of seed cotton plant^{-1} and weight of seed cotton boll^{-1} were significantly affected by various plant spacing. In general, seed cotton boll^{-1} was increased with increase in plant spacing levels from 60 × 30 cm to 60 × 75 cm, number of bolls also increased from 12.98 to 24. 65 bolls plant^{-1} with increased plant spacing. Its favorable results were observed on the yield of cotton and it was supported by Wankhade and Bathkal (1994).

The yield of seed cotton ha^{-1} is the resultant of the component of the yield contributory characters and the plant spacing. The yield component plant^{-1} at closer spacing (60 × 30 cm) was low as compared to that of wider plant spacing (60 × 75 cm). These might be due to crowding of plant under closer spacing. It was observed that when plant spacing was decreased from 60 × 75 to 60 × 60 cm ha^{-1} than the additional gain in seed cotton yield was 0.91 g ha^{-1}. When the plant spacing was further decreased

Table 33.1: Morphological Indices and Yield as Influenced by Various Treatments

Treatments	Plant Height (cm)	Functional Leaves at 140 DAS	Monopodial Branches plant^{-1} at 140 DAS	Symnpodial Branches plant^{-1} at 140 DAS	Dry Matter Accumulation Plant^{-1} (g) at 140 DAS	Days to 50% Flowering	Soil Moisture (%) at 160 DAS	No. of Bolls Picked per Plant	Yield of Seed Cotton Plant^{-1} (g)	Weight of Seed Cotton Boll^{-1} (g)	Seed Cotton Yield (q ha^{-1})	Biological Yield (q ha^{-1})
A. Effect of Spacing Levels (cm)												
S_1 (60 × 30)	84.18	97.42	1.88	21.85	48.25	72.29	18.42	12.98	25.57	2.03	13.34	26.54
S_2 (60 × 45)	88.39	99.78	1.99	24.06	75.26	73.03	18.66	21.84	43.41	2.06	15.74	27.84
S_3 (60 × 60)	85.09	102.17	2.14	22.55	81.53	73.43	18.90	22.27	44.21	2.12	11.39	22.00
S_1 (60 × 70)	71.85	104.64	2.36	18.86	89.78	73.66	20.06	24.65	49.02	2.17	10.48	19.74
S.E. (m) ±	0.64	0.07	0.04	0.36	0.71	0.08	0.02	0.35	0.88	0.01	0.42	0.21
C.D. at 5%	1.86	0.21	0.14	1.04	2.05	0.24	0.07	1.02	2.55	0.04	1.23	0.60
B. Effect of Nitrogen Level (kg ha^{-1})												
N_0 0 kg N	76.90	96.41	1.98	19.93	6.26	72.81	18.82	17.16	34.24	2.03	10.55	20.44
N_1 25 kg N	80.60	98.54	2.06	21.11	70.53	73.01	18.95	20.02	39.85	2.06	12.43	23.01
N_2 60 kg N	84.05	102.80	2.13	22.13	79.04	73.20	19.07	21.86	43.75	2.11	14.04	25.78
N_3 75 kg N	87.95	106.26	2.21	24.15	82.00	73.39	19.20	22.70	44.36	2.18	13.92	26.89
S.E. (m) ±	0.64	0.07	0.04	0.36	0.71	0.08	0.02	0.35	0.88	0.01	0.42	0.21
C.D. at 5%	1.86	0.21	0.14	1.04	2.05	0.24	0.07	1.02	2.55	0.04	1.23	0.60
C. Interaction Effect												
S.E. (m) ±	1.28	0.14	0.09	0.72	1.42	0.17	0.04	0.71	1.76	0.03	0.85	0.42
C.D. at 5%	–	–	–	–	–	–	–	2.05	5.10	–	2.46	1.21
Mean	82.38	101.0	2.09	21.83	73.71	73.10	19.01	20.43	40.55	2.09	12.74	24.03

from 60 × 60 to 60 × 45 cm. The additional increase in plant spacing reaching to 60 × 30 cm, seed cotton yield was reduced by 2.40 g ha^{-1} due poor growth in dense population. Similar results were reported by Brar and Singh (2000).

Yield of lint and biological yield ha^{-1} was increased with decreased plant spacing. However, maximum lint yield and biological yield observed under 60 × 45 cm spacing as compared to 60 × 30 cm spacing.

Quality characters like ginning percentage earliness index, seed index and harvest index were not significantly affected by the plant spacing. Similar findings were recorded by Kubde and Lakhadive (1993).

Table 33.2: Yield and Quality Attributes of Cotton as Influenced by Various Treatments

Treatments	Lint Yield (q ha^{-1})	Ginning Percentage (%)	Earliness Index (%)	Seed Index (%)	Harvest Index (%)
A. Effect of Spacing Levels (cm)					
S_1 (60 × 30)	5.03	38.36	0.65	6.18	0.53
S_1 (60 × 45)	6.05	38.51	0.67	6.17	0.57
S_1 (60 × 60)	4.37	38.40	0.64	6.16	0.54
S_1 (60 × 75)	4.02	38.40	0.62	6.18	0.54
S.E. (m) ±	0.17	0.14	0.06	0.02	0.01
C.D. at 5%	0.50	–	–	–	–
B. Effect of Nitrogen Levels (kg ha^{-1})					
N_0 0 kg N	4.05	38.43	0.65	6.16	0.54
N_1 25 kg N	4.69	38.50	0.64	6.16	0.56
N_2 50 kg N	5.39	38.43	0.64	6.18	0.54
N_3 75 kg N	5.33	38.32	0.65	6.17	0.53
S.E. (m) ±	0.17	0.14	0.06	0.02	0.01
C.D. at 5%	0.50	–	–	–	–
C. Interaction Effect					
S.E. (m) ±	0.35	0.28	0.01	0.02	0.02
C.D. at 5%	–	–	–	–	–
Mean	24.03	38.42	0.64	6.17	0.54

Effect of Nitrogen Level

The beneficial effect of nitrogen could be noticed on the number of boll harvested plant^{-1}. The number of picked bolls plant^{-1} was increased by 2.86, 4.7 and 5.54 with the application of 25, 50 and 75 kg N ha^{-1}, respectively over no nitrogen level. The weight of seed cotton boll^{-1} also showed more or less similar effect of nitrogen levels as that of number of boll picked plant^{-1}. Seed cotton yield per plant^{-1} was significantly increased with increase in nitrogen levels from 0 to 50 kg ha^{-1}. Beyond 50 kg N ha^{-1} level, there was no significant response in increasing seed cotton yield plant^{-1}. It could be seen that mean increase in seed cotton yield plant^{-1} with first, second, third increment of nitrogen were of the 5.61, 3.90 and 0.61 g plant^{-1}, respectively. Similar findings were reported by and Wankhade (1992).

As the dose of nitrogen was increased to 25, 50 and 50 kg N ha^{-1}, it increases seed cotton yield ha^{-1} to 1.88, 3.49 and 3.37 q ha^{-1}, respectively over no nitrogen. But the seed cotton yield differences for 50 and 75 kg N ha^{-1} was non-significant. Similar results were reported by Wankhade and Bathkal (1994). However, increases in yield of lint was observed to be 0.64, 1.34 and 1.28 q ha^{-1} with the application of 25, 50 and 75 kg N ha^{-1} over no nitrogen. Ogunlela and Abed (1984) also observed the increased lint yield with increase in the levels of nitrogen.

Biological yield was increased with increment in nitrogen level and its value running in between 20.44 to 26.89 q ha^{-1} with the application of 0 to 75 kg N ha^{-1}. Quality characters like ginning percentage, seed index, earliness index and harvest index were not significantly affected by nitrogen level. Similar findings were reported by Bapana *et al.* (1976).

References

Bapana, J.S., Patel, C.T. and Chaudhari, V.P. 1976. Studies on varietal response to nitrogen and plant density. *J. Cotton Develop.*, 6(1): 8–10.

Brar, M.R. and Singh, V.P. 2000. Response of compact and early maturing cotton genotype to plant densities. *J. Cotton Develop.*, 14(2): 155–158.

Kubde, K.J. and Lakhdive, B.V. 1993. Effect of nitrogen, phosphorus and spacing on yield quality and nutrient uptake of *hirsutum* cotton. *P.K.V. Res. J.*, 17(12): 231–232.

Ogunlela, A.F. and Abed, M.H. 1984. Effect of nitrogen fertilizer levels and its split application on yield and earliness in Giza-69. *Agric. Res. Develop.*, 54(9): 113–115.

Wankhade, S.T. and Bathkal, B.G. 1994. Yield potential of Asiatic cotton under varying nitrogen and plant density levels under (rainfed) condition. *PKV, Res. J.*, 18(1): 53–55.

Wankhade, S.T. 1992. Effect of spacing and varieties on growth, yield and quality of Asiatic cotton (*G. arboreum*) under rainfed condition. *Indian J. Agron.*, 37(3): 523–526.

Chapter 34

A Comparative Study of Leaf Architecture of Three Plants Species of Family Solanaceae

Archana Nigam

P.G. Department of Botany, Government P.G. (Autonomous) College, Satna – 485 001, M.P., India

ABSTRACT

The family Solanaceae is one of the important family from the economic point of view. A number of species have since been introduced for their proved food, medicinal and ornamental value. Our information on cuticular and epidermal morphology of Solanaceae is however mainly confined to the plants that are used in medicine. During present observation, three species of Solananceae *i.e., Capsicum annum, Cestrum diurnum* and *Datura innoxia* have been selected for comparative morphological studies of their leaves the results show remarkable variations in different species.

Keywords: Solanaceae, Epidermal structure, Capsicum annum, Cestrum diurnum, Datura innoxia.

Introduction

The Solanaceae is a large family of flowering plants with 85 genera and over 2200 species. The cuticular and epidermal morphology of angiosperms has received scant treatment from taxonomists although Florin (1931) has strongly pleaded for such investigation, which would greatly facilitate phylogenetic consideration and assisting of generic affinities.

Works on individual epidermal characters are those of Netolitzky (1932) and Howell (1945). An investigation of the trichomes in tobacco leaves as they related to chemical changes during curing

(Akers 1975). The study of veins also suggests closeness to the observation made up Fisher (1885) distinguished principal and secondary or minor vein ending in dicotyledons. Strain (1933) studied the xylem structure of vein ending in dicotyledons and established categories of vein ending based on the number of terminal tracheids. Hooker's (1855} record 14 genera and 50 species of this family from India. Hickey (1973) classified the vein ending into simple and branched. Inamdar and Murthy (1978) studied the leaf architecture in some Solanaceae.

The present observations. which deals with the cuticular morphology of Solanaceae is therefore an attempt to fill the gap. The purpose of this study is to present an inclusive coherent classification of the architecture of dictyledonous leaves. The proposed system is based in part or portion of previous classification.

Materials and Methods

Satna is a main town of Rewa division where number of industries, historical and tourist places are situated. It is between 23° to 24° of North latitude and 80° to 81.5° of east latitude. The leaves of three species of Solanaceae, for the present investigation were collected from the different area of Satna and fixed in F.A.A. Leaves were cleared in 2.5 to 5.1 NaOH solution followed by NaCl. The preparation were stained in Saffranin and mounted in glycerin. Camera lucida drawing were made, the size of leaf was measured.

Epidermal peels of fixed mature leaves were taken by scraping with scalpel. For difficult materials, the double treatment method of Leelavathi and Ramayya (1975) was employed the peels specimens were studied from base middle and apex regions covering from mid vein to the margin-where leaves are small, the peels represented the entire leaf surface. These were stained and mounted in Saffranin glycerin (Ramayya and Rajgopal, 1968) and observed at low and high magnifications.

Results

From the comparative study of the leaf morphology and epidermal features of the three species of Solanaceae suggests that different genera show some morphological variations in different characters where leaves in the vegetative part are mostly alternate. In the floral regions they become opposite. They are simple rarely pinnately divided and are always exstipulate. The comparative morphological studies of selected genus of Solanaceae family are as follow.

Capsicum annum L. (*Eng*: Red pepper. Chillies, *Verna*: Lal mirch)

The *Capsicum annum* is a cultivated herb. The fruits are extremely pungent and used as a spice and condiment eaten raw and prickled. Leaves are ovate, tapering to a sharp point, entire up to 15 cm long, dark green on the upper surface and pale on the upper surface and pale on the lower surface.

The epidermal cells of both the surface are almost similar, sinuous walled, polygonal and irregularly displaced. In lower epidermis cells of basal portion are irregular, nearly 100–130 μm long × 25–55 μm wide, anticlinal wall 15 μm–15 μm thick. Sinousity varies from 15–35 μm (25 μm). The cells of middle portion are also similar size nearly 105–125 μm long, 25–35 μm wide, anticlinal wall 1.5 μm thick, nearly 20–15 μm (25 μm) sinuosity present. The cells of apical region of same size of middle part but comparatively more sinuous. Basal middle and apical cells are almost of the same size as 70–140 μm long × 20–70 μm broad, anticlinal wall 1.5–2 μm thick with variation of sinousity 15–40 μm (25 μm).

Stomata are irregularly dispersed on both surface having 3–4 subsidiaries, shape and size similar to epidermal cells nearly 30–90 μm long and 25–40 μm wide, one is comparatively smaller than the other, guard cells nearly 35–48 m long × 25–32 µm wide and stomatal pores are 15–28 µm long × 2-4 µm wide. Frequency of stomata is more on the lower surface than that of the upper. In some of the stomata common subsidiary cells are present which suggests contiguous nature of stomata. As Solereder (1908) define it is pair of stomata lying side by side and touching one and other by one of their longitudinal walls. Thus cruciferous type of stomata present.

Cestrum diurnum L. (*Eng*: Day Jasmine, *Verna*: Din ka Raja)

The *Cestrum diurnum* is grown in the garden as an ornamental shrub for its fragment flowers. The epidermal cells of both surfaces are almost similar, sinuous walled with rectangular and irregularly displaced. In lower epidermis the cells of basal portion are irregularly nearly 90–108 µm long × 30–60 µm wide, anticlinal wall 1.8–3 µm thick. Sinuosity is 25 µm. The cells of middle portion are also of the similar size nearly 95–120 µm long × 25–55 µm thick and 15–35 µm (25 µm) sinuosity present. The cells of apical region are also similar in shape nearly 100–120 µm long × 50–55 µm wide. Anticlinal wall and sinuosity is similar than middle and apex of leaf. Hairs are glandular but very short.

In upper epidermis the basal middle and apical cells are almost of the same measurement as 60–130 µm long × 30–80 µm wide, anticlinical wall 1.7–2.8 µm thick with sinuosity 15–35 µm (25 µm).

The stomata are irregularly dispersed on both surface having 3-4 subsidiaries, shape and size similar to epidermal cells nearly 35–90 µm long, and 25–40 µm wide. One is smaller than the other. Guard cell nearly 30–40 µm long × 25–30 µm wide and stomatal pores is 15–30 µm long × 2–5 µm wide. Frequency of stomata is more as the upper surface that that of the lower. Thus Rannunculaceous types of stomata are present.

Datura innoxia MILL. (*Verna*: Sadah dhatura)

The plant of *Datura innoxia* is annual wild herb up to 2 meter height with petiolate, exstipulate simple, alternate, ovate, and glabrous leaves. The petiole with a few large coarse teeth on the margin and base is oblique. The epidermal cells of two surfaces of leaves are almost similar, both are sinuous called of polygonal and irregular shape and size. In lower epidermis the cells of leaf base are almost sinuous walled, polygonal and irregularly shaped or elongated with 10–20 µm (15 µm) sinuosity while epidermal cells measuring 85–120 µm long × 35 × 60 µm wide and thickness of anticlinal wall 1.8–2 µm. The cells of apical region are also irregularly arranged and comparatively smaller. The basal and middle region cells measuring 50–90 µm long × 35–55 µm wide. Thickness of anticlinal wall varies from 2–4 µm and sinuosity of cells also less than the other two region which is nearly 2--6 µm (4 µm).

The upper epidermal cells in the basal region of leaves are also less, sinuous, or straight, irregularly arranged, measuring 40–95 µm long × 36–60 µm wide. Anticlinal wall 5–8 µm thick, cells are rectanguloid, Oval in shape, sinuosity varies from 1.5–2 µm. While the cell of middle portion are measuring 90–140 µm long × 36–48 µm wide. Thickness of anticlinal wall is 1.5–2 µm. Cell ware irregularly oriented, cell wall is sinuous and sinuosity varies from 6–10 µm but the cells of apical region are 70–110 µm long × 36–50 µm wide with 2–3 µm thick anticlinal walls. Cells are irregular, sinuous. Sinuosity varies from 6–10 µm. Hairs present on the upper epidermis with cylindrical pointed apex.

The stomata are generally present on the both surface, irregularly distributed, its frequency is comparatively more on the lower surface, 3–4 subsidiaries are present. Size of subsidiary cells are 40–75 µm long × 20–60 µm wide, while one is slightly smaller than the other. The size of guard cells also variable from 35–45 µm long × 22–28 µm wide. Stomatal pore varies from 18–28 µm long × 3–6 µm wide.

The subsidiary cells of upper epidermis are nearly 20–40 µm long × 30–55 µm wide with 35–45 µm long × 30–35 µm wide guard cells while size of stomatal pores are generally measuring is 15–30 µm long × 5–7 µm wide in different parts of the upper epidermis. In some of the stomata common subsidiary cells are present which suggests the contiguous nature of stomata, thus Cruciferous type of stomata present.

Discussion

In addition, a more sophisticated knowledge of leaf architecture has permitted a start to be made in discerning phylogenetic trends from leaves (Hickey, 1971b). In Solanaceae venation pattern is similar to that or majority of dicotyledons. Presence of loop like structure in big areoles is of frequent occurrence where the areoles are devoid of free veins, which probably may help in the transport system. Simple and branched vein ending are found.

Coleman and Greyson (1976) reported the presence of multiseriate fimbriate vein in tomato. Multiseriate fimbriate vein is observed in *Cestrum diurnum*. Patel and Shah (1974) reported the extension cells in bringal and chilli. Similar types of elongated parenchymatous cells are noticed in *Datura innoxia*.

Comparative morphological studies of different genera and species suggest that in the genera *Cestrum diurnum*. Rannunculaceous stomata are found while in *Cupsicum annum* and *Datura innoxia* cruciferous types of stomata are found which denote that there are morpho-ecological variations in leaf architecture. Similarly, the comparative observation of hairs also suggests that in genera *Capsicum annum*, *Cestrum diurnum*, *Datura innoxia* glandular hairs are found, they are almost glandular and two celled lonted.

Capsicum shows close affinity in the presence of identical glandular and non-glandular hairs, cruciferous or partially cruciferous type of stomata. High stomatal frequency, very small stomata and epidermal cells and in the presence of typical small glandular hiars. Among the non-glandular hairs, usually stellate or peltate, types of hairs are found as well as the simple uniseriate ones, but usually not together in the same species of *Solanum*.

References

Akers, C.P. 1975. I Tobacco leaf trichomes, ultra structure of glandular trichoesm. II. Chemical changes in trichomes during flue caring. *Ph.D. Dissertation*. Department of Botany, North California State University, Raleigh.

Coleman W.K. and Greyson. R. 1976. The growth and development of leaf in tomato (*Lyco. esculentum*) II, Leaf ontogeny. *Can. J. Bot.*, 54: 2704–2717.

Fisher, A. 1885. Studien uben die Sicbrohren der dicotylenblatter. Ber. Sac. Ges. Wiss leipzig. Math. *Phy. Cl.*, 37: 245–290.

Florin, R. 1931. Untersuchungen Zur stamniesgeschichte der Coniferales and Cordaitales. Kgl. Svenst a Veten Skapsakad. Handl. Stockholm. Ser. 3 B and 10.

Hickey L.J. 1971b. Evolutionary significance of the leaf architectural features in the woody dicots. (Abst.). *Amer J. Bot.*, 58: 469

Hickey L.J. 1973. Classification of the leaf architecture of dicotyledons leaves. *Amer. J. Bot.*, 60: 17–33.

Hooker, J.D. 1855. *The flora of British India*. L. Reeve and Co., Kent, England, as mentioned in *Can. J. Bot.*, 42(1964): 803.

Howell, J.T. 1945. Concerning stomata on leaves in Arctostaphylos. Was Mann Collector, 6: 57–63.

Inamdar, J.A. and Murthy, G.S.R. 1978. Leaf architecture in some Solanaceae. *Flora*, 167: 265–272.

Leelavathi, A. and Ramayya, N. 1975. Rapid isolation in leaf epidermis by double treatment method. *Geobios*, 2: 117–119.

Netolitzky, F. 1932. Handbuch. d. Pflanzen antomit. IV. Die pflanzenhaare. Berlin. as mentioned in *Can. J. Bot.*, 42 (1964): 803.

Patel, J.D. and Shah, J.J. 1974. Development studies on leaf of brinjal (*S. melongens* L.) and chilli (*Cap. annum* L.). *Proc. Indian Acad. Sci.*, 80(B): 197–206.

Ramayya, N. and Rajgopal, T. 1968. Foliar epidermis as taxonomic aid in the "*Flora of Hyderabad*". Part I: Portulacaceae and Aizoaceae. *J. Osmania Univ.*, 5: 147–160.

Solereder, H. 1908. *Systematic anatomy of dicotyledons I and II*. Oxford. *Geophytology*, 14(1): 68.

Strain, R.W. 1933. A study of vein ending in leaves. *Am. Midland Naturalist*, 14: 367–375.

Chapter 35

Association of Stability Parameters in Pigeon Pea

C. Vanniarajan, P. Rangasamy and T. Nepolean

Centre for Plant Breeding and Genetics, Tamil Nadu Agricultural University, Coimbatore – 641 003, T.N., India

ABSTRACT

Fifty three genotypes of pigeon pea comprising 40 F_1 hybrids and 13 parents were raised under four different environments. Stability analysis and association between the stability parameters *viz.*, mean (x), regression coefficient (b_i), and deviation from regression ($s^2 d_i$) were carried out as per Eberhart and Russel and Goulden methods. The results revealed that relationship between mean and non-linear responses and between linear and non-linear responses could not be generalized due to character specific.

***Keywords**: Pigeon pea, Stability, Stability parameters, Linear, Non-linear.*

Introduction

Association analysis between stability parameters of yield and yield components of crop provides information on the responsiveness, stability and genetic control of stability parameters subjected to environmental variations (Jatsara and Paroda, 1981). An ideal genotype in any practical situation is one with high mean performance, desired linear response and low deviation from linearity. If these aspects controlled by independent genetic system, then in that situation such an ideal genotype can be developed through standard breeding procedures. It is thus important to know whether these components of adaptation for the character of interest is independently inherited or not. Since information on the association of stability parameter among themselves is meager in pigeon pea, an investigation was carried out to compute correlation coefficient among stability parameters.

Materials and Methods

Fifty three genotypes of pigeon pea which comprises of 40 F_1 hybrids and 13 parents were raised in a randomised block design with three replications under four different environments during 1992 and 1993 in Agricultural college and Research Institute, Madurai and Agricultural Research Station, Kovilpatti. Stability analysis and association between the stability parameters *viz.*, mean (x), regression coefficient (b_i) and deviation from regression ($s^2 d_i$) were carried out as per the method of Eberhart and Russel (1996) and Goulden (1952) respectively. The correlation coefficients between stability for ten biometrical traits are presented in Table 35.1.

Table 35.1: Correlation Coefficient Between Stability Parameters for Different Characters in Pigeon Pea

Character	*x Vs b_i*	*x Vs s^2d_i*	*B_i Vs s^2d_i*
Days to 50% flowering	-0.336*	-0.028	0.066
Days to maturity	-0.190	0.069	0.109
Plant height	0.596*	0.242	0.114
Branches per plant	0.330*	0.152	0.213
Clusters per plant	0.122	0.562*	-0.365*
Pods per plant	0.418*	0.545*	-0.130
Pod length	0.234	0.152	0.189
Seeds per pod	-0.206	0.150	-0.083
100 seed weight	0.356*	-0.024	0.348*
Seed yield per plant	0.794*	0.432*	0.246

x: Mean; b_i: Regression coefficient; s^2d_i: deviation from regression coefficient. Significant at 5% level.

Results and Discussion

Significant and positive correlation between mean performance (X) and linear regression coefficient (b_i) for plant height, branches per plant, pods per plant, 100 seed weight and seed yield per plant indicate that the genotypes with high mean values for these characters were responsive to favourable environments. Therefore mean values for these characters may be used to judge the responsiveness. This confirms the findings of Shoran (1985). Significant and negative correction was also noticed between mean performance and linear regression coefficient for days to 50 per cent flowering.

A significant positive association between mean performance and deviation from linear regression ($s^2 d_i$) for clusters per plant, pods per plant and seed yield per plant reflected that the genotypes with high mean values would be highly unstable under the influence of changing environments. This has already been reported by Shoran (1985) in pigeon pea and Tomar and Singh (1992) in cotton. Non-significant association between mean and deviation from linear regression for all the other characters studied indicate that non-linear component of genotype x environment interaction ($s^2 d_i$) are independent of the mean performance. It also indicated the involvement of different genetic system in the control of these stability parameters.

Association between linear and deviation from linear regression from linear regression coefficient for clusters per plant and 100 seed weight indicate that the genotypes showing high responsiveness would be unstable. Therefore the relationship between mean and non-linear responses on one hand and between linear and non-linear response on the other, could not be generalized in this crop because these relationships were observed to be character specific. This is in accordance with the findings of Shoran (1985) in pigeon pea.

Conclusion

The relationship between the stability parameters in pigeon pea cannot be generalized and it is character specific.

References

Eberhart, S.A. and Russel, W.A. 1966. Stability parameters for comparing varieties. *Crop Sci.*, 6: 36–40.

Goulden, C.H. 1952. *Methods of Statistical Analysis*. John Wiley and Sons, New York, p. 53–55.

Jatsara, D.S. and Paroda, R.S. 1981. Genetype x environment interaction in segregating generations of wheat. *Indian J. Genet.*, 41: 12–17.

Shoran, J. 1985. Role of plant characters in pigeon pea. *Int. Pigeon Pea Newst.*, 4: 15–18.

Tomar, Y.S. and Singh, V.P. 1992. Phenotypic stability of hybrids and their parents for seed cotton yield and its components in desi cotton. *Indian J. Genet.*, 52: 238–244.

Chapter 36

Woody Vegetation Structure in a Sacred Grove of Pudukottai District, Tamil Nadu, South India

B. Muthukumar#, S. Dhanasekar*, S. Soosairaj**,*
*N. Nagamurugan***, B. Balaguru*****,

***Department of Botany, National College, Tiruchirappalli – 620 001, T.N.*
***Department of Botany, St. Joseph's College (Autonomous), Tiruchirappalli – 620 002, T.N.*
****Department of Biotechnology, Kurinji College of Arts and Science, Tiruchirappalli – 620 002, T.N.*
*****National Innovative Foundation, Ahmedabad – 380 015, Gujarat, India*

ABSTRACT

The inventory area (1 ha) yielded a total of 38 woody species (> 10 gbh) coming under 35 genera and 23 families. Of these 15 species were trees, 15 shrubs and 8 straggler lianas and vines. The subfamily Mimosoideae contributed maximum number of species (6 species) followed by Rubiaceae (5 species) and Euphoribiaceae (3 species). *Chloroxylon swietenia* (216) followed by *Wrightia tinctoria* (96) were the highest contributors to the stand density. Species richness, stand density and diversity indices consistently decreased with increasing size classes of woody species from 10cm to 202 cm girth. *Chloroxylon swietenia, Wrightia tinctoria* and *Albizia anara* have showed the highest density values and highest frequency values.

Keywords: Sacred groves, Dry deciduous forest, Species richness, Conservation.

Corresponding Author, E-mail: bmkumar64@yahoo.com.

Introduction

Sacred groves are small patches of native vegetation traditionally protected and managed by local communities. They are vestiges of an ancient practice in which people protect the forest to avoid the wrath of its resident god (Gadgil and Vartak, 1975, 1976). Social protection allows the groves to harbour rich biological diversity and provides socio environmental beliefs to the society (Singh, 1997) and it also plays an important role in protection of particular ecosystem and serve the function of preserving the genetic diversity of the common species of trees (Harikrishnan Nair *et al.*, 1997). Visalakshi (1995) and Parthasarathy and Karthikeyan (1997) quantified woody species diversity of a few sacred groves of Pondicherry. The sacred grove studies in Tamil Nadu was done by Seth (1993), Swamy (1998), Britto *et al.* (2001; 2004) and Natarajan *et al.* (2004). The present study aims at analysis of woody vegetation and structure in one such sacred grove in Pudukottai district of Tamil Nadu.

Materials and Methods

Study Area

The sacred grove is situated between Kotthamangalapatti and Pasumalaipatti villages of Pudukottai district of Tamil Nadu lying at 10° 30′ 54″ N and 78° 46′ 53″ E longitude. It occupies an area of 2 ha. The area is drought prone and climate is hot and humid with temperatures ranging from 23°C (min) to 38°C (max) respectively. The area receives an annual average rainfall of 700 mm. The forest is classified as tropical dry mixed deciduous after Champion and Seth (1968).

Methods

One-hectare (100 × 100 m) plot was subdivided into 10x10 m workable units. All living trees having ≥ 10 cm gbh were identified and their girth were measured about 1.3 m from ground level. For multi-stemmed trees, girth was measured separately and its basal area was calculated and summed. For species diversity and dominance Shannon-Wiener (H′) = $-\Sigma pi \ln pi$ and Simpson (cd) indices = Σpi^2 were calculated as per Magurran (1988). Importance Value Index was calculated according to the formulae used by Balsev *et al.* (1987).

Results and Discussion

The inventory area (one ha) in the sacred grove yielded a total of 38 woody species (≥ 10 cm gbh) coming under 35 genera and 23 families. Of these 15 species were trees, 15 shrubs and 8 either straggler or liana and vine. The family Mimosoideae contributed 6 species followed by Rubiaceae with 5 species and Euphorbiaceae (3 species). Frequency, abundance and density of the sampling plot is given in Table 36.1. Highest density is reported by *Chloroxylon swietenia* followed by *Wrightia tinctoria* and *Albizia amara*. *Chloroxylon swietenia* followed by *Albizia amara* and *Wrightia tinctoria* have contributed highest frequency values. The abundant species encountered in study area are *Chloroxylon swietenia* followed by *Wrightia tinctoria* and *Strychnos nuxvomica*. The sampling site has stand density of 667 stems per hectare. This is lower when compared to the Coromandel Coast (1367,974 stem per hectare) by Parthasarathy and Karthikeyan (1997) however, it is within the range as per Kalakad National Park of the Western Ghats (574–915 stems), which harbours a tropical wet evergreen virgin forest (Parthasarathy *et al.*, 1992). The site has with higher stand density, when compared to that of Uppangle sites 610 and 635 stems per hectare (Loffeier, 1998).

Chloroxylon swietenia (216) followed by *Wrightia tinctoria* (96) were the higest contributors to the stand density. Nine species were found to contribute only one individual to the total density. They are *Acacia horrida, Cassine glauca, Cissus quadrangularis, Cissus vitigenea, Commiphora berryi, Ficus benghalensis, Jatropha gossypiifolia, Premna serratifolia* and *Pavetta indica*.

Table 36.1: Consolidated Details of Quantitative Biodiversity Analysis

Parameters	*Values*
Species richness	38
Number of Genera	35
Number of families	23
Diversity Indices	
Shannon-Wiener Index (H')	2.503
Simpson index (λ)	0.152
Stand diversity (ha^{-1})	667
Total Basal area (m^2ha^{-1})	15.005

Table 36.2: Stand Density, Basal Area and Importance Value Index (IVI) of the Woody Species in the Study Area

Species Name	*Stand Density*	*Basal Area*	*IVI*
Acacia caesia	5	0.725	2.088
Acacia horrida	1	0.016	0.508
Acacia leuopholea	8	0.509	3.738
Albizia amara	82	5.559	2.088
Azadirachta indica	6	0.469	5.533
Benkara malbarica	28	0.033	11.430
Canthiun parviflorum	4	0.006	1.645
Capparis sepiaria	17	0.225	8.314
Cassine glauca	1	0.003	0.419
Chloroxylon swietenia	216	2.829	68.750
Cissus quadrangularis	1	0.001	0.406
Cissus vitigenea	1	0.003	0.419
Clausena dendata	5	0.012	1.833
Cleistanthus collinus	3	0.010	1.270
Commiphora berryi	1	0.005	0.436
Cordia oblique	14	0.108	5.831
Derris scandens	14	0.113	5.373
Dichorastachus cinerea	2	0.020	0.940
Diospyros ebenum	13	0.120	5.810
Euphorbia antiquorum	20	0.250	9.190
Flacourtia indica	2	0.005	0.840
Ficus benghalensis	1	0.170	1.350
Gmelina asiatica	4	0.034	1.800

Contd...

Table 36.2–Contd...

Species Name	*Stand Density*	*Basal Area*	*IVI*
Holoptelea integrifolia	3	0.130	2.130
Jatropha gossypiifolia	1	0.009	0.400
Jasminum angustifolium	3	0.004	1.220
Lannea coromandelica	4	0.110	2.370
Premna serratifolia	1	0.002	0.410
Paveta indica	1	0.001	0.400
Pterolobium hexapetalum	5	0.005	2.030
Prosopis juliflora	24	0.190	8.140
Ranida dumetorum	5	0.010	2.070
Reissantia indica	3	0.021	1.340
Senna auriculata	4	0.003	1.380
Strychnos nuxvomica	47	0.390	15.790
Tarena asiatica	7	0.020	2.944
Wrightia tinctoria	96	3.460	46.918
Zizyphus oenopilia	13	0.020	5.110

Stand density is indirectly proportional to the girth class *viz.* (473 stems in > 50 cm; 151 in 51–100 cm; 39 in 101–150 cm and only one stem in above 150 cm girth class. *Albizia amara* (5.559 m^2) is the highest contributor followed by *Wrightia tinctoria* (3.46 m^2) and *Chloroxylon swietenia* (2.28 m^2) to the total basal area.

High important value index was observed in *Chloroxylon swietenia* (68.75) followed by *Wrightia tinctoria* (46.9) and *Strychnos nux-vomica* (15.79). The ratio of Ab/% F shows the distribution pattern of species. *Albizia amara, Canthium parviflorum, Gmelina asiatica* and *Lannea coromandelica* have showed regular distribution while all other species were contagious in their distribution. *Pterolobium hexapetalum* has the highest Ab/% F ratio. The diversity index shows that the Shannon-Wiener index (H') = 2.503 and Simpson index (λ)= 0.152.

Conservation

This study identifies the woody species diversity in a sacred grove. Natives and indigenous peoples in this region have an intimate knowledge of nature, wild life and their environment on ground of religious believes, has become important for considering and realizing this practices model for conservation of biodiversity. Imposition of religious restriction not to bring out any items from this grove would virtually help in maintaining soil fertility coupled with biological diversity through nutrient release. Thus this sacred grove is one of the finest examples of traditional conservation practices.

References

Balsev, H., Luteyn, J., Oliguard, B. and Holm-Nielson, L.B. 1987. Composition and structure of adjacent unflooded and floodplain forest in Amazonian Ecuador. *Opera Bot.*, 92: 37–57.

Champion, H.G. and Seth, S.K. 1968. *A Revised Survey of the Forest Types of India.* Government of India Publications, New Delhi.

Debal Deb, Kaushik Deuti and Malhotra, K.C. 1997. Sacred Grove relics as bird refugia. *Curr. Sci.*, 73(10): 815–817.

Gadgil, M. and Vartak, V.D. 1975. Sacred groves of India. A plea for continued conservation. *J. Bombay. Nat. Hist. Soc.*, 72(2): 315–320.

Gadgil, M. and Vartak, V.D. 1976. Sacred groves of western ghats of India. *J. Bombay Nat. Hist. Soc.*, 77: 623–647

Ghosh, S. 2001. Sacred groves and their role in plant conservation. *Science and Culture*, 67(11–12): 347–349.

Harikrishnan Nair, G., Gopikumar, K., Pramond, K., Krishnan, G. and Sunilkumar, K.K. 1997. Sacred groves of India: Vanishing greenery. *Curr. Sci.*, 72(10): 697–698.

Induchudhanan. 1996. Ecology of the sacred groves of Kerala. *Ph.D. Thesis*, Pondicherry University, Pondicherry.

John Britto, S., Balaguru, B., Natarajan, D. and Arockiasamy, D.I. 2001. Comparative analysis of tree diversity and its population density in a sacred grove at Malliganatham, Pudukkottai district of Tamil Nadu. *Ad. Pl. Sci.*, 14(11): 327–330.

John Britto, S., Balaguru, B., Natarajan, D., Soosairaj, S. and Nagamurugan, N. 2004. Ethnobotany of certain Sacred Groves in Tamil Nadu, South India. *J. Trop. Med. Plants*, 5(1): 55–61.

Kalam, M.A. 1996. Sacred groves of Kodaya districts Karnataka (South India): A socio historical study. Francis de Pondicherry, Pondicherry.

Kulkarni, D.K., Brave, J.B., Jagdale, R.P. and Inamdar, A.C. 1993. Floristics of a sacred forest patch from Sudergad district Orissa state. *J. Econ. Taxon. Bot.*, 17(2): 415–419.

Loffeier, M.E. 1998. Reconstitution après exploitation selective et forest simpervirente du Coorg (Inde). *Acta oecological Genaralis*, 9: 69–87.

Magurran, A.E. 1988. *Ecological Diversity and its Measurements.* Princeton University Press, New Jersey.

Mitra, A. and Pal, S. 1994. The spirit of the sanctuary. *Down to Earth*, 31: 21–36.

Natarjan, D., John Britto, S., Balaguru, B., Nagamurugan, N. and Soosairaj, S. 2004. Distribution of sacred grove forests in the Chitteri hills of the Eastern Ghats of Tamil Nadu. *Ind. J. Environ. Ecoplan.*, 8(2): 375–379.

Parathasarathy, N. and Karthikeyan, R. 1997. Plant biodiversity inventory and conservation of two tropical dry evergreen forests on the Coromandal coast, South India. *Biodiv. Conser.*, 6: 1063–1083.

Singh, G.S. 1997. Socio-cultural evaluation of sacred groves for biodiversity conservation in north western Himalaya. *J. Hill Res.*, 10(1): 43–50.

Visalakshi, N. 1995. Vegetation analysis of two tropical dry evergreen forests in South India. *Trop. Ecol.*, 36: 117–127.

Chapter 37

Diversity, Relative and Rank Abundances of Predatory Coccinellids in an Irrigated Rice Ecosystem, Madurai, Tamil Nadu

M. Kandibane**, *S. Rahuraman and *N. Ganapathy*****

**Department of Agricultural Entomology, Tamil Nadu Agricultural University, Coimbatore – 641 003, Tamil Nadu*

***Department of Agricultural Entomology, Agricultural College and Research Institute, Madurai – 625 104, Tamil Nadu*

ABSTRACT

A total of seven taxa of predatory coccinellids were recorded in rice ecosystem during *kharif* 2000. In partially weeded plot, all the seven taxa were present, while in weeded plot, five taxa were observed. The coccinellids showed more abundance in partially weeded plot than in weeded plot. Among the coccinellids, *Micraspis discolor* (Fab.) was the dominant predator in both the ecosystems. Weed plants present in partially weeded rice ecosystem provided pollen and nectar and favourable climates for the survival of coccinellids. The coccinellids present on the weed hosts in partially weeded plot, migrated to weeded plot, when the population of herbivores increased. Hence, natural control could be possible.

Keywords: *Predatory coccinellids, Diversity, Relative abundance, Rank abundance, Rice ecosystem.*

Introduction

The ladybird beetles (Coleoptera: Coccinellidae) are brightly coloured, oval, convex insects, comprising of about 5000 species. The greater numbers of these species are predaceous both as grubs and adults, they feed on wide variety of soft-bodied insects such as leafhoppers, planthoppers, mealybugs, aphids and on the eggs of various insects. The most common and widely distributed species in rice ecosystem are *Microspis discolor* (Fabricius) and *M. crocea* (Mulsant) (Pathak and Khan, 1991). The immature stage (grubs) of coccinellid, *Coccinella repanda* Thumb was noticed feeding on nymphs of brown planthopper, where as the adults feed on the scrapping of the rice leaves (Samal and Misra, 1982). The grubs of the beetle, *Varania discolor* Fabricius were found predating on nymphs and adults of brown planthopper, *Nilaparvatha lugens* (Stal) a major pest of rice, where as the adult beetles feed on rice pollen (Samal and Misra, 1982). In Tamil Nadu, the inventory, diversity, relative and rank abundances of coccinellids in weeded and partially weeded rice ecosystems had not clearly been studied earlier. Keeping the above points in mind, the present investigation to study the diversity and relative abundance of predatory coccinellids was taken up in rice ecosystem.

Materials and Methods

The field trial was conducted in irrigated rice ecosystem at the wetlands of Agricultural College and Research Institute, Madurai, Tamll Nadu during *Kharif* 2000. Four ruling rice varieties *viz.*, MDU 5, ADT 36, ADT 39 and ADT 43 were used and each variety replicated into two treatments namely weeded plot (all the weeds removed) and partially weeded plot (10 weeds allowed/m^2 with rice plants).

The size of the experimental plots was 8 × 6 m. The study area receives water from the vaigai dam. Monthly minimum and maximum temperatures varied between 28° and 38° May being the warmest (40°C maximum temperature) and January the coolest (28°C minimum temperature) months during 2000. The average rainfall of Madurai was 893 mm during 2000. The collection of arthropods was done with sweep net at weekly intervals and the collected insects sorted out into respective taxa based on taxonomic characters. A total of seven samplings were taken during the season. The number of taxa in weeded and partially weeded rice ecosystems were recorded in each week. The species of weed plants allowed in partially weeded plots were collected and identified. In the present investigation, Jaccard coefficient index of similarity (Jaccard, 1908) was used to study the similarity of coccinellid fauna in weeded and partially weeded rice ecosystems.

$$C_j = j/(a + b - j)$$

where,

j is the number of taxa occurring in both samples A (weeded) and B (partially weeded)

a is the number of taxa in sample A (weeded) and

b is the number of taxa in sample B (Partially weeded)

Results and Discussion

The present study revealed the occurrence of seven taxa of predatory coccinellids in rice ecosystem during *kharif* 2000 (Table 37.1). The relative abundance of coccinellids showed that among the seven species, *Micraspis discolor* was the dominant species and exhibited more abundance in weeded plot (8.0) than in partially weeded plots (6.0). Sheltan and Edwards (1988) stated that common species of insects always had greater abundance only in undiversified ecosystem and also had ability to survive in existing maximum and minimum environmental conditions. *Coccinella transversalis, Hormonia octomaculata, Cheilomenes sexmaculatus, Brumoides sutralis, Chilocorus nigritus, M. discolor* and M. sp. Showed 2, 4, 3, 5, 7, 1 and 6 rank, respectively and had more abundance in partially weeded plot. All the seven taxa were recorded only in partially weeded rice ecosystem. The rare species like *C. sexmaculatus, B. suturalis* and *C. nigritus* exhibited more abundance in diversified ecosystem due to availability of alternate resources like pollen and nectar as stated by Pimm (1998). The diversity of predatory coccinellids exhibited less similarity values of 0.60, 0.50, 0.40 and 0.33 in the third, fourth fifth and sixth week, respectively (Table 37.2). During tillering and flowering stages of the crop, the diversity of coccinellids was more due to availability of favourable microclimates and abundance of alternate resources as indicated by Ohgushi (1992). The coccinellids exhibited 0.80 similarity in the first and the last week of the crop growth possibly due to the presence of common species of coccinellids (*M. discolor*, *M.* sp) at this stages as expressed by Odum (1971). The diversity of coccinellids did not show more variation between the rice varieties. In partially weeded rice ecosystem, a total of 18 weed species were recorded. Among them, *Echinochloa colonum*, *Cyperus rotundus*, *C. iria*, *C. difformis*, *Panicum repens* and *Bracharia mutica* were dominant.

Table 37.1: Relative and Rank Abundances of Coccinellids in Irrigated Rice Ecosystem During *kharif* 2000

Coccinellid Fauna	*Weeded*		*Partially Weeded*	
	*	**	*	**
Coccinella transversalis (Fab.)	1.2	3	3.0	2
Hormonia octomaculata (Fab.)	1.0	4	2.0	4
Cheilomenes sexmaculatus (Fab.)	1.0	4	2.8	3
Brumoides sutralis (Fab.)	0	5	1.6	5
Chilocorus nigritus (Fab.)	0	5	1.0	7
Micraspis discolor (Fab.)	8.0	1	6.0	1
Micraspis spp.	3.2	2	1.2	6
Total number of taxa	**5**		**7**	

*: Values in the columns are average of 20 sweeps.

**: Numbers in the columns are rank abundance values.

Table 37.2: Taxonomic Similarity of Predatory Coccinellids in Irrigated Rice Ecosystem During *kharif* 2000

Fauna	*I Week*				*II Week*			
	MDU 5	*ADT 36*	*ADT 39*	*ADT 43*	*MDU 5*	*ADT 36*	*ADT 39*	*ADT 43*
Coccinellids	0.80	0.71	0.83	0.80	0.80	0.71	0.80	0.80
Fauna	*III Week*				*IV Week*			
	MDU 5	*ADT 36*	*ADT 39*	*ADT 43*	*MDU 5*	*ADT 36*	*ADT 39*	*ADT 43*
Coccinellids	0.71	0.66	0.60	0.60	0.50	0.57	0.60	0.50
Fauna	*V Week*				*VI Week*			
	MDU 5	*ADT 36*	*ADT 39*	*ADT 43*	*MDU 5*	*ADT 36*	*ADT 39*	*ADT 43*
Coccinellids	0.40	0.50	0.40	0.40	0.33	0.33	0.42	0.42
Fauna	*VII Week*							
	MDU 5	*ADT 36*	*ADT 39*	*ADT 43*				
Coccinellids	0.80	0.80	0.71	0.71				

*: Values in the columns are indices.

Conclusion

The present study was concluded that number of species and abundance were more in partially weeded plots than in weeded plots. The species of weed plants present in the partially weeded plots provided pollen and nectar for the survival of coccinellids, when the sucking insects were absent in weeded plot. The coccinellids always maintained the abundance of sucking insects in partially weeded plots. When the population of sucking insects increased in weeded plot, the coccinellids present on the weed hosts in partially weeded plot, migrated to weeded plot. Hence the diversity of coccinellids was found to be more in partially weeded plots.

References

Jaccard, P. 1908. Nouvelles recherctus surla distribution floral. *Bulletin Societe Sciences Naturale*, 44: 226–270.

Odum, E.P. 1971. *Fundamentals of Ecology*, 3rd edn. Philadelphia Saunders. pp. 574.

Ohgushi, T. 1992. Resource limitation on insect herbivore populations. In: *Effects of Resource Distribution on Animal-plant Interaction*, (Eds.) M.D. Hunter, T. Ohgushi and P.W. Price. Academic press, New York, p. 287–325.

Pathak, M.D. and Khan, M.Q. 1991. Control of rice insect pests. In: *Major Insect Pests of the Rice Plants.* Proc. Symp. Intern. Rice. Res. Inst. Philippines, 1990: 238–256.

Pimm, S.L. 1984. The complexity and stability of ecosystems. *Nature*, 307: 321–326.

Samal, P. and Misra, B.C. 1982. *Coccinella repanda* Thumb. A predatory coccinellid beetle of rice brown planthopper, *Nilaparvata lugens* (Stal.). *Oryza*, 19: 212.

Shelton, M.D. and Edwards, C.R. 1983. Effect of weeds on the diversity and abundance of insects in soybeans. *Environ. Entomol.*, 12: 296–298.

Chapter 38

Storability of *Corcyra cephalonica* Stainton (Galleridae: Lepidoptera) Eggs for Trichogrammtid Production

J.R. Kadam, S.R. Parate, A.P. Chavan, D.B. Kadam and B.M. Mhaske

Department of Entomology, Mahatma Phule Krishi Vidyapeeth, Rahuri Maharashtra, India

ABSTRACT

The results revealed that there was no deterioration of the *Corcyra* eggs when stored at 0, 5, and 10°C with 85, 80 and 75 per cent relative humidity for 20, 10 and 4 days respectively. These were cent per cent accepted by the test parasitoid. *Trichogramma chilonis* for parasitization. However, the eggs stored still more longer up to 36, 22 and 14 days under similar conditions respectively, also showed the satisfactory acceptability of 70 per cent.

Keyword: *Corcyra cephalonica, Eggs, Trichogrammtid production.*

Introduction

The eggs of rice moth, C. *cephalonica* are used as a factitious host for the multiplication of some predators and parasitoids. Among *Trichogrammatids, Trichogramma minutum* Riley (Krishnamurty, 1940; Abbas ând Anwar, 1963), *T. australicum* Girault (Breniere, 1965), *T. erosicornia* (King, 1930), *T exiguum* (Maninder and Verma, 1982) were reared on eggs of rice moth. The egg larval parasitoid, *Chelonus blackburni* Cameron (Manoharan and Balsubramanian, 1982 and Prasad *et al.*, 1982) and the larval parasitoids. *Apanteles angaleti* Muesebeck (Subbarao and Gopinath, 1961), *Bracon hebetor* Say (Krishnamurty and Seshagiri Rao, 1944), *Aphibolus venater* (Pingale, 1954) are also reared on *Corcyra*.

Mass production of rice moth eggs and their storability is the bottleneck in economic production and efficient utilization of Trichogrammatids. The studies on storability of mass produced eggs of the fictitious host are also necessary to reduce wastage in no demand situation, so as stored host eggs could be used for production of the bioagents immediate after the demand. Therefore, an attempt was made to study eggs storability of the fictitious host for a Trichogrammatid production to find out most suitable storability approach leading to rationalising supply of the rice moth eggs for Trichogrammatid production.

Materials and Methods

A laboratory experiment was conducted in the Department of Agricultural Entomology, Mahatma Phule Krishi Vidyapeeth, Rahuri, Dist. Ahmednagar (Maharashtra, India). Eggs of *C. cephalonica* were stored in three separate conditions by adjusting 0°C with 85 per cent relative humidity in freezer and 5°C and 10°C with 80 and 75 per cent relative humidity in B.O.D. incubators, respectively. For this purpose 1.8 c.c. fresh eggs samples were taken on alternate days and irradiated with ultra-violet rays for 30 minutes keeping at 15 cm distance in laminar flow cabinet. These eggs were divided in to nine equal (0.2 c.c.) parts which were kept separately in 7 cm × 5 cm size polythene bags. Further, the nine parts were grouped in to three equal sets. The three packets each of 0.2 c.c. in a set were treated as three replications and the set was placed in freezer for storage at 0°C with 85 per cent relative humidity for deterioration (browning and blackening) of eggs. From remaining two sets one was incubated at 5°C with 80 per cent relative humidity and the another at 10°C with 75 per cent relative humidity in B.O.D. incubators separately during the same period. At initiation of deterioration of the eggs samples in each of the storage conditions the samples were removed from storing devices. From each of the samples 1000 eggs were separated and pasted on a strips of post card thick paper with help of 10 per cent gum arabic following usual procedure of preparation of eggs cards. The sample strip of each of the replication so prepared were subjected to the parasitoids for parasitization by placing 1 part of 6 days old parasitised egg cards with the unparasitised egg strips in the ratio of 1 : 6. Likewise eggs in three replication of each storage condition were tested for their acceptability for the parasitization by the parasitoid. Observations on the acceptability by *T. chilonis* as judged from blackened eggs with exit holes were recorded under compound microscope at 10th days after releasing of parasitoid. From this percentage acceptability of all samples were determined. The experimental data suitably arranged for tabulation.

Results and Discussion

Storability of *C. cephalonica* Eggs

The results of the study on egg storability are presented in Table 38.1. The storability was judged from extent of deterioration of the eggs stored for different periods and at different temperature. The mean deterioration of the eggs at 0, 5, and 10 °C with 85, 80 and 75 per cent relative humidity was 40.68, 58.22 and 71.12 per cent respectively, on 78th day of storage. At the same temperature and per cent relative humidity condition there was no deterioration of the eggs when stored for 20, 10, and 4 days; while cent per cent deterioration was observed when these stored for 78, 60, and 36 days, respectively. The results also showed that the eggs could be stored at these temperatures and per cent relative humidity up to 35, 25, and 15 days respectively bearing marginal deterioration up to 30 per cent. Thus, the results indicated that with decrease in storage temperature, increase in relative humidity there was increase in storability. Similarly, with increase in storage period there was increase in deterioration of the eggs. Bigler (1986) reported maximum storage of 4 weeks for irradiated *Ephestia* eggs at 2°C and 90 per cent relative humidity. The *Corcyra* eggs stored for 4 weeks at 0°C temperature and 85 per cent relative humidity in present study. The narrow difference in the storability of the eggs of *Ephestia* and *Corcyra* was due to variability in species in host insect storage temperature and relative humidity.

Table 38.1: Influence of Temperature on Storability of *Corcyra cephalonica* Eggs

Ageing (Days)	*Deterioration (%) of Eggs Stored at*		
	0°C	*5°C*	*10°C*
0–4	0	0	0
5–10	0	0	1.00–12.80
11–15	0	3.50–12.30	16.20–30.00
16–20	0	12.70–24.20	34.50–45.80
21–25	1.00–9.25	26.50–30.00	47.60–60.65
26–30	11.70–19.90	33.50–47.90	67.90–82.29
31–35	21.75–29.35	49.85–59.95	86.10–98.14
36–40	30.70–40.90	63.20–71.90	100
41–45	42.85–52.00	74.40–81.85	100
46–50	54.60–60.10	83.90–90.70	100
51–55	62.30–70.30	92.10–97.60	100
56–60	71.90–78.00	98.40–100.00	100
61–65	79.10–86.50	100	100
66–70	88.70–92.30	100	100
71–75	94.05–99.00	100	100
76–78	99.90–100.00	100	100
Mean	40.68	58.22	71.12

Acceptability Stored Eggs by Egg Parasitoid *T. chilonis*

The data on acceptability of the eggs by the egg parasitoid *T. chilonis* are presented in Table 38.2. Cent per cent of the eggs stored at 0, 5, and 10 °C with 85, 80 and 75 per cent relative humidity up to 20, 10 and 4, days respectively, were parasitized by the *Trichogramma*. Thereafter, the acceptability of the eggs decreased. However, considerable (70 per cent) acceptability by the parasitoid was observed when these stored up to 36, 22 and 14 days at these temperature and per cent relative humidity, respectively. Mean acceptability of the eggs stored up to 5 weeks at 0, 5, and 10 °C temperature with 85, 80 and 75 per cent relative humidity was 93.11, 77.79 and 59.60 per cent, respectively. *Corcyra* eggs stored at 10°C for 5 and 10 days showed 100 and 89 per cent parasitization by *T. chilonis* (Anonymous, 1996). However, in present study eggs stored at 10°C with 75 per cent relative humidity for 5 and 10 days showed 99 and 87.2 per cent parasitization by same parasitoid. The narrow difference in acceptability after 5 and 10 days of storage was due to variation in the temperature and relative humidity.

Table 38.2: Effect of Storability of *C. cephalonica* Eggs at Different Conditions on Acceptability by Eggs Parasitoid, *Trichogramma chilonis*

Ageing (Days)	*Percentage Acceptability of the Eggs Stored at Different Temperature*		
	0°C	*5°C*	*10°C*
0	100.00	100.00	100.00
1	100.00	100.00	100.00
2	100.00	100.00	100.00
4	100.00	100.00	100.00
6	100.00	100.00	98.00
8	100.00	100.00	92.40
10	100.00	100.00	87.20
12	100.00	93.00	80.40
14	100.00	88.10	72.90
16	100.00	83.30	65.50
18	100.00	79.80	60.40
20	100.00	75.80	54.20
22	98.00	71.20	50.20
24	93.20	66.70	44.60
26	88.30	62.50	32.10
28	84.80	54.80	23.20
30	80.10	52.20	17.70
32	76.40	48.20	10.10
34	72.00	43.30	3.20
36	69.30	36.80	0.00
Mean	93.11	77.79	59.60

References

Abbas, H.M. and Anwar, M.S. 1963. Breeding of *Corcyra cephalonica* Stainton for mass production of *Trichogramma minutum* Riley. *Agriculture Pakist.*, 14(2): 209–214.

Anonymous, 1996. Efficiency in parasitization of *Trichogramma chilonis* at different period of storage. Report of Res. Rev. of Sub-Committee in Agril. Entomology, M.P.K.V., Rahuri (M.S.).

Bigler, F. 1986. Mass production of *Trichogramma maidis* Pint. and its field application against *Ustrinia nubilalis* Hbn. in Switzerland. *J. Appl. Entomol.*, 101: 23–29.

Breniere, J. 1965. Bionomics and technique of mass rearing of *Trichogramma australicum. Entomophaga*, 10(10): 83–96.

King, C.B.R. 1930. Report of Entomologist. *Bull. Tac. Res. Int.*, Ceylon, 4: 9.

Krishnamurty, B. 1940. *Trichogramma minutum* Riley in relation to *Sitotroga cerealella.* Mysore Curr. Sci., 9(12): 544.

Krishnamurty, B. and Seshagiri Rao, D.S. 1944. A preliminary note on the breeding of a beneficial ectophagus larval parasite (Braconidae) on laboratory host. *Curr. Sci.*, 13(3): 81–82.

Maninder, S. and Verma, G.C. 1982. The growth of population of *Trichogramma exiguum* Perkins. *Indian J. Ent.*, 44(2): 179–181.

Manoharan, V. and Balsubramanian, M. 1982. Most parasitoid density response to fecundity of *Chelonus blackburni*. *Proc. Ind. Nation. Acad.*, 48 (2): 214–217.

Pingale, S.V. 1954. Biological control of some stored grain pests by the use of a big predator, *Amphibolus venator*. *Indian J. Ent.*, 16(3): 300–302.

Prasad, J., Pawar, A.D. and Sharma, R.K. 1982. Biocontrol trials with an exotic egg larval parasitoids, *Chelonus blackburrni* Cameron on *Heliothis armigera*. *J. Adv. Zoology*, 3(2): 160–161.

Subbarao, B.R. and Gopinath, K. 1961. The effect of temperature and humidity on the reproduction potential of *Apanteles angalati*. *Ent. Exp. Appl.*, 4(2): 119–122.

Chapter 39

Stability Analysis in Sorghum [*Sorghum bicolor* (L.) Moench]

S.P. Patil,* M.R. Manjare,** S.R. Kamdi and A.M. Dethe******

M.Sc. (Agri.) Student, **Associate Professor of Agricultural Botany, * Ph. D. Scholars Cytogenetics and Plant Breeding, Department of Agricultural Botany, Mahatma Phule Krishi Vidyapeeth, Rahuri – 413 722*

ABSTRACT

Nineteen genotypes of sorghum comprising varieties, hybrids and elite lines were evaluated under three environments (sowing dates) for ten quantitative traits including grain yield. The significant value of G × E interactions revealed differential response of the genotypes to varying environmental conditions. Stability parameters revealed that the genotype SPV–1592 possesses average stability for grain yield, suggesting its suitability for inclusion in breeding programme for the development of stable variety.

Keywords: ***Sorghum bicolor, Stability, G × E interaction.***

Introduction

The ultimate aim of any plant breeding is to develop cultivars with high-yielding potential with consistent performance over diverse environments. Productivity of population is the function of its adaptability, while later is the compromise of fitness, stability and flexibility. Thus, the predictability of performance *i.e.*, stability depends upon the adaptability of the genotype.

Sorghum [*Sorghum bicolor* (L.) Moench] is one of the most important food and fodder crops of the world in general and of semi-arid tropics in particular. In the Maharashtra State as a main cereal crop, *rabi* sorghum is grown in varying conditions as residual moisture and major area under this is on medium to light soils. The present investigation was planned to identify well buffered stable genotypes among varieties, hybrids and elite populations of sorghum.

Materials and Methods

Nineteen genotypes involving the existing varieties, hybrids and newly developed lines were evaluated during *rabi* 2002 in three environments *i.e.*, sowing dates, (*i*) 27th September 2002, (*ii*) 11th October 2002, (*iii*) 26th October 2002 at Post Graduate Institute Farm, Mahatma Phule Krishi Vidyapeeth, Rahuri (M.S.). Each entry was sown in a 4-row plot of 4.5 m length with 45 cm interrow and 15 cm distance between the plants. All the treatments received recommended package of practices to ensure satisfactory crop growth. Ten randomly selected plants per plot in each replication were used to record data for the characters *viz.*, days to 50 per cent flowering, days to maturity, plant height, earhead length, earhead girth, number of grains per earhead, 1000 grain weight, grain yield per plant, grain yield per hectare and dry fodder yield per hectare. Data from three environments as well as pooled data were subjected to analysis of variance (Panse and Sukhatme, 1995). The traits, which showed significant genotype-environment (G × E) interaction, were subjected to stability analysis as per Eberhart and Russell (1966). The three stability parameters:

1. Overall mean performance of each genotype across the environments (X)
2. The regression of genotype on the environmental index (bi), and
3. Squared deviation from regression coefficient (S^2di) were worked out for all the genotypes.

The significance of stability parameter (bi) and its deviation from unity were tested by Students' 't' test.

Results and Discussion

The analysis of variance for phenotypic stability (Table 39.1) indicated that the mean differences due to genotypes were significant for all the characters except days to maturity when tested against G × E interaction. Significant differences were also observed for all the characters when tested against pooled deviation. Except for earhead length and earhead girth, environmental variances were found significant for all the characters. The highly significant G × E interaction observed for all the characters indicated that genotypes showed varied response to different environments. Bakeit (1990) and Reddy *et al.* (2004) also reported significant G × E interaction in most of the traits studied.

Table 39.1: Analysis of Variance for Stability with Three Environments (Eberhart and Russell, 1966)

Sl.No.	*Source*	*d.f.*	*Mean Sum of Squares*				
			Days to 50% Flowering	*Days to Maturity*	*Plant Height (m)*	*Earhead Length (cm)*	*Earhead Girth (cm)*
1.	Genotype	18	15.47+@	11.802@	0.208+@	3.729+@	2.785+@
2.	Environment (E)	2	197.484+@	328.656+@	0.605+@	1.423	0.554
3.	G × E	36	5.028*	6.366*	0.0261*	0.583*	0.783*
4.	E + (G × E)	38	15.157	23.328	0.0566	0.627	0.771
5.	Environment (linear)	1	394.95	657.26	1.211	2.847	1.107
6.	G × E (linear)	18	8.305@	9.023@	0.0395@	0.606	0.889
7.	Pooled deviation	19	1.659	3.517*	0.0120*	0.530*	0.641*
8.	Pooled error	54	1.017	1.0202	0.00316	0.170	0.239

Contd...

Table 39.1–Contd...

Sl.No.	Source	d.f.	Mean Sum of Squares				
			Number of Grains per Earhead	1000 Grain Weight (g)	Grain Yield per Plant (g)	Grain Yield per ha (kg)	Dry Fodder Yield per ha (kg)
1.	Genotype	18	101560.90+@	31.824+@	346.63+@	878195.60+@	5058631.00+@
2.	Environment (E)	2	132128.00+@	6.531+@	15.593+@	193200.00+@	1112192.00+@
3.	G × E	36	4509.77*	0.263*	0.884*	8056.00*	46435.56*
4.	E + (G × E)	38	11226.53	0.593	1.659	1?800.42	102528.00
5.	Environment (linear)	1	264257.80	13.073	31.170	386421.40	2224072.00
6.	G × E (linear)	18	7124.46@	0.412@	1.718@	15652.31@	90158.31@
7.	Pooled deviation	19	1795.01*	0.107*	0.0516	443.475**	2612.72*
8.	Pooled error	54	177.33	0.035	0.0833	139.11	801.185

+ = Significant at 5 per cent and 1 per cent level respectively against the G × E interaction;
@ = Significant at 5 per cent and 1 per cent level respectively against the pooled deviation;
* = Significant at 5 per cent and 1 per cent level respectively against the pooled error

On partitioning the variances of G × E into linear and non-linear components, the linear component of G × E interaction was significant for all the traits except earhead length and girth suggesting that a large portion of G × E interaction was accounted for by the linear regression. Also, non-linear component (pooled deviation) was found highly significant for most of the characters except days to 50 per cent flowering and grain yield per plant, indicating that the genotypes differed considerably with respect to their stability for traits under study. The predominance of linear component noticed would help in predicting the performance of the genotypes across the environment (Table 39.1). Several workers reported importance of both linear and non-linear components of G x E interaction.

The environmental indices depicted that E_1 environment (first sowing date) was observed to be favourable for most of the traits; while E_2 was only favourable for earhead girth. However, E_3 environment was not found favourable for any character.

Different measures of stability have been used by various workers. Finlay and Wilkinson (1963) explained linear regression as a measure of stability; whereas Eberhart and Russell (1966) emphasized that both linear (bi) and non-linear (S^2di) components of G × E interaction be considered while judging the phenotype stability of a genotype; whereas deviation from regression (S^2di) should be considered as measure of stability. Accordingly, the mean (X) and deviation variance (S^2di) of each genotype were considered for stability and linear regression (bi) was used for testing varietal response. Genotypes with lowest or non-significant mean squared deviation being the most stable and *vice-versa*. The three parameters X, bi and S^2di together gave an idea of adaptability of genotypes across the environments. An ideal and stable genotype, according to Eberhart and Russell (1966) may be characterised as with high performance, unit regression and zero or minimum deviation from regression. Thus, taking into account all these points in the present study, the genotypes were categorized accordingly. The mean (X), regression coefficient (bi) and deviation from regression (S^2di) for the traits studied are presented in Table 39.2. In the present investigation, the magnitude of regression coefficient and deviation from regression varied from genotype to genotype.

Table 39.2: Stability Parameters (Eberhart and Russell, 1966)

Sl.No.	Genotype	Days of 50% Flowering			Days to Maturity			Plant Height (cm)			Earhead Length (cm)			Earhead Girth (cm)		
		Mean (X)	Bi	S²di	Mean (X)	Bi	S²di	Mean (X)	Bi	S²di	Mean (X)	Bi	S²di	Mean (X)	Bi	S²di
1.	SPV–1411	76.16	1.23	1.87	124.16	1.20	– 0.58	2.28	– 0.01*	– 0.0031	13.67	2.15	0.008	18.50	2.43	– 0.23
2.	SPV–1504	67.33	0.31	– 0.86	120.66	0.60	– 0.47	2.05	2.18*	– 0.0015	15.11	1.67	0.94*	16.07	9.72	3.64*
3.	SPV–1546	74.00	1.31	– 0.55	123.33	1.39	3.25	2.38	1.82	– 0.0002	15.27	0.02	0.56	17.73	– 0.26	0.15
4.	SPV–1587	71.83	0.01	– 0.85	123.16	1.20	– 0.58	2.51	0.14	– 0.0032	14.68	3.63	0.10	17.34	1.00	– 0.24
5.	SPV–1588	74.66	– 0.23	– 0.01	121.83	– 0.11**	– 0.83	2.61	0.02*	– 0.0030	15.21	2.23	1.49*	16.91	– 0.32	– 0.23
6.	SPV–1589	75.33	1.08	1.46	122.66	1.27	5.18*	2.82	0.05*	– 0.0031	15.99	0.48	– 0.15	15.83	– 4.30	– 0.11
7.	SPV–1590	73.83	1.15	6.35*	122.16	0.29*	4.19	2.12	1.43	0.0436*	15.22	– 0.60	0.71*	14.78	– 2.76	– 0.22
8.	SPV–1591	74.16	1.86**	– 0.50	122.83	1.44	1.07	2.20	1.72	– 0.0007	16.30	0.19	0.97	16.69	5.84	0.81
9.	SPV–1592	73.50	1.78*	– 0.51	122.00	1.50	– 0.37	2.19	1.09	– 0.0052	16.36	0.27	– 0.16	16.92	– 0.70	0.54
10.	RSV–117	75.16	1.39	– 0.46	122.50	1.32	– 0.99	2.07	1.35	– 0.0004	15.31	4.40	– 0.11	16.14	1.31	1.34*
11.	RSV–143	74.83	1.46	1.23	122.83	1.19	3.80	2.16	2.13*	0.0238*	15.18	– 3.44*	0.12	16.73	1.15	– 0.16
12.	RSV–268	76.83	1.31	– 0.94	121.00	0.47	0.63	2.33	1.62	– 0.0031	17.46	– 0.79	– 0.09	17.65	3.34	– 0.19
13.	RSV–271	77.16	1.31	0.17	126.33	1.36	15.42*	2.31	1.91	0.0865*	14.54	1.88	0.12	17.19	– 2.61	– 0.17
14.	RSV–272	77.00	1.16	– 0.71	126.50	1.32	– 0.99	2.13	1.26	– 0.0016	13.37	– 0.88	0.42	16.48	– 0.13	– 0.02
15.	RSV–491	75.83	1.16	3.74	124.16	1.20	– 0.58	2.69	0.02*	– 0.0031	15.79	0.14	0.34	15.98	– 4.32	– 0.14
16.	CSV–216	72.50	0.00	– 1.01	123.00	1.39	7.29*	2.45	0.69	0.0310*	16.48	0.01	– 0.16	16.22	– 0.54	3.03*
17.	Phule Maulee	72.33	1.24	– 0.90	118.00	0.11*	0.51	2.05	0.93	– 0.0031	13.71	2.88	0.46	15.08	9.22*	0.01
18.	M–35–1	74.16	1.32	5.65*	122.33	1.33	6.19*	1.97	0.36	– 0.0028	13.61	4.56	1.43*	16.07	2.31	0.07
19.	Swati	72.66	0.07	– 0.97	121.50	0.47	5.28*	1.76	0.22	0.0041	16.17	0.15	– 0.15	15.09	– 1.24	– 0.22
	Grand Mean	74.17	–	–	122.57	–	–	0.077	–	–	15.23	–	–	16.49	–	–
	S.E. ±	0.910	0.281	–	0.319	–	–	0.442	–	–	0.515	3.003	–	0.566	3.320	–

Contd...

Table 39.2–Contd...

Sl.No.	Genotype	Number of Grains per Earhead			1000-grain Weight (g)			Grain Yield per Plant (g)			Grain Yield per ha (kg)			Grain Yield per ha (kg)		
		Mean (X)	Bi	S²di	Mean (X)	Bi	S²di	Mean (X)	Bi	S²di	Mean (X)	Bi	S²di	Mean (X)	Bi	S²di
1.	SPV–1411	1623.50	0.29	– 152.41	41.49	0.91	– 0.02	66.40	0.37**	– 0.078	2649.3	2.01**	– 139.2	6358.3	2.01**	– 695.3
2.	SPV–1504	1426.30	0.94	413.53	33.59	3.27**	0.71*	49.34	3.49**	0.163	1973.9	2.62**	682.1*	4737.5	2.62**	4050.9*
3.	SPV–1546	1840.30	2.00*	2945.3*	42.71	0.51	0.43*	78.74	0.43**	– 0.023	3365.1	0.15**	12.70	8076.2	0.15**	77.93
4.	SPV–1587	1543.60	1.13	4913.1*	40.47	– 0.17**	0.04	63.80	1.60**	– 0.003	2726.8	0.62*	– 134.5	6544.3	0.62*	– 735.11
5.	SPV–1588	1803.50	1.55	4142.7*	44.24	0.56	– 0.02	86.53	0.33**	– 0.066	3997.9	0.13**	– 135.1	8875.2	0.13**	– 780.53
6.	SPV–1589	1936.50	0.52	– 165.58	39.41	0.50	– 0.04	77.67	0.45**	– 0.073	3319.4	0.17**	– 143.8	7966.5	0.17**	– 804.83
7.	SPV–1590	1750.00	0.42	6.68	35.14	1.59	0.18*	63.94	0.95	– 0.003	2732.9	0.37**	– 99.42	6559.1	0.37**	– 566.53
8.	SPV–1591	1586.80	0.44	– 140.3	37.77	1.57	– 0.04	62.70	0.97	0.003	2561.6	1.25	6346.5*	6147.9	1.25	3688.4*
9.	SPV–1592	1576.30	2.57**	8503.5*	39.46	0.64	– 0.030	66.19	1.04	– 0.044	2640.3	0.37*	– 138.2	6336.6	0.37**	– 771.97
10.	RSV–117	1519.30	1.30	3521.3*	38.22	0.87	– 0.031	57.59	3.70**	– 0.048	2297.9	1.33*	228.7	5514.5	1.25	1419.1
11.	RSV–143	1417.20	2.03**	1606.6*	37.40	0.94	– 0.030	58.78	0.70	– 0.074	2178.6	1.89**	– 140.1	5228.5	0.37**	– 714.81
12.	RSV–268	1633.00	0.20	– 165.25	36.68	0.46	– 0.031	61.06	0.25**	– 0.073	2261.6	1.82**	– 133.1	5427.8	1.33*	– 668.02
13.	RSV–271	1833.50	0.30	– 40.51	42.83	0.30	– 0.035	79.33	0.51*	– 0.058	3390.4	0.19**	– 73.37	8137.1	1.89**	– 69.70
14.	RSV–272	1482.20	1.44	3472.1*	36.29	0.68	0.01	57.97	0.49*	– 0.077	2148.1	1.80**	– 139.6	5155.4	1.82**	– 715.56
15.	RSV–491	1883.80	0.87	– 43.84	38.00	0.68	0.07	79.39	0.21**	– 0.060	3392.9	0.08**	– 87.87	8142.9	0.08**	– 487.57
16.	CSV–216	1763.30	0.17*	93.18	32.71	0.95	0.04	59.13	0.42**	– 0.052	2527.2	0.16**	– 115.9	6065.3	0.16**	– 662.88
17.	Phule Maulee	1177.10	1.62	1223.6*	34.63	1.56	– 0.009	45.90	0.67	– 0.078	1831.9	1.53**	– 136.5	4396.8	1.53**	– 720.31
18.	M–35–1	1672.20	0.74	775.8*	34.60	2.19**	– 0.011	59.31	1.89**	0.121	2199.8	2.28**	266.1	5279.6	2.28**	1630.31
19	Swati	1632.80	0.35	– 173.46	36.44	0.91	0.17*	62.17	0.37**	– 0.079	2657.1	0.14**	– 136.5	6377.1	0.14**	– 756.20
	Gram Mean	1636.90	–	–	37.98	–	–	65.04	–	–	2660.7	–	–	6385.6	–	–
	S.E. ±	29.96	0.36	–	0.231	0.395	–	0.160	0.176	–	14.89	0.150	–	36.143	0.149	–

* and ** Significant at 5 and 1 per cent level, respectively.

The genotypes *viz.*, SPV–1546, SPV–1591 and SPV–1592 for days to 50 per cent flowering; SPV–1592 for days to maturity, SPV–1546 and RSV–268 for plant height; RSV–117 for earhead length and SPV–1591 for earhead girth exhibited below average stability indicating their suitability for favourable environment.

Whereas, the genotypes *viz.*, SPV–1504, SPV–1587, CSV–216 and Swati for days to 50 per cent flowering; SPV–1504, SPV–1588, RSV–268 and Phule Maulee for days to maturity; SPV–1411, SPV–1587, SPV–1588, SPV–1589 and RSV–491 for plant height; RSV–271 for earhead girth; SPV–1589, SPV–1590, RSV–271 and CSV–216 for number of grains per earhead; SPV–1587, SPV–1588, SPV–1589 and RSV–271 for 1000 grain weight; SPV–1411, SPV–1546, SPV–1588, SPV–1589, RSV–271 and RSV–491 for grain yield per plant; SPV–1546, SPV–1587, SPV–1588, SPV–1599, SPV–1590, RSV–491 for grain yield per hectare as well as dry fodder yield per hectare depicted above average stability suggesting their suitability for poor environment.

None of the genotype was found stable for all the characters under study. For days to 50 per cent flowering Phule Maulee; for days to maturity RSV–117; for earhead length SPV–1546, SPV–1589, SPV–1592, RSV–268, RSV–491, CSV–216 and Swati; for earhead girth SPV–1411, SPV–1546, SPV–1587, SPV–1588, SPV–1592, RSV–143 and RSV–268; for number of grains per earhead RSV–491; for 1000 grain weight SPV–1411, SPV–1592, RSV–117 and RSV–491 and for grain yield per plant SPV–1592 showed general adaptability *i.e.*, average stability. No genotype was found average stable for plant height, grain yield per hectare and dry fodder yield per hectare.

'SPV–1592' had stability (general adaptability) for major traits including grain yield per plant; while M–35–1 was found to be unstable for major characters (Table 39.3). 'Phule Maulee' and 'SPV–1504' were observed with good stability for earliness. Thus, present results are in conformity with those of Heinrich *et al.* (1983), Bakeit (1990) and Narkhede *et al.* (1997).

Table 39.3: Nature of Stability of Genotypes under Different Environment for Ten Characters in Sorghum

Sl.No.	Character	Genotypes Showing Stability			Unstable Genotypes
		Average	Above Average	Below Average	
1.	Days to 50% flowering	Phule Maulee	SPV–1504, SPV–1587, CSV–216, Swati	SPV–1591, SPV–1592, SPV–1546	SPV–1590, M–35–1
2.	Days to maturity	RSV–117	SPV–1504, SPV–1588, RSV–268, Phule Maulee	SPV–1590	SPV–1590, M–351, Swati
3.	Plant height	–	SPV–1411, SPV–1587, SPV–1588, SPV–1589, RSV–491	SPV–1546, RSV–268	RSV–271, CSV–216
4.	Ear head length	SPV–1546, SPV–1589, SPV–1592 RSV–268, RSV–491, CSV–216, Swati	–	RSV–117	SPV–1591
5.	Earhead girth	SPV–1411, SPV–1546, SPV–1587, SPV–1588, SPV–1592, RSV–143, RSV–268	RSV–271	SPV–1591	–

Contd...

Table 39.3–Contd...

Sl.No.	Character	Genotypes Showing Stability			Unstable Genotypes
		Average	Above Average	Below Average	
6.	Number of grains per earhead	RSV–491	SPV–1589, SPV–1590, RSV–271, CSV–216	–	SPV–1546, SPV–1588, M–35–1
7.	1000-grain weight	SPV–1411, SPV–1592, RSV–117, RSV–491	SPV–1587, SPV–1588, SPV–1589, RSV–271	–	SPV–1546
8.	Grain yield per plant	SPV–1592	SPV–1411, SPV–1546, SPV–1588, SPV–1589, RSV–271, RSV–491	–	–
9.	Grain yield per ha	–	SPV–1546, SPV–1587, SPV–1588, SPV–1589, SPV–1590, RSV–271, RSV–491	–	–
10.	Dry fodder yield per ha	–	SPV–1546, SPV–1587, SPV–1588, SPV–1589, SPV–1590, RSV–271, RSV–491	–	–

Present investigation illustrated the most stable and promising sorghum genotypes *viz.*, SPV–1592, SPV–1546, SPV–1587, SPV–1588, SPV–1504 and Phule Maulee; which can be recommended for multi-location evaluations to assess their suitability across locations or they can be used as the donor parents for generating the breeding material suited for development of new varieties with wider adaptability over seasons/locations or with specific adaptability to a particular environment.

References

Bakeit, B.R. 1990. Stability of grain yield and its components of grain sorghum genotypes [*Sorghum bicolor* (L.) Moench] as affected by different irrigation regimes. *Cereal Res. Comm.*, 18(1–2): 117–124.

Eberhart, S.A. and Russell, W.A. 1966. Stability parameters for comparing varieties. *Crop Sci.*, 6: 36–40.

Finlay, K.W. and G.N. Wilkinson. 1963. Analysis of adaptation in a plant breeding programme. *Aust. J Agrjc. Res.*, 14: 742–754.

Heinrich, G.M., Francis, C.A. and J.D. Eastin. 1983. Stability of grain sorghum yield components across diverse environments. *Crop Sci.*, 23(2): 209–212.

Narkhede, B.N., Shinde, M.S. and S.P. Patil. 1997. Stability performance of *rabi* sorghum hybrids. *J. Maharashtra Agric. Univ.*, 22(2): 174–175.

Panse, V.G. and P.V. Sukhatme. 1995. *Statistical Methods for Agricultural Workers*, Revised Edn. by Sukhatme V.G. and V.N. Amble. ICAR Publication, New Delhi.

Reddy Raghu Ram, P., Maruthi Sankar, G.R. and N.D. Das. 2004. Genotype X Environment interactions in *rabi* sorghum. *J. Maharashtra Agric. Univ.*, 29(1): 021–024.

Chapter 40

Effect of Salinity on the Growth of *Rhizobium* and their Interrelationship Between Different Osmotica Like Aspartate, Calcium Chloride and Trehalose

S. Manju, K. Vijayalakshmi, C. Sheela Sasikumar

Post Graduate Department of Microbiology, D.G. Vaishnav College, Arumbakkam, Chennai – 600 106

ABSTRACT

Salinity is a serious threat to agriculture in arid and semi arid regions. The improvement of crop production by biofertilizers in this area requires first, the selection of effective *Rhizobial* strain well adapted to the conditions of this region. For this purpose, we have isolated *Rhizobium* from five different leguminous plants with well developed nodules. When inoculated at increasing salt concentration the growth of the organisms decrease. The present study shows the osmolytes like calcium chloride, aspartate and treholose relieved the growth inhibition under saline conditions. These strains might use various osmolytes during the osmoprotection process depending upon the availability in the environment. Among these osmolytes aspartate is more efficient than other two. But they should also be tested in strain cultivar interaction experiments and in field inoculating trails using cultivars previously selected under typical arid and semi arid regions before use.

Keywords: *Osmotolerant, Rhizobium, Osmolytes, Calcium chloride, Trehalose, Aspartate.*

Introduction

The saline soils may adversely affect the plant growth and microbial population. Not very much is known concerning the effect of salinity on soil microorganisms. Particular interest, of course, would be attached to the nitrogen fixing organisms, because symbiotic nitrogen fixation by root nodule bacteria of legumes alone accounts for approximately 20 per cent global nitrogen. (Somani and Totawat, 1993). In general osmotolerant bacteria accumulate some substances called compatible solutes, either by absorption or by synthesis. It is assumed that these substances confer protection against the deleterious effect of the low water activity, maintain the appropriate cell volume, and protect intracellular macromolecules against the destabilizing effect of the salt (Csonka and Hanson, 1991). Compatible solutes identified so far include amino acid (glutamate and proline), amino acid derivatives (ecotine, betaines and analogous molecules) and carbohydrates (saccharose, trehalose, glycerol and arabitol) (Kempf and Bremer, 1998).

Balanced fertilizers use has always been the concern in India and there have been group discussion and campaigns to implement it. Optimum desirable NPR ratio given is 4 : 2 : 1 one of the major concerns in today's world is pollution and contamination of soil. The use of chemical fertilizer and pesticides has caused tremendous harm to the environment. An answer to this is the biofertilizer now used in most countries. Biofertilizer are organisms that enrich the quality of soil. Plants have number of relation with bacteria, fungi and algae, the common of which are with *Mycorrhiza, Rhizobium* and *Cyanophyceae.* These are known to deliver a number of benefits including plant nutrition, disease resistance and tolerance to adverse soil and climatic condition.

The present study focuses on the role of aspartate, calcium chloride and trehalose, when externally provided in osmoprotection of various *Rhizobium* species which displays a variation in salt tolerance.

Materials and Methods

Isolation and Identification of *Rhizobium*

Rhizobium was isolated from *Arachis hypogea,* fenugreek, *Mimosa pudica, Ciltoria ternetae* and *Desmodium heterophyllum* in yeast extract mannitol agar and identified by colony morphology, gramstaining, motility, nitrate reduction and sugar utilization test (sugars such as glucose, fructose, maltose, sucrose, lactose and mannitol were used. Colonies from 5 different isolates were inoculated into 10 ml YEM broth and incubated in RT shaker overnight. Then O.D. was taken at 594 nm (1 O.D = 10^7 cells/ml).

Salinity Tolerance

YEM broth with NaCl concentration ranging between 1 per cent–5 per cent w/v were inoculated with each isolates at a concentration of 10^7 cells/ml. then the corresponding O.D. values was measured at 594 nm based on turbidity in liquid media after 3 days incubation at 28°C.

Effect of Aspartate, Calcium Chloride and Trehalose

1 per cent of aspartate, calcium chloride and trehalose were sterilized by filtration and incorporated separately into YEM broth ranging between 1 per cent–5 per cent NaCl and incubated at RT for 3 days. The growth was estimated by the measurement of O.D. at 594 nm after 48 hours of incubation and the results were tabulated (Ilham Bouhmouch *et al.*, 2001)

Results and Discussion

Result shows increasing salt concentration may have detrimental effect on soil microbial population as a result of direct toxicity or through osmotic stress, thereby decreases the growth of the

Without osmolyte | With aspartate
With calcium chloride | With trehalose

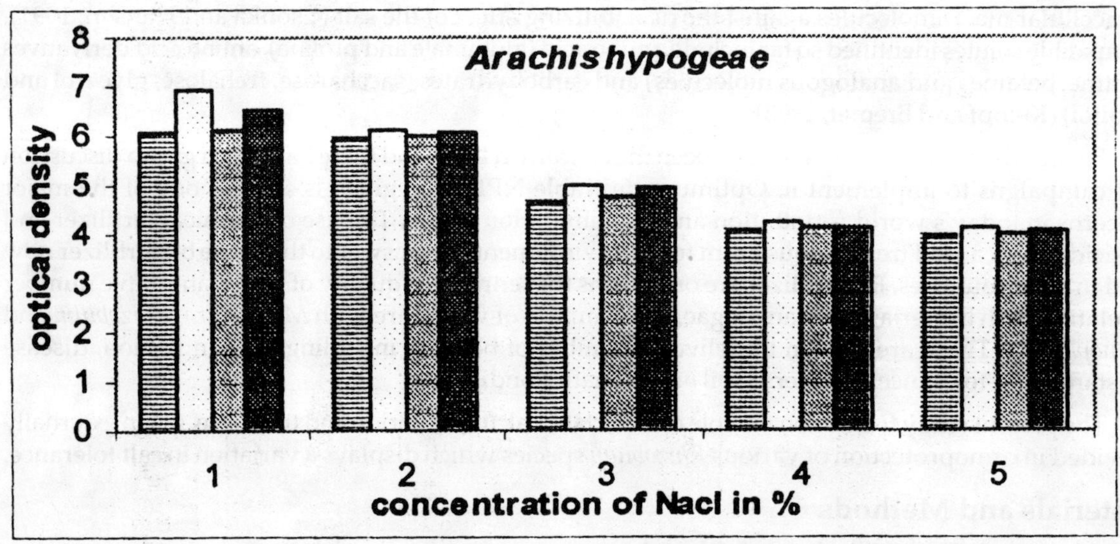

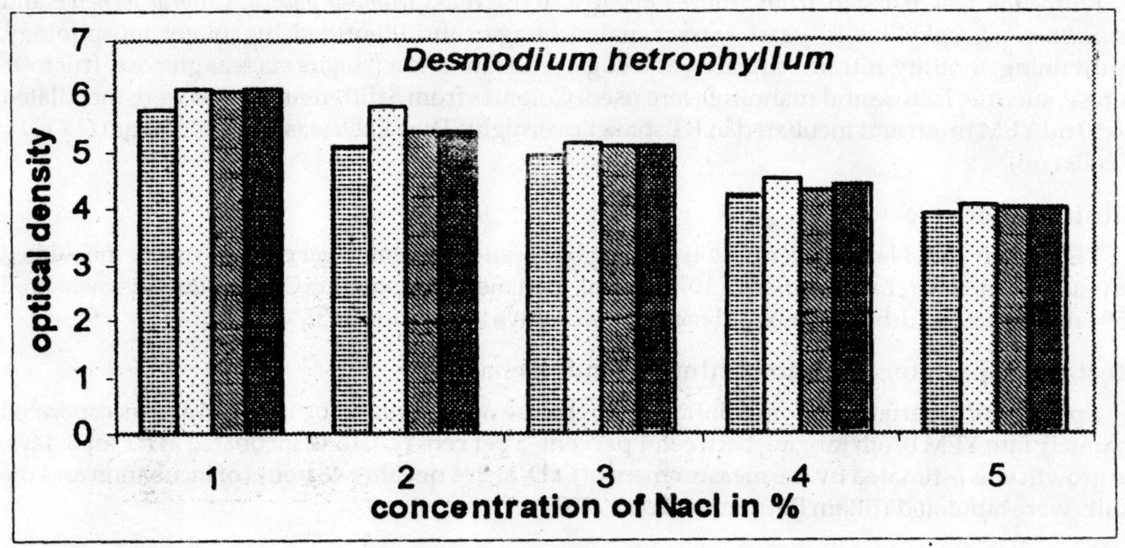

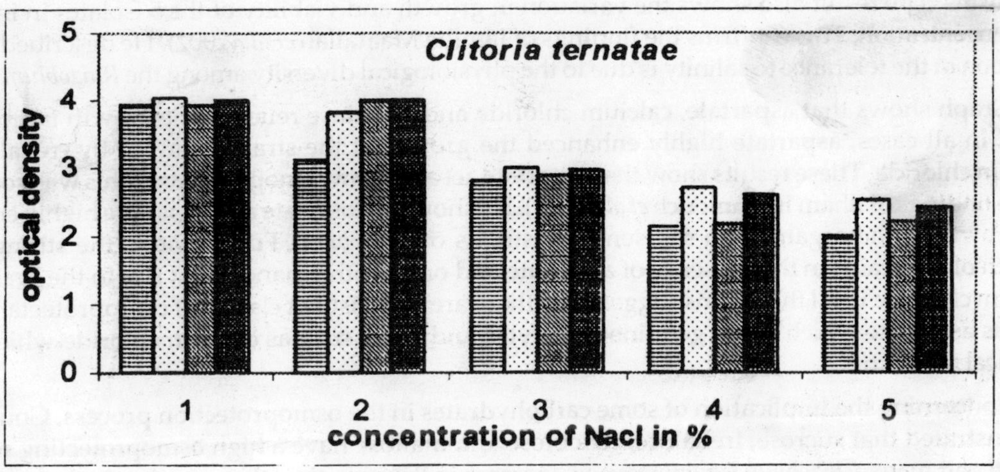
Clitoria ternatae
optical density
0
1
2
3
4
5
1
2
3
4
5
concentration of Nacl in %

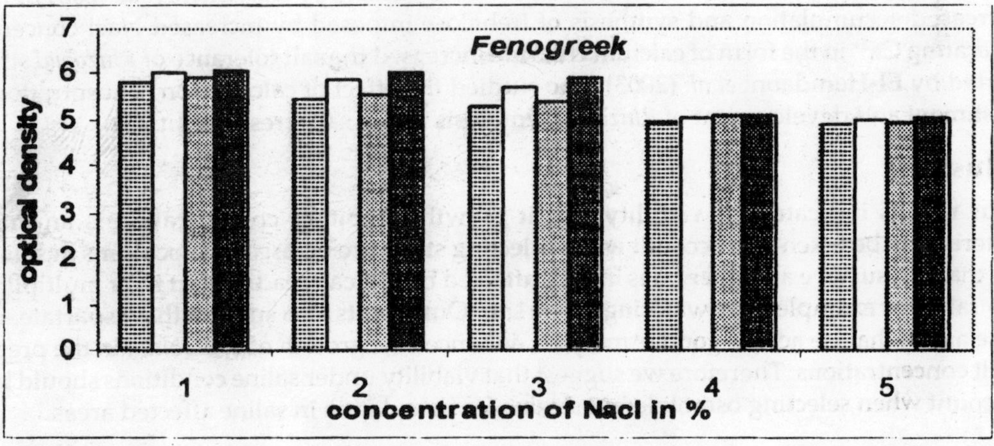
Fenogreek
optical density
0
1
2
3
4
5
6
7
1
2
3
4
5
concentration of Nacl in %

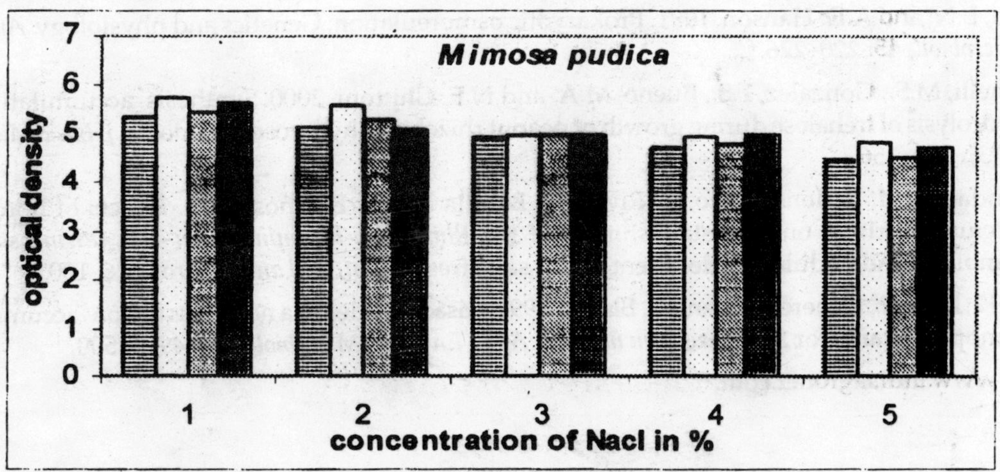
Mimosa pudica
optical density
0
1
2
3
4
5
6
7
1
2
3
4
5
concentration of Nacl in %

organisms. The result also shows the variation in growth and viability of the 5 isolates in increased salt concentration. This confirms the findings of Jamilia Maatallah *et al.* (2002). He described that the variation in the tolerance to salinity is due to the physiological diversity among the *Rhizobium* isolates.

Graph shows that aspartate, calcium chloride and trehalose relieved the growth inhibition by NaCl. In all cases, aspartate highly enhanced the growth of the strain followed by trehalose and calcium chloride. These results show that aspartate act as potent osmoprotectant. This was confirmed by the findings of Ilham Bouhmouch *et al.* (2001). He showed glutamate and aspartate highly enhanced the growth of the organism in salt sensitive strains of *Rhizobium*. Furthermore, the strong rise in growth of all strains in the presence of aspartate and on the other hand slight rise in the presence of calcium chloride in all the strains suggest that there are two distinct classes of osmoprotectant: those such as aspartate which act as genuine osmolytes and those such as calcium chloride which act as chemical mediators.

Concerning the implication of some carbohydrates in the osmoprotection process, Gouffi *et al.*, demonstrated that sucrose; trehalose, cellobiose and maltose have a high osmoprotecting role in *S. meliloti* and many *Rhizobium leguminosarum*. Increase in the growth of the organism in the presence of trehalose encountered in our data supports the results indicated by Dardanelli *et al.* (2001). He shows the increased accumulation and synthesis of trehalose imposed by increased Nacl concentration. Incorporating Ca^{2+} in the form of calcium chloride increased the salt tolerance of *Rhizobial* strain was supported by EI-Hamdaoni *et al.* (2003) who studied the effect of calcium ion concentration on the establishment and development of *Rhizobial* symbiosis under salt stress conditions.

Conclusion

Our results indicate that viability at salt growth inhibitory concentration is an important characteristic to be taken into account when selecting strains to be used as inoculums in saline soils. Strains that can survive at higher rates in soils affected by salt can readily start their multiplication if salinity falls, for example after watering of the soil. Our results also suggest that aspartate, calcium chloride and trehalose acts as good osmolytes, enhances the growth of *Rhizobium* in the presence of high salt concentrations. Therefore we suggest that viability under saline conditions should be taken into account when selecting osmotolerant strains for inoculation in saline affected areas.

References

Csonka, L.N. and A.D. Hanson. 1991. Prokaryotic osmoreulation: Genetics and physiology. *Ann. Rev. Microbiol.*, 45: 220–226.

Dardanelli, M.S., Gonzalez, P.S., Bueno, M.A. and N.E. Ghittoni. 2000. Synthesis, accumulation and hydrolysis of trehalose during growth of peanut rhizobia in hyperosmotic media. *J. Basic Microbiol.*, 40(3): 149–156.

EI-Hamdaoui, M. Redondonieto, R. Rivilla, I. Bonilla and L. Bolanos. 2003. Effects of boron and calcium nutrition on the establishment of the *Rhizobium leguminosarum*–pea (*Pisum sativum*) symbiosis and nodule development under salt stress. *Plant, Cell and Environ.*, 26: 1003.

Gouffi, K., Pica, N., Pichereau, V. and C. Blanco. 1999. Disaccharides as a new class of non–accumulating osmoprotectants for *Sinorhizobium meliloti*. *Appl. Environ. Microbiol.*, 65: 1491–1500.

http://www.indiaagronet.com.

Ilham, Bouhmouch, Fatiha Brhada, Abdelkarim Filli-Maltouf and Jamal Aurag. 2001. Selection of osmotoerant and effective strains of Rhizobiaceae for inoculation of common bean (*Phaseolus vulgaris*) in Moroccan saline soils. *Agronomie*, 21: 591–599.

Maatallah, Jamila, El Bekkay Berraho, Juan Sanjuan and Carmen Lluch. 2002. Phenotypic characterization of rhizobia isolated from chickpea (*Cicer arietinum*) growing in Moroccan soils. *Agronomie*, 22: 321–329.

Kempf, B. and E. Bremer. 1998. Uptake and synthesis of compatible soutes as microbial stress response to high osmolarity environments. *Arch. Microbiol.*, 170: 319–330.

Somani, L.L. and K.L. Totawat. (Eds.) 1993. *Management of Salt Affected Soils and Waters*. Agrotech Publishing Academy, Udaipur, p. 27–36.

Chapter 41

Integrated Weed Management in *Kharif* Groundnut (*Arachis hypogaea* L.)

A.P. Chaudhari, T.K. Tiwari, C.B. Gaikwad and A.V. Solanke

Department of Agronomy, Mahatma Phule Krishi Vidyapeeth, Rahuri, M.S., India

ABSTRACT

A field experiment was conducted during the rainy (*kharif*) season of 2000–2001 on integrated weed management in groundnut (*Arachis hypogaea*, L.) on clay loam soils of Rahuri, Maharashtra. Weed density and weed dry matter was significantly less in weed free check. Pre-emergence application of oxyfluorfen @ 0.125 kg a.i.ha^{-1} with hand weeding 30 DAS or glyphosate (POE) @ 1.5 kg a.i.ha^{-1}, soil solarization for 30 days and oxyfluorfen (PE) @ 0.25 kg a.i.ha^{-1} effectively controlled dry matter accumulation in weed.

Maximum values of shelling per cent, 100 kernel weight and yield were recorded in weed free check followed by two hand weedings, oxyfluorfen (PE) @ 0.125 kg a.i.ha^{-1} + glyphosate (POE) @ 1.5 kg a.i.ha^{-1} and oxyfluorfen (PE) @ 0.125 kg a.i.ha-l + 1 HW 30 DAS, maximum net returns and B:C ratio were observed under weed free treatment (Rs. 14.766/- ha^{-1} and 2.8 respectively followed by two hand weedings, oxyflorfen (PE) @ 0.125 kg a.i ha^{-1} + HW, 30 DAS and oxyfluorfen (PE) @ 0.125 kg a.i.ha^{-1} + Glyphosate (POE) @ 1.5 kg ha^{-1}.

Keywords: *Groundnut, Weed biomass, Herbicide, Weed management, Yield, Economics.*

Introduction

Groundnut is extensively grown in India during the *kharif* season. Initial slow growth combined with prostate nature of its growth and hot humid climate prevailing during the *kharif* season permit early and severe crop weed competition resulting in loss of yield to the tune of 75 per cent (Gnanamurthy and Balasubramania, 1998). Chemical control of weeds form an excellent alternative to manual weeding. However, pre-emergence application of herbicides may allow the emergence of weeds after some time. Under such situation integration of pre-emergence herbicidal treatments with hand weeding or post-emergence herbicides may help in reducing the losses caused by weeds. Weed control can also be achieved by mulching moistened soil with transparent polyethylene sheet during hot season. This technique of soil solarization in India was reviewed by Yaduraju (1993). The present study was, therefore, initiated to find out an effective and economical weed control method in groundnut.

Materials and Methods

A field experiment was conducted in clay loam soil during *kharif* 2000 at Post Graduate Institute Research Farm, Rahuri, Maharashtra. The soil was slightly alkaline in reaction (pH 7.8), containing 113.82, 15.18 and 421.70 kg ha^{-1} available nitrogen, phosphorus and potassium, respectively. The experiment comprised 10 treatments, the details of which are furnished in Table 41.1. The experiment was laid out in randomized block design and replicated three times. The groundnut variety SB-XI was sown as 30 × 10 cm apart during first week of July. A basal dose of 25 kg N and 50 kg P_2O_5 ha^{-1} was applied through urea and single super phosphate respectively. All the recommended package of practices were followed to raise the crops.

For soil solarization, in summer months, the plots were covered with transparent polyethylene sheet (0.014 mm) for 30 days following the procedure of Chauhan *et al.* (1988). All such plots were uniformly irrigated to field capacity. One day prior to imposing the treatments. Fifteen days prior to sowing of groundnut, the finally prepared plot was uniformly irrigated to field capacity. A flush of young weed seedlings appeared was removed by harrowing in stale bed technique.

Weed samples were collected 1 m^2 quadrate at harvest for weed dry matter. The weed index (WI) and weed control efficiency (WCE) were computed as per Somani (1992).

Results and Discussion

Weed Density and Weed Dry Matter

The predominant weed species observed were *Cyperus rotundus, panicum ischmi, Cyanotis auxillaris, Cynodon dactylon, Commelina benghalensis* among monocots and *Digera arvensis, Parthenium hysterophorus, Acalypha indica* L., *Phyllanthus niruri, Lactuca runcinata, Euphorbia geniculata, Portulaca oleracea, Tridex procumbens* among the dicot weeds.

All the weed control treatments effectively controlled the weeds. Significantly lower weed density and weed dry matter and significantly more weed control efficiency was observed in all weed control treatments over the unweeded control. The weed density was significantly lowest in weed free treatment and was at par with 2 hand weedings and pre-emergence application of oxyfluorfen @ 0.125 kg a.i. + glyphosate (POE) @ 1.5 kg a.i.ha^{-1}. The highest weed weight (16.45 q ha^{-1}) was recorded in unweeded control (T_{10}). Weed free check recorded significantly lowest weed weight. It was followed by pre-emergence application of oxyfluorfen @ 0.125 kg a.i.ha^{-1} + glyphosate (POE) @ 1.5 kg a.i.ha^{-1}, oxyfluorfen (PE) @ 0.125 kg a.i.ha^{-1} + 1 HW at 30 DAS, soil solarization for 30 days and early post application of oxyfluorfen @ 0.25 kg a.i.ha^{-1} which were at par with each other. The results are in conformity with the reports of Patel *et al.* (1996) and Biradar *et al.* (1997). The weed index was the maximum with unweeded control compared with other treatments (Table 41.1). Under the present investigation, effectiveness of herbicidal application in reducing weed competition upto early crop-growth stage seem to be due to their adverse effects on physiology of weed growth and development.

Yield Attributes

Weed free check, two hand weedings at 20 and 40 DAS, oxyfluorfen (PE) @ 0.125 kg a.i. ha^{-1} + glyphosate (POE) @ 1.5 kg a.i.ha^{-1} and oxyfluorfen (PE) @ 0.125 kg a.i.ha^{-1} + 1 HW 30 DAS recorded

Table 41.1: Effect of Weed Control Methods on Weed Intensity, Weed Control Efficiency, Weed Index and Weed Dry Weight in Groundnut

Treatment	*Weed Intensity* (m^{-2})	*Weed Control Efficiency (%)*	*Weed Index (%)*	*Weed Dry Weight* ($q\ ha^{-1}$)
T_1: Oxyfluorfen 23.5% EC (PE) @ 0.25 kg a.i.ha^{-1}	7.00	89.12	2039	3.17
T_2: Oxyfluorfen 23.5% EC (POE) @ 0.125 kg a.i.ha^{-1}	21.33	66.84	33.66	4.44
T_3: Oxyfluorfen 23.5% EC (PE) @ 0.125 kg a.i.ha^{-1} + 1 HW 30 DAS	5.33	91.71	7.66	2.79
T_4: Glyphosate 41% EC (POE) @ 1.5 kg a.i.ha^{-1}	17.00	73.57	28.28	4.29
T_5: Oxyfluorfen 23.5% EC (PE) @ 0.125 kg a.i.ha^{-1} + Glyphosate 41% EC (POE) @ 1.5 kg a.i. ha^{-1}	3.67	94.30	6.36	2.16
T_6: Soil solarization for 30 days	6.00	90.67	12.26	3.02
T_7: Stale bed technique (Irrigated plot 15 days before sowing than harrowing)	56.67	11.90	45.06	12.06
T_8: Two hand weedings (20 and 40 DAS)	2.33	96.38	4.67	3.94
T_9: Weed free	1.33	97.93	–	0.77
T_{10}: Unweeded control	64.33	–	47.05	17.13
CD (P = 0.05)	3.77	–	–	1.05

PE: Pre-emergence; POE: Post-emergence; DAS: Days after sowing; HW: Hand weeding.

Table 41.2: Effect of Weed Control Methods on Yield Attributes, Net Return and Benefit : Cost Ratio of Grounduut

Treatment	Shelling (%)	100 kernel Weight (g)	Yield (q ha^{-1})			Additional Income Over Control (Rs ha^{-1})	Benefit : Cost Ratio
			Pod	Haulm	Kernel		
T_1: Oxyfluorfen 23.5% EC (PE) @ 0.25 kg a.i.ha^{-1}	70.79	49.04	10.39	13.41	7.26	3884	2.34
T_2: Oxyfluorfen 23.5% EC (POE) @ 0.125 kg a.i.ha^{-1}	69.94	48.02	8.63	11.11	5.98	1766	2.22
T_3: Oxyfluorfen 23.5% EC (PE) @ 0.125 kg a.i.ha^{-1} + 1 HW 30 DAS	70.92	50.32	12.05	15.81	8.42	7016	2.78
T_4: Glyphosate 41% EC (POE) @ 1.5 kg a.i.ha^{-1}	70.37	48.71	9.37	12.76	6.54	2541	2.27
T_5: Oxyfluorfen 23.5% EC (PE) @ 0.125 kg a.i.ha^{-1} + Glyphosate 41% EC (POE) @ 1.5 kg a.i. ha^{-1}	70.97	50.87	12.22	16.24	8.50	6538	2.55
T_6: Soil solarization for 30 days	70.78	49.07	11.45	14.35	7.98	4733	2.30
T_7: Stale bed technique (Irrigated plot 15 days before sowing than harrowing)	69.52	47.23	7.18	10.08	5.26	31	2.07
T_8: Two hand weedings (20 and 40 DAS)	71.01	50.55	12.45	16.18	8.68	7926	2.07
T_9: Weed free	71.35	51.11	13.05	16.99	9.24	8183	2.80
T_{10}: Unweeded control	69.48	47.15	6.92	9.83	4.80	–	2.14
CD (P = 0.05)	0.57	1.66	1.25	1.73	0.89	–	–

PE: Pre-emergence; POE: Post-emergence; DAS: Days after sowing; HW: Hand weeding.

higher shelling (per cent) and 100 kernel weight compared with unweeded (Table 41.2). Soil solarization for 30 days and oxyfluorfen (PE) @ 0.25 kg a.i.ha^{-1} were at par with these treatments in respect of shelling (per cent). Bhagat (1997) also reported the beneficial effect of herbicides in control of weeds in groundnut crop.

Crop Yield and Economics

Significantly higher dry pod, haulm and kernel yields were obtained with weed free treatment which was at par with that observed with two hand weedings (20 and 40 DAS), pre-emergence application of oxyfluorfen @ 0.125 kg a.i.ha^{-1} + glyphosate (POE) @ 1.5 kg a.i.ha^{-1} and oxyfluorfen @ 0.125 kg a.i.ha^{-1} + 1 HW 30 DAS. Bhagat (1997) observed that highest yield in 2 hand weedings (13.8 q ha^{-1}) followed by oxyfluorfen (PE) @ 0.1 kg a.i.ha^{-1} + HW (12.5 q ha^{-1}). This might be due to the combined action of pre-emergence application of herbicides which supressed the initial weed growth and 1 HW 30 DAS or post emergence herbicidal application resulted in removal/killing of weeds which emerged later. Further, the high yield with these treatments might also be utilization of available nutrients by the crop otherwise due to utilised by the weeds. This was supported by the lower dry weight of weeds and higher yield attributing character under these treatments resulting in higher pod yields.

Weed free check, 2 hand weeding (20 and 40 DAS) and pre-emergence application of oxyfluorfen @ 0.125 a.i.ha^{-1} with 1 HW 30 DAS or post emergence application of glyphosate @ 1.5 kg a.i.ha^{-1} gave higher net returns and benefit : cost ratio.

Thus, weed control with either two hand weedings (20 and 40 DAS) or pre-emergence application of oxyfluorfen @ 0.125 kg a.i.ha^{-1} + 1 HW 30 DAS or oxyfluorfen (PE) @ 0.125 kg a.i.ha^{-1} + glyphosate (POE) @ 1.5 kg a.i.ha^{-1} is economically viable in groundnut fields.

References

Bhagat, R.S. 1997. Integrated weed management in groundnut (*Arachis hypogaea*, L.). *M.Sc. (Agri.) Thesis*, Dr. PDKV, Akola, India.

Biradar, I.B., Hosmani, M.M. and Chitapur, B.M. 1997. Soil solarization as a new tool for weed control in groundnut. *Indian J. Weed Sci.*, 20(1): 82–84.

Chauhan, Y.S., Nene, Y.L., Johansen, C., Haware, M.P., Saxena, N.P., Singh, S., Sharma, S.B., Sahrawat, R.L., Burford, J.R., Rupela, O.P., Kumar Rao, J.V.D.K. and Sithanantham, S. 1988. Effects of soil solarization of pigeonpea and chickpea. *Res. Bull II*, ICRISAT, Patancheru, pp. 16.

Gnanamurthy, P. and Balasubramanian, P. 1998. Weed management practices and their influence on weed growth and yield of groundnut. *Indian J. Agron.*, 43(1): 122–125.

Patel, S.R., Nageshwar Lal and Thakur, D.S. 1996. Weed control efficiency of different herbicides in rainfed groundnut (*Arachis hypogaea*, L.). *Indian J. Agric. Sci.*, 67(9): 399–402.

Somani, L.L. 1992. *Dictionary of Weed Science*. Agrotech Publishing Academy, Udaipur, Rajasthan, pp 175.

Yaduraju, N.T. 1993. Role of soil solarization in weed management. In: *Proc. Int. Symp. Indian Soc. Weed Sci.*, Hissar (India), Nov. 18–20, 1: 343–349.

Chapter 42

Response of Rice (*Oryza sativa* L.) Crop to Integrated Nutrient Management

K. Hema, G. Swarajya Lakshmi, B.V. Seshaiah and B. Venkateswarlu

Department of Soil Science and Agricultural Chemistry,
Agricultural College, Bapatla – 522 101

ABSTRACT

A pot culture experiment was conducted with rice (*Oryza sativa* L.) Var Vijetha, (MTU 1001) to study the effect of organic manures and biofertilizer alone and in combination with inorganic nitrogen on growth, yield and uptake of macronutrients during *rabi*, 2002, at net house, Agricultural College, Bapatla on clayey soil. The treatments consisted of control (without manure or fertilizer) 50, 75 and 100 per cent RDFN through inorganic fertilizer, application of entire N through organics (PM, GM and FYM) and biofertilizer (*Azolla*) alone and also 50 and 75 per cent substitution of inorganic N through organics (PM, GM and FYM) and biofertilizer (*Azolla*). In total there were sixteen treatments laid out in a completely randomized design replicated thrice. The treatment T_{15} (75 per cent RDFN + 2.0 t PM ha^{-1}) recoded the highest plant height, drymatter production, grain and straw yields and uptake of macro nutrients (N, P and K) at all the stages of Crop growth.

Introduction

However, the present fertilizer Consumption ratio ($N : P_2O_5 : K_2O$) is 8.9 : 2.8 : 1.0 as against the optimal ratio of 4 : 2 : 1. This leads to much greater drain on native soil fertility and soil will not support high production levels in future. It has been demonstrated that inorganic N fertilizers, generally are not efficiently utilized by rice and are prone to heavy losses as N gases (Dedatta and Buresh, 1989). The continuous use of high analysis fertilizers increase the crop yield in the initial years, but adversely affects the sustainability at a later stage (Virmani, 1994). Indiscriminate use of chemical fertilizers also results in the deficiency of nutrients other than the applied and decline in organic carbon (Singh *et al.*, 1999).

The current decline in organic matter levels in many potential areas of productivity is a matter of great concern for Indian agriculture. To compensate this, there is a need to develop integrated nutrient management (INM) system. Hence, proper blending of chemical fertilizers with locally available organics not only improves soil health but also helps to maximize the sustainable production.

Materials and Methods

A pot culture experiment was conducted during rabi 2002 at nethouse of Agricultural College, Bapatla. It is located at 80°3′ East longitude, 15°84′ North latitude at an elevation of 5.4 m above mean sea level. The pots are uniformly filled with twelve kgs of soil. The soil is clay in nature having pH of 8.2, E.C 0.42 dSm^{-1}, Organic carbon 0.432 per cent, CEC 39.79 C mol (p^{+}) kg^{-1}, available N, P, K are 282.50, 39.20 and 504.0 kg ha^{-1} respectively and available Fe, Mn, Cu and Zn are 2.079, 5.22, 2.995 and 0.858 ppm respectively. Farm yard manure contained 1.22 per cent N, 0.62 per cent P_2O_5, 0.90 per cent K_2O, 308.8 ppm Fe, 62.2 ppm Zn, 83.9 ppm Cu and 291 ppm Mn; Poultry manure contained 3.79 per cent N, 3.09 per cent P_2O_5, 2.12 per cent K_2O, 920 ppm Fe, 231.1 ppm Zn, 306 ppm Cu and 447 ppm Mn; Daincha contained 2.01 per cent N, 0.22 per cent P_2O_5, 2.03 per cent K_2O, 405 ppm Fe, 66.7 ppm Zn, 68 ppm Cu and 329 ppm Mn and *Azolla* contained 2.38 per cent N, 0.24 per cent P_2O_5, 1.90 per cent K_2O, 93.82 ppm Fe, 12.90 ppm Zn, 28.8 ppm Cu and 133 ppm Mn. Soil and nutrient analysis of different manures and biofertiliser was done as per the procedure suggested by Tandon (1993).

In total there were sixteen treatments laid out in completely randomized design (CRD) and replicated thrice. The experiment was conducted with rice variety MTU 1001(Vijetha). Treatmental details of the experiment were *Azolla*, farm yard manure, daincha and Poultry manure both alone and in combination with inorganic fertilizer.

T_1: Control
T_2: Recommended Dose of Fertilizer Nitrogen (RDFN)
T_3: 75 per cent RDFN
T_4: 50 per cent RDFN
T_5: 6t *Azolla* ha^{-1}
T_6: 75 per cent RDFN + 1.5t *Azolla* ha^{-1}
T_7 50 per cent RDFN + 3.0t *Azolla* ha^{-1}
T_8: 20t FYM ha^{-1}
T_9: 75 per cent RDFN + 5.0t FYM ha^{-1}
T_{10}: 50 per cent RDFN + 10.0t FYM ha^{-1}
T_{11}: 20t Daincha ha^{-1}
T_{12}: 75 per cent RDFN + 5.0t Daincha ha^{-1}
T_{13}: 50 per cent RDFN + 10.0t Daincha ha^{-1}
T_{14}: 8t Poultry manure ha^{-1}
T_{15}: 75 per cent RDFN + 2.0t Poultry manure ha^{-1}
T_{16}: 50 per cent RDFN + 4.0t Poultry manure ha^{-1}

Fertilizers were applied in the form of urea, single super phosphate and muriate of potash to supply N, P and K respectively. The entire dose of P, K and half of the N were applied basally and mixed thoroughly in soil to all treatments except control. The balance of the nitrogen was applied at active tillering and panicle initiation stages in two equal splits. Thirty days old seedlings were transplanted. Farmyard manure, poultry manure and green manure (Daincha) were applied fifteen days before transplanting for decomposition, where as *azolla* was applied fifteen days after transplantation. Three hills were maintained in each plot and all the recommended measures were followed meticulously.

Plant samples were collected at active tillering, panicle initiation stages and at harvest for computing dry matter production samples were dried at 60°C in an hot air oven to a constant weight and powdered with the help of plant sample grinder and were analysed for N, P and K. The plant N content in plant analysis was estimated by microkjeldhal method as outlined by Piper (1966). Wet digestion method as described by Piper (1966) was followed for determining phosphorus by digesting the plant samples and in the digest phosphorus was determined by vandomolybdate yellow colour method and potassium by Flame photometry.

Results and Discussion

Drymatter Production

The results showed that the drymatter production increased with crop growth in all the treatments. The rate of increase was more from panicle initiation stage to harvest than from active tillering to panicle initiation stage. The manures and biofertilizer in conjunction with inorganic fertilizer recorded significantly higher drymatter production over control at all the stages of crop growth. Combined application of organic and inorganic fertilizer N have significantly enhanced the drymatter production. This may be ascribed to the prevalence of better physical conditions and there by increased availability of nutrients which enhanced the uptake of nutrients, resulting in improved drymatter production. Similar results were reported by Pagiliai *et al.* (1987) and Madhavi *et al.* (1996).

Among all the treatments the treatment T_{15} (2.0 t PM + 75 per cent RDFN) produced highest drymatter (16.67 g pot^{-1} at active tillering stage, 27.50 g pot^{-1} at panicle initiation stage, 93.37 g pot^{-1} at harvest respectively) followed by T_{16} (4.0 t PM + 50 per cent RDFN) while T_1 (control) recorded the lowest value of drymatter production at all the stages of crop growth. The treatment T_{15} recorded highest DM this might be due to poultry manure, a rich source of nutrients and during the decomposition of the manure higher availability of the nutrients might have resulted.

Among the individual sources, the treatment T_2 (100 per cent RDFN) recorded the highest drymatter production (11.57 g pot^{-1} at active tillering, 23.50 g pot^{-1} P_1 and 79.71 g pot^{-1} at harvest stage) at all the stage of crop growth. It might be due to the fact that during early stages of crop growth, the inorganic nitrogen might have resulted in more drymatter accumulation, due to the ready availability of nutrients as compared to the application of manures and biofertilizer alone. The increase in drymatter at recommended dose of N, P and K was earlier reported by Misra *et al.* (1995).

Among all the combinations, the treatment T_{12} (5.0 t GM + 75 per cent RDFN), among the FYM combinations, the treatment T_9 (5.0 t FYM + 75 per cent RDFN) and among the *Azolla* combinations, the treatment T_6 (1.5 t Azolla + 75 per cent RDFN) recorded the highest DM at all the stages of growth. It might be due to the supply of more inorganic N as compared to the test of the combinations. Similar results were observed by Mahapatra *et al.* (1991).

Application of 50 and 75 per cent RDFN alone recorded lower values of drymatter production compared to their combinations with organics (PM, GM and FYM) and biofertilizer (*Azolla*) at all the stages of crop growth. This might be due to the synergistic effects of combined use of inorganic fertilizer with organic manures. These results were in conformity with findings of Bhandari *et al.* (1992) and Muneshwar Singh *et al.* (2001).

Grain and Straw Yields (Table 42.1)

The highest grain and straw yields (33.10 g pot^{-1} and 66.6 g pot^{-1} respectively)were recorded by the treatment T_{15} (2.0 t PM + 75 per cent RDFN), while T_1 (control) recorded the lowest grain and straw yields (15.51 pot^{-1} and 24.77 g pot^{-1} respectively) followed by T_5 (6t *Azolla*) which recorded 23.22 g pot^{-1} and 43.49 g pot^{-1} respectively. This might be due to the immediate availability of N through inorganic nitrogen, later by the mineralization of N through poultry manure resulting in steady supply of nutrients through out the crop growth period. These were in accordance with Madhavi *et al.* (1996) and Singh *et al.* (1997). The beneficial effects of poultry manure in conjunction with inorganic nitrogenous fertilizer on field parameters have also been reported by Vasanthi and Kumara Swamy (2000).

With regard to GM, FYM and *Azolla* combinations, with increasing levels of inorganic nitrogen, the grain and straw yields were also increased. Among the green manure combinations, the treatment T_{12} (5 t GM + 75 per cent RDFN) recorded the highest grain and straw yields (31.36 g pot^{-1} and 57.10 g pot^{-1} respectively). The superiority of the green manurial legume is attributed to its fast decomposition, which probably led to better availability of the nutrients as compared to other organic materials under wet land rice cultivation (Gill and Meelu, 1982).

Among the FYM combinations, the treatment T_9 (5.0 t FYM + 25 per cent RDFN) recorded the highest grain and straw yields (29.0 g pot^{-1} and 53.18 g pot^{-1}) respectively. Similar results were reported by Singh *et al.* (1999) and Anand Swarup and Yaduvanshi (2000). Among the *Azolla* combinations, the treatment T_6 (T_5 + *Azolla* + 75 per cent RDFN) recorded the highest grain and straw yields (26.16 g pot^{-1} and 48.18 g pot^{-1}) respectively. Similar results were observed by Chandra Sekharan *et al.* (1988).

Application of 50 per cent and 75 per cent N through inorganic fertilizer and rest of the N supplied through organic sources, recorded higher grain and straw yields compared to the application of inorganic fertilizer alone. Similar results were reported by Bhandari *et al.* (1992) and Yaduvanshi (2001). A positive effect of organic and inorganic sources of nitrogen alone and incombination was also reported by Jagadeswari and Kumara Swamy (2000).

Uptake of Macro Nutrients (N, P and K) (Table 42.1)

The uptake of N, P and K were significantly influenced by the type and level of manure and fertilizer either alone or in conjunction with inorganic fertilizer N. Among all the treatments, the treatment T_{15} (2.0 t PM + 75 per cent RDFN) recorded the highest uptake of N, P and K at all the stages of the crop growth. It might be due to the steady and increased availability of nutrients from poultry manure and due to rapid mineralisation which might have resulted in increased uptake of nutrients by plant. Increase in the uptake of N, P and K from soil amended with poultry manure was reported by Madhavi *et al.* (1996) and Singh *et al.* (1997). The poultry manure combinations performed better than their corresponding combinations of GM, FYM and *Azolla*. It might be due to the steady and increased availability of nutrients by plant.

Among the individual sources, the treatment T_2 (100 per cent RDFN) recorded the highest uptake of N, P and K which might be due to the immediate availability of N which might have also enhanced the availability of P and K. The individual sources of manures (PM, FYM and GM) and biofertilizer (*Azolla*) recorded low uptake which might be due to slow release of nutrients by the manures during the early stage of crop growth.

Among green manure combinations the treatment T_{12} (5.0 t GM + 75 per cent RDFN) recorded the highest N, P and K uptake. These are in conformity with findings of with Sriramachandrasekharan

Table 42.1: Effect of Conjunctive Use of Organic and Inorganic Sources of Nitrogen on Macronutrient (N, P and K) Uptake (mg pot^{-1}) of Rice at Different Stages of Crop Growth

Treatments	*Dry Matter Production (mg pot^{-1})*					*N Uptake (mg pot^{-1})*				*P Uptake (mg pot^{-1})*				*K Uptake (mg pot^{-1})*			
	Active Tillering Stage	*Panicle Initiation Stage*	*Harvest*	*Yield (g pot^{-1})*		*Active Tillering Stage*	*Panicle Initiation Stage*	*Harvest*		*Active Tillering Stage*	*Panicle Initiation Stage*	*Harvest*		*Active Tillering Stage*	*Panicle Initiation Stage*	*Harvest*	
				Grain	Straw			Grain	Straw			Grain	Straw			Grain	Straw
T_1 Control	6.52	11.23	38.27	13.51	24.77	65.37	135.88	99.61	171.73	29.30	37.30	94.53	42.68	98.88	138.50	54.93	262.55
T_2 Recommended Dose of Fertilizer Nitrogen (RDFN) 120-60-30 kgNPKha^{-1}	11.57	23.50	79.71	28.28	51.42	140.32	337.60	249.91	508.48	111.9	150.16	311.03	175.40	260.32	356.14	207.38	791.86
T_3 75% RDFN	8.12	21.83	74.73	26.39	48.35	131.64	273.59	231.35	477.07	71.6	121.53	242.76	148.23	225.73	316.53	179.54	722.83
T_4 50% RDFN	7.91	20.56	70.19	24.86	45.32	123.10	249.55	208.82	424.71	70.5	110.56	208.78	100.10	162.65	280.98	154.41	626.92
T_5 6t *Azolla* ha^{-1}	7.58	19.83	66.72	23.22	43.49	129.67	256.12	210.21	439.90	60.36	109.93	212.83	93.10	156.54	257.79	136.76	608.85
T_6 75% RDFN + 1.5t *Azolla* ha^{-1}	11.74	21.49	74.35	26.16	48.18	166.37	287.96	252.63	449.68	131.83	138.10	320.00	155.10	268.06	340.25	200.56	783.72
T_7 50% RDFN + 3.0t *Azolla* ha^{-1}	9.65	20.16	68.23	24.00	44.23	156.97	264.76	226.08	446.58	95.13	116.96	280.80	141.33	204.23	299.04	162.00	695.88
T_8 20t FYM ha^{-1}	8.48	21.16	72.14	25.37	46.76	127.60	284.95	215.01	437.98	80.53	119.00	213.30	116.06	176.66	284.24	149.53	682.69
T_9 75% RDFN + 33.0t *Azolla* ha^{-1}	13.12	24.23	82.37	29.04	53.18	184.21	334.37	287.29	513.18	155.66	170.60	379.56	193.13	301.76	434.64	232.32	916.46
T_{10} 50% RDFN + 10.0t FYM ha^{-1}	11.63	22.33	76.69	26.86	49.16	165.74	303.68	262.86	467.02	133.30	153.00	330.33	186.33	257.79	353.55	179.06	806.22
T_{11} 20t Daincha ha^{-1}	9.91	21.81	74.72	26.28	48.43	130.10	308.24	237.85	460.08	94.43	124.50	242.60	135.40	203.15	301.70	157.68	726.44
T_{12} 75% RDFN + 5.0t Daincha ha^{-1}	14.51	25.90	88.46	31.36	57.10	199.05	378.14	357.08	607.16	176.03	219.06	439.00	255.40	350.65	492.10	282.24	1056.35
T_{13} 50% RDFN + 10.0t Daincha ha^{-1}	13.72	23.26	76.55	28.16	51.71	182.67	350.14	332.05	585.50	134.06	187.20	373.53	214.73	320.13	430.31	234.66	818.73
T_{14} 8t Poultry manure (PM) ha^{-1}	10.53	22.83	78.16	27.66	56.50	139.55	303.92	249.69	486.80	101.03	142.70	280.23	153.16	244.63	334.83	193.62	774.33
T_{15} 75% RDFN + 5.0t Poultry manure ha^{-1}	16.67	27.50	93.37	33.10	60.60	238.41	412.91	460.30	711.03	231.10	297.16	517.56	431.60	416.74	591.25	308.90	1183.72
T_{16} 50% RDFN + 10.0t Poultry manure ha^{-1}	14.86	26.08	89.14	31.74	57.40	208.03	397.72	439.17	688.80	191.63	260.23	457.03	386.16	354.16	525.94	275.08	1073.38
Sem±	0.61	0.57	2.46	1.27	1.65	5.79	19.14	12.02	26.24	7.27	8.72	9.35	8.75	11.89	21.14	12.05	32.08
CD (P = 0.05)	1.70	1.59	6.83	3.53	4.50	16.05	53.05	33.32	72.74	14.26	24.19	25.93	24.27	32.97	58.62	33.41	88.93

et al., (1996). The combinations of FYM and *Azolla* recorded lower N, P and K uptake than GM combinations at all the stages of the crop growth. It might be due to higher availability of N in green manure (*Sesbania acculeata*) treated pots because of its high N content, coupled with succulent nature which might have undergone rapid decomposition in submerged soil. The lower N, P and K uptake by *Azolla* treated pots might be due to higher lignin content which known to retard mineralization as reported by Sriramachandrasekhran *et al.* (1996). By increasing the dose of inorganic N fertilizer along with organic manures not only increased the uptake of N but also P and K which are in concurrence with the results of Bhandari *et al.* (1992) and Radha Madhav *et al.* (1996).

However the treatment T_{15} (2.0 t PM + 75 per cent RDFN) appeared superior to T_{16} (4.0 t PM + 50 per cent RDFN) and T_{14} (8 t PM) which might be due to the addition of high mineral nitrogen along with low quantity of PM which reduce the C : N ratio and thus increased the rate of decomposition, resulting in faster availability of nutrients throughout the period of crop growth.

References

Anand Swarup and Yaduvanshi, N.P.S. 2000. Integrated nutrient management on soil properties and yield of rice (*Oryza sativa*) in alkali soil. *Journal of the Indian Society of Soil Science*, 48(2): 279–282.

Bhandari, A.L., Anil Sood, Sharma, K.N. and Rana, D.S. 1992. Integrated nutrient management in a Rice: Wheat system. *Journal of the Indian Society of Soil Science*, 40: 742–747.

Chandra Sekharan, V., Jhangamuthu, G.S. and Bala Subramaniyan, P. 1988. Effect of bioferilizer (*Azolla*) and urea N on growth and yield of low land rice (Co.43). *Madras Agricultural Journal*, 75 (9 and 10): 318–322.

Dedatta, S.K. and Buresh, R.J. 1989. *Advances in Soil Science*, 10: 143–169.

Gill, H.S. and Meelu, O.P. 1982. Studies on substitution of inorganic fertilizers with organic manures and their effect on soil fertility in rice: Wheat rotation. *Fertilizer Research*, 3: 303–314.

Jagadeswari, V. and Kumara Swamy, K. 2000. Long-term effects of manure fertilizer schedules on the yield and nutrient uptake by rice crop in a permanent manorial experiment. *Journal of the Indian Society of Soil Science*, 48(4): 833–836.

Madhavi, B.L., Suryanarayan Reddy, M. and Gopal Reddy, B. 1996. Effect of Poultry manure and fertilizers on yield and micronutrients uptake by maize. *Journal of Research ANGRAU*, 24(2): 154–156.

Mahapatra, B.S., Sharma, K.C. and Sharma, G.L. 1991. Effect of integrated nitrogen management on soil N and dry matter production of low land rice. *Coryza*, 28: 87–92.

Misra, A., Dash, P. and Paikaey, R.K. 1995. Yield and nutrient uptake by winter sunflower (*Helianthus annuus* L.) as influenced by nitrogen and phosphorus. *Indian Journal of Agronomy*, 40(1): 137–138.

Pagliai, M., La Marka, M. and Lucamante, G. 1987. Changes in soil porosity in remolded soils treated with poultry manure. *Soil Science*, 144(2): 128–140.

Piper, C.S. 1966. *Soil and Plant Analysis*. Hans Publishers, Mumbai, p. 137–153.

Radha Madhav, M., Ravi Kumar, A. and Venkateswarlu, B. 1996. Effect of different sources of nitrogen on growth, yield and nutrient uptake of rice. *The Andhra Agricultural Journal*, 43: 119–122.

Singh, Munheshwar, Singh, V.P. and Sammi Reddy, K. 2001. Effect of Integrated use of fertilizer nitrogen and farmyard manure or green manure on transformation of N, K, and S and productivity of Rice. wheat system on a vertisol. *Journal of the Indian Society of Soil Science*, 49(3): 430–435.

Singh, N.P., Sachan, R.S., Panday, P.C. and Bishti. 1999. Effect of a decade long term fertilizer and manure application on soil fertility and productivity of rice-wheat system in a mollisol. *Journal of the Indian Society of Soil Science*, 47: 72.

Singh, G.R., Parihar, S.S., Chaure, N.K., Choudhary, K.K. and Sharma, R.B. 1997. Integrated nutrient management in summer sesame. *Indian Journal of Agronomy*, 42(4): 699–701.

Srirama Chandrasekharan, M.V., Ramanathan, G. and Ravichandran. 1996. Effect of organic manures on the availability of nutrients in rice rhizosphere soil. *Coryza*, 33: 126–131.

Tandon, H.L.S. (Ed). 1993. *Methods of Analysis of Soils, Plants, Water and Fertilizers*. Fertilizers Development and Consultation Organization, New Delhi, India, pp. 150.

Vasanthi, D. and Kumarswamy, K. 2000. Effects of manure fertilizer schedules on the yield and uptake of nutrients by cereal folder crop and on soil fertility. *Journal of the Indian Society of Soil Science*, 48(3): 510–515.

Virmani, S.M. 1994. The twenty-first Dr. R.V. Tamhane memorial lecture UNCED Agenda 21: The new challenges for soils research. *Journal of the Indian Society of Soil Science*, 45(3): 445–449.

Yaduvanshi, N.P.S. 2001. Effect of five years of rice-wheat cropping and NPK fertilizer use with and without organic and green manures on soil properties and crop yields in a reclaimed sodic soil. *Journal of the Indian Society of Soil Science*, 49(4): 714–719.

Chapter 43

Effect of Lactational Order on the Physico-chemical Properties of Milk in Deccani Sheep

A.S. Hulge, A.P. Chavan, S.P. Kalhapure, S.V. Kale and R.B. Walujkar

Department of Animal Science and Dairy Science, Mahatma Phule Krishi Vidyapeeth, Rahuri – 413 722, Maharashtra

ABSTRACT

The present study was carried out to investigate the effect of lactational order on the physico-chemical properties of milk in Deccani sheep revealed that the lactational order had significant ($P < 0.05$) effect on daily milk yield, protein, lactose and pH of sheep milk; whereas no significant effect was noticed on total solids, fat solids not fat, ash, acidity and specific gravity of sheep milk.

Keyword: *Deccani sheep, Lactational order, Physico-chemical properties.*

Introduction

Sheep milk is quite in significant and sporadically used, it forms an important base in the economy of many rural people. Sheep milk contains 56 gm/kg of protein compared with 33 gm/kg in cow and goat milk. Milk is not a uniform article of commerce but varies largely in composition. The variation is caused due to the number of factor affecting the physiology of lactating animal. In the present investigation an attempt was made to study the variation in the physico-chemical properties of milk of *Deccani* sheep as affected by lactational order.

Materials and Methods

The research work was conducted in the laboratory of Department of Animal Science and Dairy Science, Mahatma Phule Krishi Vidyapeeth, Rahuri, District Ahmednagar (Maharashtra). For the study 15 sheep of five different lactational order were selected. The trial was conducted in FRBD with three replication. The general and uniform management system were adopted, using stall fed condition in loose housing system. The sheep once selected under each group were maintained throughout the experimental period for getting milk samples. The milk samples of 150 ml were analysed weekly for each selected animal. The morning milk samples were analysed on the same day whereas the evening amples were analysed on the next day by keeping in refrigerator at 4°C. In all total 420 milk sample vere analysed for all the five lactational order for each parameter. In each lactational order, 84 milk ample were analysed for its chemical composition *viz.*, total solids, fat, solids not fat, protein, lactose, sh, acidity, pH and specific gravity. The analysis was made as per the standard methods.

esults and Discussion

The data presented in Table 43.1, revealed that the lactational order had significant ($P < 0.05$) effect on daily milk yield, protein, lactose and pH of sheep milk; whereas no significant effect was noticed on the total solids, fat solids not fat, ash, acidity and specific gravity of sheep milk. The daily milk yield was increased up to the third lactation and then decreased in the subsequent lactation. These results were in conformity with Peeters *et al.* (1992). Reduction in total solids percentage from first lactational order to subsequent lactational order. These results were in agreement with the findings of Juarez *et al.*, 1982). There was no difference in the fat content of milk form the sheep of different lactations though decreased slightly in subsequent lactations. These results were in agreement with the results of Kozhokaru and Mogoryanu (1972). The solids not fat content in milk was near about constant for all the five lactational order. The protein content in sheep milk was significantly increased up to fourth lactation and then decreased in the fifth lactation. These results were in agreement with that of Kozhokaru and Mogoryanu (1972). The lactose content of sheep milk was decreased with advanced lactational number and found maximum in first lactation. There was no significant difference in the ash content of milk. These results were in agreement with Casoli *et al.* (1989). The acidity per cent was constant and had no decreasing or increasing trends, significantly affected due to lactational orders in all subsequent five lactation. The pH of sheep milk significantly increasing up to second lactation and then decreased in the further lactation. There was no significant difference in the specific gravity of milk of different lactation. These results are in agreement with Yaghin (1964). The effect of interaction between the stage of lactation and lactational order for all the parameter in sheep milk found non-significant except daily milk yield.

Table 43.1: Effect of Lactational Order on the Physico-chemical Properties of Milk of *Deccani* Sheep

Parameters	Mean					SE±	S × O Interaction	CD at 5% Levels	Sx Interaction
	O1	O2	O3	O4	O5				
Daily milk yield (ml)	375.86	393.11	450.48	449.52	441.21	8.19	16.38	23.45	46.91
Total solids (%)	18.38	18.59	18.55	118.37	18.46	0.095	0.190	NS	NS
Fat (%)	7.12	7.07	7.01	6.83	7.00	0.085	0.171	NS	NS
Solids not fat (%)	11.56	11.54	11.54	11.57	11.43	0.051	0.102	NS	NS
Protein (%)	5.80	5.90	5.88	5.91	5.78	0.029	0.059	0.084	NS
Lactose (%)	4.83	4.70	4.69	4.69	4.71	0.034	0.068	0.097	NS
Ash (%)	0.96	0.96	0.97	0.96	0.97	0.004	0.008	NS	NS
Acidity (% LA)	0.196	0.189	0.195	0.193	0.197	0.003	0.007	NS	NS
pH	6.51	6.58	6.57	6.56	6.55	0.011	0.022	0.031	NS
Specific gravity	1.035	1.035	1.035	1.035	1.035	0.0003416	0.0006831	NS	NS

References

Casoli, C., Duranti, E., Morbidini, L., Panella, F. and Vizioli, V. 1989. Quantitative and compositional variation of Massese sheep milk by parity and stage of lactation. *Small Ruminant Research*, 2(1): 47–62 [*Anim. Breed. Abstr.*, 58(6): 3563].

Juarez, M. Jimenez, S. Goicoechea, A. and Ramos, M. 1982. Studies on quality of ewe's milk. In: *21st International Dairy Congress*. Vol. 1. Book 2, Moscow, USSR, Mir Pubiishers, p. 625–626. [*Dairy Sci. Abstr.*, 45(6): 4144.

Kozhokaru, L.V. and Mogoryanu, I.I. (1972). Changes with age in milk production and composition of locally selected Tsigai ewes. Trudy, Kishinevskii sel skokho zyaistvennyi Institute No. 94, 66–70, 74 [*Dairy Sci. Abstr.*, 37(9): 5251].

Pecters, R., Buys, N., Robijns, L., Vanmontfort, D., Isterdael, J. Van. 1992. Milk yield and milk composition of Flemish milk sheep, suffolk and Texel ewes and their crossbreds. *Small Ruminant Research*, 7(4): 279–288 [*Anim. Breed. Abstr.*, 60(1): 6426].

Yaghin, H. 1964. Fat content and specific gravity of milk of Awassi ewes in the flock of the Department of Animal Breeding of the Agean University. *Zir. Fak. Derg.*, 1(2): 91–108. [*Dairy Sci. Abstr.*, 32(1): 384].

Chapter 44

Efficacy of Different Insecticides Against Brinjal Shoot and Fruit Borer (*Leucinodes orbonalis* Gn.)

B.M. Mhaske, A.P. Chavan, D.B. Kadam and V.C. Girhe

Department of Entomology, Mahatma Phule Krishi Vidyapeeth, Rahuri, Dist. Ahmednagar – 413 722, Maharashtra, India

ABSTRACT

The experiment were conducted during summer 2002 on brinjal crop with an object to study the efficacy of different insecticide against brinjal shoot and fruit borer. In the promising study, the treatment with azadirachtin (Econeem) 15 and 12.5 g a.i. per ha all the doses of profenofos (1000, 750 and 500 g a.i. per ha), alpha endosulfan and endosulfan 280 g a.i. per ha, deltamethrin (11 g a.i.) and cypermethrin (60 g a.i. per ha) were effective and at par with each other in controlling shoot damage and monocrotophos was moderately effective against this pest, while profenophos 1000 g a.i. per ha recorded minimum fruit damage, however, all other treatment were at par with it. The similar results were reflected on yield as that of fruit borer control.

Keyword: Fruit and shoot borer, Efficacy, Profenophos, Brinjal.

Introduction

The brinjal shoot and fruit borer *Leucinodes orbonalis* Gn is the most noxious and serious pest of brinjal and has been reported through the country. This pest is of endemic in nature and inflicts direct losses to the extent of 26.3 to 62.5 per cent (Gangwar and Sachan, 1981) and some times goes as high as 95.8 per cent (Akhtar and Khawaja, 1973). The shoot and fruit borer causing serious damage to shoots of seedlings at vegetative stage and to fruits at the time of bearing. The infested fruits are often considered unfit for human consumption. During the experiments, an attempt was made to find out the most promising insecticides for effective control of a brinjal shoot and fruit borer.

Materials and Methods

Experimental Layout

The field experiments were conducted during summer 2002 on the Instructional Farm of Post Graduate Institute, Department of Entomology, Mahatma Phule Krishi Vidyapeeth, Rahuri, Dist. Ahmednagar, Maharashtra, India. The experiments were laid out in a randomised block design with twelve treatments and three replication on brinjal variety. Mahyco 10 (Kalpataru) in gross plot size 4.5 × 3.0 ft. and net plot size was 3.60 × 1.80 ft. The treatment was three doses of profenophos 50 EC @ 500, 750 and 1000 g a.i. per ha, three doses of azadirachtin 1 per cent EC @ 10, 12.5 and 15 g a.i. per ha, alpha endosulfan and endosulfan 35 EC @ 280 g a.i., monocrotophos 36 SL 145 g a.i. per ha, deltamethrin 2.8 EC @ 11 g a.i. per ha and cypermethrin 10 EC @ 60 g a.i. per ha were tested.

Method of Sprays

The spraying was done with due care and thorough coverage of all the plant parts which are recognized the main feeding sites of the pests. Spraying were done in morning hours and care was taken to wash the pump with water while switching on from one insecticide to another. The inescticidal spray was given soon after the infestation of pests. In all three sprays were given to the experiments.

Observation

The observation were recorded on total number of shoots and infested shoots from five randomly selected tagged plant from each of the treatment plots at 15 days after each spraying. The total number of fruits and infested fruits due to the fruit borer were recorded from each plots by critically examining the fruits at each picking. Total six picking were carried out starting from 14th week after transplanting. The per cent infestation of shoots and fruits were calculated. The percentage of shoots were calculated from the infested and healthy "number of shoots" while the percentage of damaged fruits were calculated by taking in to consideration the infested and healthy fruits on the number and weight basis, the percentage were then transformed to there arc-sin values for statistical analysis. The yield of fruits was recorded at each harvest.

Results and Discussion

Effect of New Insecticides on the Incidence of Shoot and Fruit Borer of Brinjal

It is observed from the Table 44.1, that all the insecticidal treatments significantly reduced percentage of shoot and fruit damage in comparison with untreated control. The data shows that the treatment with azadirachtin 15 g a.i. per ha was significantly superior in reducing per cent shoot infestation, azadirachtin 10 and monocrotophos 145 g a.i. per ha. Fruit damage on number basis, the treatment with profenofos 50 EC @ 1000 g a.i. per ha recorded lowest fruit damage of 22.81 per cent as against 41.97 per cent in untreated control. Also the fruit damage on weight basis, profenofos 50 EC @ 1000 g a.i. per ha recorded lowest fruit damage of 24 per cent as against 39.88 per cent in untreated control and rest of the treatment were equally effective and were at par with it.

Effect on Fruit Yield of Brinjal

The data on fruit yield of brinjal as influenced by various insecticides are presented in Table 44.1. All insecticidal treatments produced significantly higher yield over untreated control (142.57 q/ha) of brinjal fruits. The treatment with profenofos 1000 g a.i. per ha recorded significantly highest yield (230.37 q/ha) of brinjal fruits. The rest of the treatments were equally effective giving good yield and were at par with it, except monocrotophos 145 g a.i. per ha (186.55 q/ha).

Table 44.1: Effect of New Insecticides on the Incidence of Shoot and Fruit Borer of Brinjal

Treatment	Dose (g a.i. per/ha)	Av. % Shoot Damage	Av % Fruit Damage		Yield (q/ha)
			Number Basis	Weight Basis	
Profenofos 50 EC	500	4.43	24.59	26.76	198.58
		(12.11)	(29.73)	(31.18)	
Profenofos 50 EC	750	3.99	23.64	26.71	207.37
		(11.54)	(29.06)	(31.11)	
Profenofos 50 EC	1000	3.56	22.81	24.00	230.37
		(10.94)	(28.52)	(29.33)	
Azadirachtin 1% EC	10	7.19	26.88	28.77	201.05
		(15.56)	(31.24)	(32.46)	
Azadirachtin 1% EC	12.5	4.66	25.33	27.78	216.02
		(12.52)	(30.20)	(31.82)	
Azadirachtin 1% EC	15	3.42	24.78	24.87	221.27
		(10.63)	(29.87)	(29.93)	
Alpha endosulfan 35 EC	280	3.80	26.10	27.61	219.88
		(11.24)	(30.72)	(31.69)	
Monocrotophos 36 EC	145	6.86	26.18	27.24	186.55
		(15.23)	(30.79)	(31.44)	
Endosulfan 35 EC	280	3.85	25.65	25.84	230.21
		(11.24)	(30.40)	(30.53)	
Deltamethrin 2.8 EC	11	4.05	26.35	26.23	208.92
		(11.54)	(30.85)	(30.79)	
Cypermethrin 10 EC	60	3.86	24.04	25.37	216.48
		(11.39)	(29.33)	(30.26)	
Untreated control		10.41	241.97	239.88	142.57
		(18.81)	(40.40)	(39.17)	
S.E.±		1.36	0.96	1.19	11.01
C.D. at 5% level		3.99	2.82	3.48	32.28

*: Figures in parantheses are arc-sin transformed values.

Gangwar and Sachan (1981) reported that *L. orbonalis* controlled by three applications of deltamethrin and endosulfan. Agnihotri *et al.* (1990) found cypermethrin and deltamethrin highly effective against shoot and fruit borer and increasing the yield of brinjal fruits. Srinivas and Peter (1993) evaluated nine insecticide *viz.*, pyraclofos, acephate, cartaphydrocloride, cypermethrin, Polytrin-C, alphamethrin, deltaphos, etofenprox and profenofos against *L. orbonalis* on brinjal. Singh *et al.* (1996) reported significant reduction of shoot and fruit damage by endosulfan. Hole *et al.* (1997)

reported that, deltamethrin and cypermethrin reduced per cent infestation of shoot and fruit borer with increasing yield of brinjal. Ghongale *et al.* (2002) tested some ready mix insecticides containing cypermethrin, deltamethrin, triazophos and profenofos and these were quite effective in controlling shoots and fruit borer. The present results are in corroboration with these workers.

References

Agnihtri, N.P., S.N. Sinha, H.K. Jain and A.K. Chakraborti. 1990. Bio-efficacy of some synthetic pyrethroid insectcide against *L. orbonalis* and their residues on brinjal fruits. *Indian J. Ent.*, 52(3): 373–378.

Akhtar, M. and Z.S. Khawaja. 1973. Quantitative estimation of damage to brinjal by brinjal borers. *J. Agric. Res. Pakistan*, 11(1): 78–83.

Gangwar, S.K. and J.N. Sachan. 1981. Seasonal incidence and control of insect pest of brinjal with special reference to shoot and fruit borer *L. orbonalis* in Meghalaya. *J. Res. Assam Agric. Univ.*, 2(2): 187–192.

Ghongale, P.R., N.S. Satpute, S.N. Tikar and N.R. Naitam. 2002. Field evaluation of ready mix insecticides against shoot and fruit borer of brinjal *L. orbonalis*. *Pestology*, 26(3): 26–28.

Hole, U.B., A.R. Walunj and A.B. Tambe. 1997. Bio-efficacy of Decis against shoot and fruit borer of brinjal. *Pestology*, 21(5): 31–33.

Singh, I.P., R.K. Prasad and U. Mukharjee. 1996. Efficacy and relative toxicity of some insecticide against *L. orbonalis*. *J. Res. Birsa Agric. Univ.*, 8(1): 71–75.

Srinivas, S.N. and C. Peter. 1993. Efficacy of certain new insecticides to the brinjal shoot and fruit borer *L. orbonalis*. *Pestology*, 17(6): 36–38.

Chapter 45

Somatic Embryogenesis in Bamboo (*Bambusa arundinacea*)

M.P. Meshram*, Priti A. Sonkamble**, A.M. Sonkamble and D.A. Joshi******

**JRF, DAE-BRNS Research Project, **Assistant Professor, Department of Agricultural Botany, ***Assistant Professor, Department of Horticulture, Anand Niketan College of Agriculture, Warora, District Chandrapur, Dr PDKV, Akola, Maharashtra*

ABSTRACT

An efficient protocol for *in vitro* propagation of *Bambusa arundinacea*, an evergreen temperate bamboo has been developed. Seeds cultured on solidified Murashige and Skoog (MS) medium supplemented with 2,4-D 2 mg/l produced embryogenic callus with globular shape embryos. Somatic embryos thus formed were transferred to MS medium containing BAP 2 mg/l for germination of somatic embryos. Upon transferring to MS containing only 2 mg/l BAP, the dark green embryos developed into healthy plantlets along with well developed root system.

***Keywords:** Somatic embryogenesis, Bamboo (Bambusa arundinacea)*

Introduction

One of the most unique and versatile group of plants known to mankind, the bamboos occupy a special place in the lives of the rural poor, especially in Asia.

Bamboos are the most important items of forest produce used by rural communities in Asia and the Pacific. Over 75 genera 1250 species of bamboos occur in the world (Anonymous, 1978) of which about 130 reported in India (Sharma, 1987).

The largest forest area under bamboos is in India with 9.57 million hectares of bamboo forests or 12.8 per cent of the total forest area. The principle bamboos are *Dendrocalamus strictus* and *Bambusa arundinacea*.

Bambusa arundinacea commonly known as Thorny bamboo or Spiny bamboo is a tall woody, stems thorny, numerous, tuffed, up to 40 m tall, curving at top, branches numerous, internodes 30–45 cm long, prominent stout spines, lowest nodes rooting, stem sheaths leathery, orange yellow when young, hairy outside, shining and ribbed inside 30–45 cm long blade, triangular, glabrous covered with a brownish belt of bristly hairs inside, leaves thin linear up to 20 cm long, glabrous above hair beneath, leaf sheaths hairy small. Beside being extensively used as a raw material for good quality paper it is also used as a food in some parts of China and India, wood used by Chinese in household carpentry, furniture, boxes etc. and leaves used as a fodder (Duke, 1983).

In India, industrial production of paper was started in the 1930s. Since independence, however, there has been a rapid expansion of paper factories using bamboo pulp. At one time the supply of bamboo was thought to be perpetual but with rapidly increasing population; associated with demands for fuel and farmland, resulted in significant decrease of land area under forest including bamboo. In addition to the decreasing forest cover, overexploitation, bad management, lack of adequate state control over natural forest lands and vested interests have resulted in a situation where the forest stock of bamboo and the total annual net growth has decreased and will continue to decrease significantly in Asia (Barney, 1980).

Similarly, most of the larger economic bamboos flower only once every 30–60 year resulted in few viable seed from large mass of empty florets and viable seeds also suffer much damage due to rodents, insect attack and rapid loss of viability due to poor storage. Death of bamboo clumps occur following seedling. The resultant regeneration takes time to establish and has to face factors such as intense grazing pressure and forest fires (Saxena and Bhojwani, 1993).

Tissue culture has tremendous potential in this context and could be useful in overcoming the above difficulties. Micropropagation is being attempted using nodal cuttings from the minor branches of bamboo but early flowering of such micropropagated bamboo plants is a major problem in this regard.

Somatic embryogenesis offers easy method for mass propagation. The juvenile period and growth habit of plantlets from somatic embryos are similar to those of *in vivo* seedlings. Thus bamboo plants from somatic embryos can be relied upon to last for their full vegetative life span.

There are several bamboo species for which regeneration of complete plantlets through somatic embryogenesis has been obtained. These includes *Bambusa bamboos* (Mehta *et al.*, 1982), *B. beecheyana* (Yeh and Chang, 1986b), *B. oldhamii* (Yeh and Chang, 1986a), *Dendrocalamus strictus* (Rao *et al.*, 1985), *D. strictus* (Saxena and Dhawan, 1999) however none of these reports deal with mass propagation of *B. arundinacea*.

In the present study, an attempt has been made to develop an efficient *in vitro* micropropagation method in *B. arundinacae* through somatic embryogenesis.

Materials and Methods

Sterilization of Explants and Initiation of Culture

Seeds of *B. arundinacea* were obtained from forests in Alibag (Maharashtra), India. They were kept in a polyethylene bag and stored in a freezer until use. At the time of culture, after dehusking, relatively large and healthy looking seeds were selected as explant for initiating culture.

The mature healthy seeds were washed in 2 per cent Tween-20 solution for 10 minutes. The Tween-20 solution was removed by washing in running tap water for 3–4 times-followed by rinse in distilled water and later on surface sterilization were carried out in aseptic condition under laminar flow hood.

The seeds were surface sterilized with 70 per cent ethyl alcohol solution for 5 min. followed by washing in sterilized double distill water. The seeds were further sterilized by treating them with 0.1 per cent solution of Mercuric chloride for 7 min. followed by three washing with sterilized DDW. Then the seeds were cultured on semisolid Murashige and Skoog (1962) medium containing 2,4-D 2 mg/l, 3 per cent sucrose and 0.8 per cent agar in 25 × 150 mm glass tubes (Borosil, India). The pH was adjusted to 5.8 with 1 N HCL and 1N NaOH prior to autoclaving at 121°C and 15 psi pressure for 20 min. after 6 weeks of incubation at 25±2°C under 12 hours photoperiod with a light intensity of 1000 lux individual or clusters of embryo were dissected out and transferred to germination media.

Germination of Embryos

For the germination of embryos, semisolid MS medium containing 3 per cent sucrose and 0.2 per cent gelrite was used. The basal media was supplemented with 6-benzyl amino purine (BAP) at different concentration.

Results and Discussion

Sterilization of Explants and Initiation of Culture

The sterilization procedure used yielded 70 to 80 per cent aseptic cultures. However in certain seed lots, the percentage of microbial contamination was as high as 75 per cent. An increase in the concentration of Mercuric chloride and/or in the duration of sterilization treatment (15 to 20 min.) helped to obtain a higher frequency of aseptic cultures but adversely affect the ability of seed to form callus. The similar results were reported by Saxena and Dhawan (1999) while studying somatic embryogenesis in *D. strictus*.

In the present study MS medium was used as a basal medium and found to be the best for somatic embryogenesis. Rao *et al.* (1985) and Rout and Das (1994) used B_5 and ½ MS media respectively, for somatic embryogenesis whereas Niraula and Raj Bhandary (1987), Saxena and Dhawan (1999) achieved regeneration of *D. strictus* on MS medium through somatic embryogenesis. In other studies concerning somatic embryogenesis of bamboos such as *B. oldhamii* (Yeh and Chang, 1986a), *O. acuminate aztecorum* (Woods *et al.*, 1992) and *S. latiflora* (Yeh and Chang, 1987) MS medium was used.

Of the various concentrations of 2,4-D used, 2 mg/l was found to be best for callus induction. Saxena and Dhawan (1999) reported the 3mg/l concentration of 2,4-D found optimum for *D. strictus*. Although callus became visible within 8–10 days of culture (Plate 45.1) but a reasonable amount of callus obtained only after 5–6 weeks of culture. Generally two types of callus were formed *i.e.*, yellow, compact and white friable callus. On 2 mg/l 2, 4-D, 30 to 60 per cent of the cultures formed yellow and compact callus with globular shape embryo (Plate 45.2).

At the end *i.e.*, after 5–6 weeks, mature embryos having dark green colour were separated from callus and were transferred to germination media.

Germination of Embryos

For the germination of the embryos, BAP were tested at different concentration. Of the various concentrations, MS+BAP 2 mg/l proved to be optimal for embryo germination whereas Saxena and

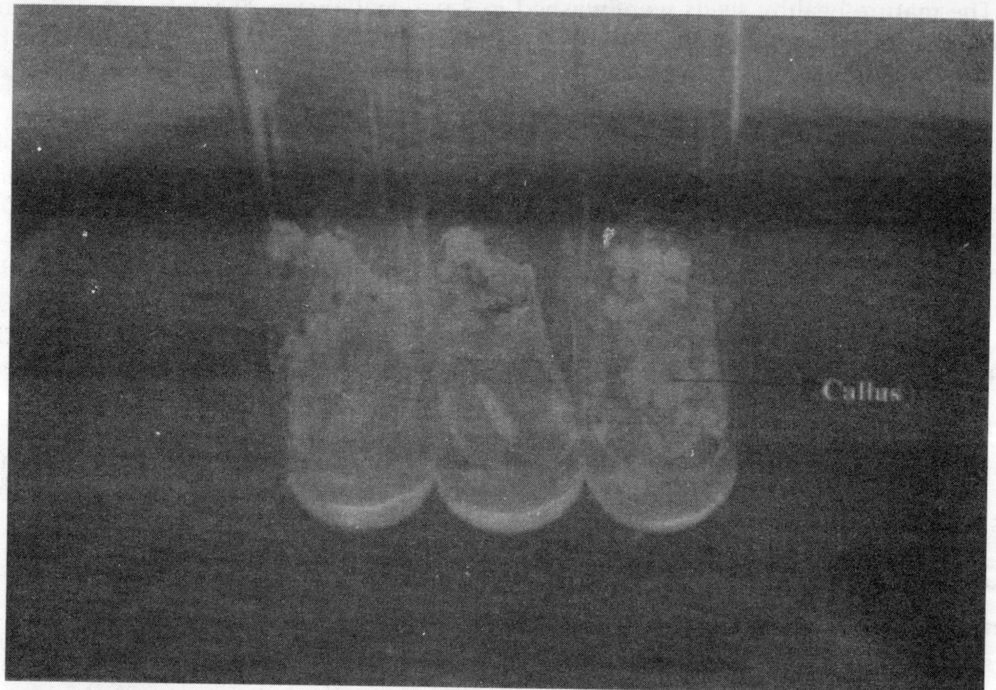

Plate 45.1: Initiation of Callus

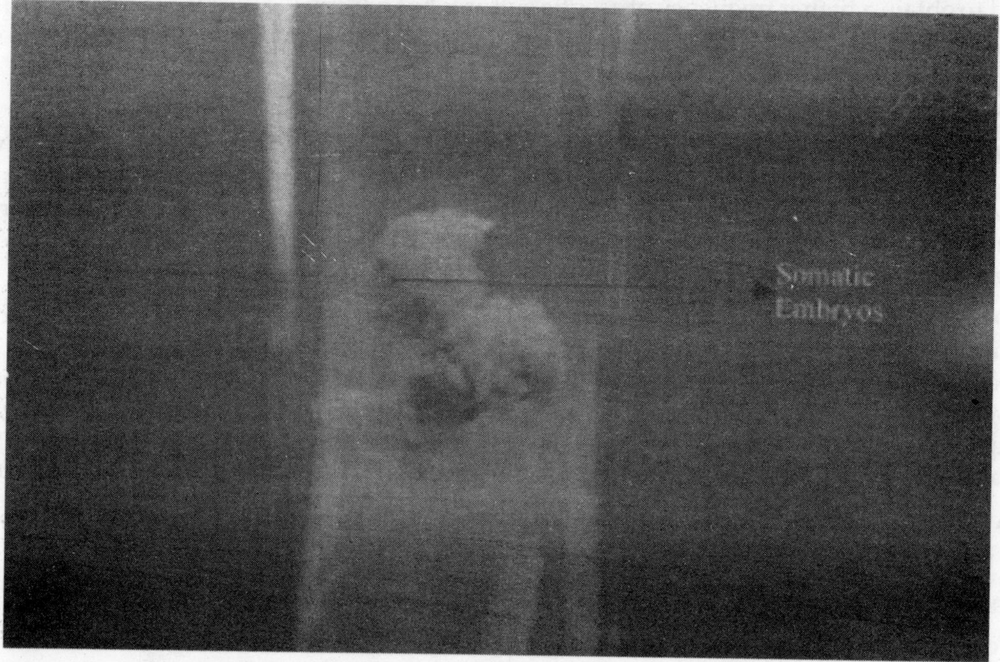

Plate 45.2: Development of Somatic Embryos from Callus

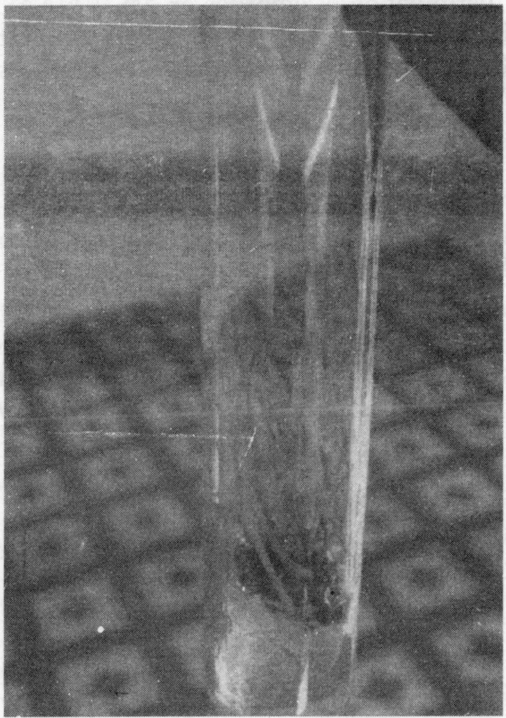

Plate 45.3: Germination of Somatic Embryos

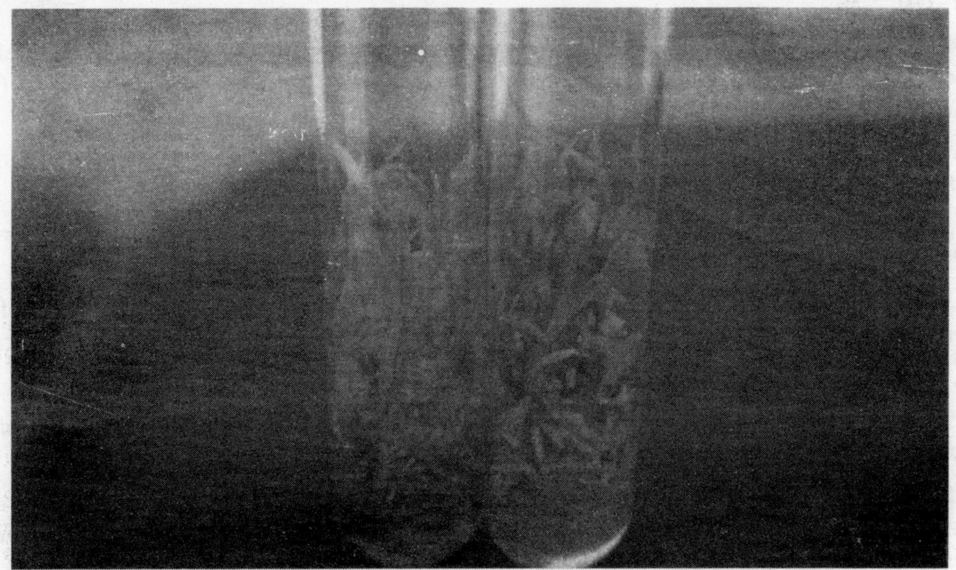

Plate 45.4: Plantlets Developed from Somatic Embryos

Dhawan (1999) achieved germination of embryos on MS + NAA 5 mg/l + Kn per cent mg/l in *Dendrocalamus strictus*. After 8 to 10 days of incubation embryos starts to develop shoots (Plate 45.3) and after 4 weeks of incubation, each test tube contains 15–25 shoots (Plate 45.4) Most of the shoots were 5–7 cm long and well established root system such plantlets were transferred to polythene bags containing 1 : 1 : 1 sand, soil and FYM respectively.

The regeneration protocol of *Dendrocalamus strictus* developed by Rao *et al.* (1985) prescribes a separate rooting stage but this additional stage not only prolongs the production time but also increases the cost of plantlet production. In the present study, the shoots formed well developed roots on germination medium itself.

In the above experiments, the MS medium was solidified with 0.2 per cent gelrite instead of agar. In early investigation pertaining to somatic embryogenesis in bamboos, only agar was used for solidification, but in present study the germination frequency of embryos and plant growth was significantly better on gelrite than on agar. According to George (1993), there are many reports that suggest the superiority of gelrite over agar as a solidifying agent.

References

Anonymous. 1978. *Bamboos Forest News for Asia and Pacific*, FAO-Bangkok, 2(4).

Barney, G.O. 1980. *Global 2000: Report to the President*. US Government Printing Office. Washington, D.C.

Duke, J.A. 1983. *Handbook of Energy Crops* (Unpub.)

Murashige, T. and F. Skoog. 1962. A revised medium for rapid growth and bioassays with tobacco tissue cultures. *Physiol. Plant*, 15: 473–497.

Mehta, U., Ramanuja Rao, I.V. and H.Y. Mohan Ram. 1982. Somatic embryogenesis in bamboo. In: *Plant Tissue Culture*, (Ed.) A. Fujiwara. Japanese Association of Plant Tissue Culture, Tokyo, p. 109–110.

Niraula R., Raj Bhandary S.B. 1987. *In vitro* propagation of *Dendrocalamus strictus* and its establishment in soil. In: *Proceedings Regional Workshop on Tissue Culture of Tropical Crop Plants, Dhaka*, (Eds.) A.S. Islam, M.M. Haque, p. 17–20.

Saxeana S. and S.S. Bhojwani. 1993. *In vitro* clonal Multiplication of 4 year old plants of the bamboo, *Dendrocalamus longispathus* Kurz. *In Vitro Cell Dev Biol.*, 29: 135–142.

Rao, I.U., Ramanuja Rao, I.V. and V. Narang. 1985. Somatic embryogenesis and regeneration of plants in the bamboo *Dendrocalamus strictus. Plant Cell Rep.*, 4: 191–194.

Rout, G.R. and P. Das. 1994. Somatic embryogenesis and in vitro flowering of 3 species of bamboo. *Plant Cell Rep.*, 13: 683–686.

Sharma, Y.M.L. 1997. *Inventory and Resource of Bamboos*, p. 4–17.

Saxena, S. and V. Dhawan. 1999. Regeneration and large scale propagation of bamboo (*Dendrocalamus strictus* Nees) through somatic embryogenesis. *Plant Cell Rep.*, 18: 438–443.

Woods, S.H., Phillips G.C., Woods J.E. and G.B. Collins. 1992. Somatic embryogenesis and plant regeneration from zygotic embryo explants in Mexican weeping bamboo *Otatea acuminata aztecorum*. *Plant Cell Rep.*, 11: 257–261.

Yeh, M.L. and W.C. Chang. 1986a. Plant regeneration through somatic embryogenesis and paint regeneration from zygotic embryo explants in Mexican weeping bamboo (*Bambusa oldhamii* Munro). *Theor. Appl. Genet.*, 73: 161–163.

Yeh, M.L. and W.C. Chang. 1986b. Somatic embryogenesis and subsequent plant regeneration from inflorescence callus of *Bambusa beecheyana* var. beecheyana. *Plant Cell Rep.*, 5: 409–411.

Yeh, M.L. and W.C. Chang. 1987. Plant regeneration via somatic embryogenesis in mature embryo-derived callus culture of *Sinocalamus latiflora* (Munro) McClure. *Plant Sci.*, 51: 93–96.

Chapter 46

Stability of Yield and its Components in Urdbean (*Vigna mungo* L. Hepper)

Ch. Mallikarjuna Rao, Md. Mujeeb Khan and Y. Koteswara Rao

Regional Agricultural Research Station, Lam, Guntur – 522 034

ABSTRACT

Twelve genotypes of Urdbean (*Vigna mungo* L. Hepper) were studied under four r environments for seven different characters for their stability using the Eberhart, and Russell method. The G × E (linear) and pooled deviation (for days to maturity) were significant. The magnitude of G × E (linear) was greater than pooled deviation revealing the preponderance of linear component over non-linear component for all the characters studied. The genotypes LBG 719 and LBG 729 are found to be stable over all environments for seed yield. These genotypes could be used in breeding programmes to obtain high yielding stable segregants. The association was found to be positively significant with days to maturity and number of pods per plant at Ragolu.

***Keywords:** Urdbean, Stability, G × E, Eberhart and Russell.*

Introduction

Urdbean is a major pulse crop grown in rabi season both in uplands and rice fallows under different climatic zones of Andhra Pradesh. It is being grown in 5.6 lakh ha with a total production of 3.9 lakh tonnes with the productivity of 700 kg/ha in Andhra Pradesh. Genotype × Environment not only underlies the very success of stability of genotypes but also the post-breeding adaptive evaluation of improved strains before they being released for commercial cultivation. Hence an attempt is made to study the stability of urdbean genotypes evaluated over four different locations.

Materials and Methods

Twelve urdbean genotypes were grown in Randomized block design with three replications under four locations (E_1: Lam, E_2: Munipalle (Rice fallows), E_3: Jagityal, E_4: Ragolu) during rabi 2002–2003. Each genotype was sown in 6 rows of 4 m length plot. The inter and intra row spacing of 30 and 10 cm was adopted. Recommended package of practices were adopted through out the crop season. Data recorded in each plot for days to 50 per cent flowering, days to maturity, plant height (cm), number of pods per plant, number of seeds per pod, 100 seed-weight (g) and seed yield (kg/ha). Stability analysis was carried out as suggested by Eberhart and Russell (1966). Character association between seed yield and different yield components was worked out according to Dewey and Lu (1959).

Results and Discussion

The pooled variances due to genotypes are highly significant fulfilling the requisite of stability analysis (Table 46.1). Insignificante (linear) variance except for number of seeds per pod and 100 seed weight indicates variation among environments is not linear. However, variance due to G × E (linear) was highly significant indicating differential performance of genotypes under diverse environments. Significant pooled deviation indicate performance of different varieties fluctuated from their linear path of response to environments for days to maturity. The magnitude of G × E (linear) interaction was greater than pooled deviation revealing the preponderance of linear component over non-linear component for all the characters studied. The similar results were reported by Manivannan *et al.* (3) in mungbean.

Table 46.1: Analysis of Variance for Stability in Urdbean

Source of Variation	*Degrees of Freedom*	*Days to 50% Flowering*	*Days to Maturity*	*Plant Height*	*Number of Pods per Plant*	*Number of Seeds per Pod*	*100 Seed Weight*	*Seed Yield*
Genotypes	11	9.577**++	17.731**++	164.990**++	313.941**++	1.455**++	15.146**++	1759.703**
Env+ (G × E)	36	18.828**++	41.532**++	42.522**	64.555**	0.511**	1.369**++	560.736**++
Environment (linear)	1	1	1	1	1	1*	1**++	1
G × E (linear)	11	166.652**++	366.870-++	132.461**++	299.283**++	1.400**++	11.341**++	4888.380**++
Pooled deviation	24	0.426	1.112++	7.836	16.91	0.191	0.122	26.374
Pooled Error	88	0.290	0.410	22.047	45.616	0.316	0.118	21.240

+: Significant at 5 per cent level (tested against pooled error)

++: Significant at 1 per cent level (tested against pooled error)

*: Significant at 5 per cent level (tested against pooled deviation)

**: Significant at 1 per cent level (tested against pooled deviation)

The environmental index values were positive and high for days to 50 per cent flowering, days to maturity and seed yield in E_3 (Jagityal) indicating favourable environment for these traits (Table 46.2). The environment E_1 (Lam) is favorable for plant height, while E_2 (Munipalle) for number of pods per plant and E_4 (Ragolu) for 100 seed-weight were found to be suitable. According to Eberhart and Russell, a stable genotype is one with gi (Genotype mean) > X (General mean), $b_1 = 1$ and $S_2d_i = 0$.

Among the twelve genotypes studied, all the entries recorded non-significant S2di for all characters except for days to maturity (Table 46.3).

Table 46.2: Environmental Index Values for Yield and Yield Components of Urdbean in Different Environments

Sl.No	Character	E_1	E_2	E_3	E_4
1.	Days to 50 per cent flowering	– 4.139	– 2.556	4.888	1.806
2.	Days to maturity	– 6.458	– 3.292	6.597	3.153
3.	Plant height	3.409	0.548	1.298	– 5.256
4.	Number of pods per plant	0.464	– 0.053	– 0.381	– 0.030
5.	Number of seeds per pod	0.097	6.736	– 1.208	– 5.625
6.	100 seed weight	– 1.629	0.216	0.494	0.919
7.	Seed Yield	– 0.426	– 0.184	1.083	– 0.472

E_1: Environment–1 (Lam); E_2: Environment–2 (Munipalle); E_3: Environment–3 (Jagityal); E_4: Environment–4 (Ragolu).

Grouping of these genotypes into different environments is done based on stability parameters. For seed yield the genotypes LBG 719 and LBG 728 performed better in all environments studied, while LBG 693 and LBG 17 performed well under favourable environments (Table 46.4). The genotype LBG 693 is found to be stable in overall environments for days to maturity, plant height and 100 seed weight. The genotypes LBG 714 and LBG 402 are stable in favourable environments for days to maturity, number of pods per plant and number of seeds per plant. The genotype LBG 728 is stable in poor environments for traits like days to 50 per cent flowering, days to maturity, number of pods per plant and number of seeds per plant.

Table 46.3: Estimates of Stability Parameters for Yield and Yield Component Characters in Urdbean

Sl.No.	Genotype	Days to 50% Flowering			Days to Maturity		
		x	b_i	S_2d_i	x	b_i	S_2d_i
1.	LBG 709	47.67	0.943	0.121	83.58	1.087	0.967*
2.	LBG 693	48.50	0.765	–0.122	89.33	1.075	0.286
3.	LBG 712	49.50	1.730	0.176	85.50	0.407	0.493
4.	LBG 714	46.50	0.280	0.211	89.08	1.178	–0.009
5.	LBG 716	49.25	1.462	0.362	90.17	1.242	–0.254
6.	LBG 719	50.33	1.059	0.486	88.17	1.838	2.123**
7.	LBG 728	48.83	0.831	0.658	87.00	0.719	0.128
8.	LBG 731	45.17	0.484	–0.101	83.83	1.056	1.745**
9.	LBG 734	48.17	1.895	0.070	87.67	1.627	1.377*
10.	LBG 402	50.25	1.181	–0.199	88.17	0.192	0.694
11.	LBG 17	42.17	0.972	0.033	87.83	1.709	–0.270
12.	LBG 685	49.33	1.258	–0.069	88.83	0.948	1.130*
	Grand mean	47.97	1.07	–	87.43	1.08	–

Contd...

Table 46.3–Contd...

Sl.No.	Genotype	Plant Height (cm)			No. of Pods per Plant		
		x	b_i	S_2d_i	x	b_i	S_2d_i
1.	LBG 709	30.17	0.591	–16.528	30.00	0.367	–45.471
2.	LBG 693	39.50	1.089	–17.643	28.30	2.344	–28.962
3.	LBG 712	31.25	1.694	–14.080	28.42	0.887	–15.263
4.	LBG 714	37.08	1.483	–12.588	29.83	1.330	–31.969
5.	LBG 716	34.67	0.804	–20.442	23.58	0.263	–33.429
6.	LBG 719	35.25	0.771	1.605	24.33	2.160	–29.779
7.	LBG 728	32.67	0.874	–14.878	31.83	0.140	–40.883
8.	LBG 731	31.92	0.793	–20.326	24.83	2.247	–42.355
9.	LBG 734	28.75	1.150	–5.852	23.92	0.353	–21.220
10.	LBG 402	35.25	0.829	–12.003	27.50	1.851	–1.020
11.	LBG 17	32.42	1.189	–19.653	27.17	0.956	–23.440
12.	LBG 685	29.17	1.301	–18.152	28.75	1.356	–30.655
	Grand mean	33.18	1.05	–	27.37	1.18	–

Sl.No.	Genotype	No. of Seeds per Pod			100 Seed-weight (g)		
		x	b_i	S_2d_i	x	b_i	S_2d_i
1.	LBG 709	7.07	0.384	–0.074	4.77	1.018	–0.077
2.	LBG 693	7.18	1.774	–0.019	4.80	1.061	0.098
3.	LBG 712	7.02	0.391	–0.070	4.69	0.714	–0.045
4.	LBG 714	7.63	2.639	–0.167	4.56	1.198	–0.116
5.	LBG 716	7.19	0.221	–0.271	4.79	1.067	0.107
6.	LBG 719	7.30	2.438	0.039	4.46,	0.793	0.188
7.	LBG 728	7.44	0.305	–0.263	4.69	1.320	–0.369
8.	LBG 731	7.15	1.229	–0.199	4.79	1.131	–0.052
9.	LBG 734	6.91	1.515	–0.245	4.58	0.870	–0.117
10.	LBG 402	7.18	1.230	0.075	4.83	1.099	0.149
11.	LBG 17	7.10	0.596	–0.164	5.02	0.757	0.176
12.	LBG 685	6.87	1.449	–0.199	5.26	1.483	0.170
	Grand mean	7.17	1.18	–	4.77	1.04	–

Sl.No.	Genotype	Seed Yield (kg/ha)		
		x	b_i	S_2d_i
1.	LBG 709	1036.33	0.881	–16.896
2.	LBG 693	1065.83	1.139	–7.772
3.	LBG 712	833.17	1.057	27.947
4.	LBG 714	933.42	0.999	–16.216
5.	LBG 716	931.58	0.899	–15.384

Contd...

Table 46.3–Contd...

Sl.No.	Genotype	Seed Yield (kg/ha)		
		X	b_i	S_2d_i
6.	LBG 719	1069.33	1.052	32.754
7.	LBG 728	1067.25	1.077	–11.696
8.	LBG 731	932.83	0.990	30.125
9.	LBG 734	873.83	0.903	18.464
10.	LBG 402	1050.50	0.834	–0.980
11.	LBG 17	981.42	1.502	–17.481
12.	LBG 685	958.67	0.769	–11.254
	Grand mean	977.85	1.01	–

X: Mean; b_1: Regression coefficient; S_2d_i: Deviation from regression; *: Significant at 5 per cent level; **: Significant at 1 per cent level.

Table 46.4: Grouping of Urdbean Genotypes Suitable for Different Environments

Sl.No.	Character	Genotypes Stable Over all Environments ($g_i > X$, $b_i = 1$, S^2d_i deviates from zero)	Genotypes Stable for Favorable Environments ($g_i > X$, $b_i = 1$, S^2d_i deviates from zero)	Genotypes Stable for Poor Environments ($g_i > X$, $b_i = 1$, S^2d_i deviates from zero)
1.	Days to 50% flowering	LBG 719, LBG 709	LBG 402, LBG 712, LBG 685, LBG 716, LBG 734	LBG 728, LBG 693
2.	Days to maturity	LBG 693	LBG 716, LBG 714, LBG 17, LBG 402	LBG 728
3.	Plant height	LBG 693	LBG 714	LBG 719, LBG 402, LBG 716
4.	Number of pods per plant	LBG 17, LBG 712	LBG 714, LBG 685, LBG 693, LBG 402	LBG 728, LBG 709
5.	Number of seeds per pod	–	LBG 714, LBG 719, LBG 402, LBG 693, LBG 731	LBG 728, LBG 716
6.	100 seed weight	LBG 709, LBG 693, LBG 716, LBG 402	LBG 685, LBG 731	LBG 17
7.	Seed yield	LBG 719, LBG 728	LBG 693, LBG 17	LBG 709, LBG 402

X: Mean; b_1: Regression coefficient; S_2d_i: Deviation from regression; g_i: Genotypic mean.

The seed yield is positively and significantly associated with days to maturity (0.802**) and number of pods per plant (0.512*) at E_4 (Ragolu) location. Similarly Manivannan *et al.* (1999) also reported that seed yield exhibited significant positive correlation with number of pods per plant in mungbean. For the trait days to maturity, the association with seed yield is negative and non-significant at E_2 (Munipalle). In E_1 (Lam), the trait 100 seed weight recorded positive significant correlation with

Table 46.5: Phenotypic Correlation Coefficients of Seed Yield and its Components in Urdbean Over Environments

Character	Environment	Days to 50% Flowering	Days to Maturity	Plant Height	No. of Pods per Plant	No. of Seeds per Pod	100 Seed Weight	Seed Yield
Days to 50% flowering	E_1	1.000						
	E_2	1.000						
	E_3	1.000						
	E_4	1.000						
Days to maturity	E_1	0.074	1.000					
	E_2	0.360	1.000					
	E_3	–0.003	1.000					
	E_4	0.001	1.000					
Plant height	E_1	0.313	–0.037	1.000				
	E_2	0.498	0.238	1.000				
	E_3	0.165	–0.116	1.000				
	E_4	0.086	–0.145	1.000				
No. of pods per plant	E_1	–0.013	0.076	–0.206	1.000			
	E_2	0.127	–0.133	–0.002	1.000			
	E_3	–0.363	0.139	–0.015	1.000			
	E_4	0.046	0.256	–0.003	1.000			
No. of seeds per pod	E_1	0.302	–0.140	0.425*	–0.161	1.000		
	E_2	0.201	0.155	–0.173	–0.190	1.000		
	E_3	–0.117	0.183	–0.048	0.042	1.000		
	E_4	–0.239	–0.155	–0.198	–0.086	1.000		
100 seed weight	E_1	0.404*	–0.102	0.416	–0.206	0.425*	1.000	
	E_2	0.234	0.047	0.065	–0.002	–0.173	1.000	
	E_3	–0.067	0.283	0.194	–0.015	–0.048	1.000	
	E_4	0.011	0.274	0.505*	–0.003	–0.198	1.000	
Seed yield	E_1	–0.120	0.042	–0.005	–0.352	0.313	–0.286	1.000
	E_2	0.508*	–0.055	–0.179	0.160	0.368	0.301	1.000
	E_3	0.209	0.384	0.095	–0.043	–0.182	0.003	1.000
	E_4	–0.009	0.802**	0.290	0.512*	–0.139	–0.194	1.000

E_1: Environment–1 (Lam); E_2: Environment–2 (Munipalle); E_3: Environment–3 (Jagityal); E_4: Environment–4 (Ragolu).

Hence, the genotypes LBG 719 and LBG 728 were recorded to be superior to other genotypes with respect to responsiveness to environment and stability over environments. These genotypes could be used as parents in breeding programmes to obtain superior stable genotypes.

References

Dewey, O.R. and Lu, K.H. 1959. A correlation and path coefficient analysis of components of crested wheat grass and production. *Agronomy Journal*, 51: 515–518.

Eberhart, S.A. and Russell, W.A. 1966. Stability parameters for combining ability. *Crop Science*, 6: 36:40.

Manivannan, N., Murugesan, S., Ramamoorthi, N. and Nadarajan, N. 1996. Stability analysis for seed yield in mungbean. *Indian Journal of Pulses Research*, 9(2): 149–152.

Manivannan, N., Viswanathan P.L. and Murugan, E. 1999. Association of yield and yield components in mungbean. *Indian Journal of Pulses Research*, 12(1): 107–109.

Chapter 47

Detection of Tomato Virus on Cucumber

Rama Murthy, D.M Sawant and G.P. Deshmukh

Department of Plant Pathology, Mahatma Phule Krishi Vidyapeeth, Rahuri – 413 722, Maharashtra, India

ABSTRACT

A virus disease of cucumber causing slight clearing of veins and crumpling of the young leaves followed by light or dark green mottle, together with blistering and distoration of the leaves and stunting of the plant. The fruit is usually unmarked though it may be slightly mottled. The virus was detected on the basis of host range, symptomatology, transmission, physical properties, electron microscopy. The virus has limited host range distributed in three families *viz.*, cucurbitaceae, chenopodiaceae and solanaceae. The virus is easily mechanically transmissible and could be recovered at the greatest extent from leaves of the diseased plant. The virus was seed borne to an extent of 1.6 per cent and was found in maximum concentration in the seed coat and not detected from embryo of diseased seeds. No transmission was observed after 4 months of harvest of diseased cucumber seeds. The virus was not transmitted by 3 species of aphids and whitefly. The thermal inactivation point was between 90–95°C, dilution end point between 10^{-5}–10^{-6} and longevity *in vitro* of 120 days. Electron microscopy of leaf dip preparation from infected cucumber leaves showed rod shaped particles of 280–300 nm in length and 14–18 nm in breadth. Based on host range, symptomatology, transmission, physical properties and particle morphology the causal virus under investigation belonging to tomato virus group.

***Keywords:** Cucumber green mottle mosaic virus, Electron microscopy, Host ranged, Transmission.*

Introduction

Cucumber is known to succumb by diseases of fungal, bacterial and viral origin. Among the viral diseases, the viruses affecting cucumber are cucumber mosaic virus (CMV) (Uppal, 1934), bottle gourd mosaic virus (Vasudeva and Lal, 1943) cucumber green mottle mosaic virus (Vasudeva and Nariani,

1952), watermelon mosaic virus (Bhargava and Joshi, 1960), melon mosaic virus (Jaganathan and Ramakrishnan, 1971). Out of these cucumber green mottle mosaic virus seems to be of much prevalence in recent years and becoming a problem for the successful cultivation of cucumber (Capoor and Verma, 1948).

Materials and Methods

The virus isolate collected from commercial field and maintained on cucumber by mechanical sap inoculation. For mechanical transmission of the virus, the leaves were gently washed in tap water, immediately wiped off the excess water with the help of blotting paper and weighted. Then those leaves were macerated in sterilized and chilled mortar and pestle by adding equal quantity of 0.01 M phosphate buffer of pH 7.00 and diseased leaves on 1 : 1 (W/v) basis. The crude sap was then obtained by squeezing the pulp through double layered musclin cloth.

Seed Transmission

For seed transmission studies, seeds of cucumber plants which were inoculated with the virus when young and allowed to grow to maturity, were collected and the possibility of virus transmission through seed was determined by sowing 1000 seeds.

The presence of virus in various parts *i.e.*, testa (seed coat), cotyledons, embryo and plumule of four grades of infected seeds was determined by separating these parts of dry mature seeds as well as of the germinating seeds by sterile forceps. Inoculum was prepared separately from each of the separated parts and tested by inoculating ten leaves of *C. amaranticolor.*

Aphid Transmission

Insects were employed for transmission *viz., A. craccivora* (Koch), *A. gossypii* (Glover) and *Myzus persicae* (Sulz) and the whitefly (*Bemisia tabaci*) (Genn).

The aphids were given pre-acquisition fasting for 3 hrs. This was followed by acquisition feeding for 10 min and inoculation threshold for 1 hr. Ten aphids were released per plant for transmission studies. The plants were observed for symptoms development.

Host Range

Plant species belonging to 8 families *viz.*, Cucurbitaceae, Chenopodiaceae, Solanaceae, Leguminosae, Amranthaceae, Cruciferae, Graminae and Malvaceae were mechanically sap inoculated for host range study of the virus. Those plants which did not show any symptoms were back indexed on *C. amaranticolor.*

Physical Properties

The standard methods suggested by Noordam, 1973 were employed for studying the dilution end point, thermal inactivation point, longevity *in vitro* of the present virus.

Electron Microscopy

Leaf dip methods described by Brandes (1957, 1964) were used for electron microscopy.

Results and Discussion

Transmission by Mechanical Sap Inoculation

Experiments were carried out to recover the virus from different parts of diseased plants.

The results given in Table 47.1 showed that the extract prepared from the leaves proved to be the most infectious containing the highest concentration of virus as compared to the extracts obtained from other plant parts. Subsequently, the inoculum was prepared from the leaves of diseased cucumber plants in all future tests.

Table 47.1: Infectivity of Inocula Prepared from Different Plant Parts of Diseased Cucumber

Plant Part	*Number of Lesions per Leaf, a*	*Infection Index, b*
Leaf lamina	34.4	–
Petiole	14.0	0.41
Stems	8.5	0.25
Tendril	6.6	0.19
Root (Primary)	4.1	0.12
Root (Secondary)	1.2	0.04
Flower calyx	2.9	0.08
Corolla	1.1	0.03
Stamens	3.0	0.09
Pollens	1.9	0.06
Pistils	2.1	0.06

a: Average of 10 inoculated leaves of *C. amaranticolor.*

b: Infection index = treatment/control

Control is the number of lesions resulted from inoculation by leaf extract (Joshi and Thomberry, 1970).

Seed Transmission

The virus was transmitted in the seeds of cucumber at a minimum percentage of 1.6. These results are in agreement with Subbaiah and Gopal 1997; Van Koot and Van Dotst, 1959.

Insect Transmission

The virus was not transmitted by any aphid or whitefly. These findings are similar with those reported by several workers like Capoor and Verma, 1948; Raychaudhari and Verma, 1975, Antignus *et al.*, 1990, Subbaiah and Gopal, 1997.

Host Range

Host range of CGMMV was restricted to cucurbitaceae (Table 47.2). These results are similar with other workers Ainsworth, 1935, Vasudeva *et al.*, 1949; Rao and Verma, 1984, Vani and Verma, 1993 but the virus induced local lesion on *Chenopodium amaranticolor* and *Datura stramonium*. Subbaiah and Gopal, 1997 reported that (GMMV strain) produced local lesions present findings are addition to this study.

Physical Properties

Physical properties of virus isolate revealed dilution end point to be between 10^{-5}–10^{-6}, thermal inactivation point between 90–95°C and longevity *in vitro* of 120 days, it was clearly evident that physical properties of present virus was similar with those of tomato virus of cucumber reported by Capoor and Verma, 1948.

Table 47.2: Reaction of Host Plant to Inoculation with the Virus

Sl.No.	*Plant*	*Symptoms*	
		Local	*Systemic*
1.	Cucurbitaceae		
	Cucumis sativus L.	–	VC, SM, MO, B, D, P
	C. melo L.	–	VC, SM, MO, D, B
	C. melo var. utillissine	–	VC M_1, Mo, P, D, B
	Lagenaria siceraria cv. Long	–	VC, MM, Mo, B
	L. siceraria cv. Round	–	VC, MM, Mo, B
	Lutta acutangula Roxb	–	VC, M, Mo, P, D, B
	Trichosanthus anguinea	–	VC, M, Mo, P, D, B
	Cucurbita moschata Deshesne	–	VC, M, MO, P, D, B
	Citrullus vulgaris Scharad	–	VC, MM, Mo, B
2.	Chenopodiaceae		
	Chenopodium amaranticolor Costle and Reyn	CLL, NLL	–
3.	Solanaceae	CLL, NLL	–
	Datura stramonium L.		

VC: Vein clearing; MM: Mild mosaic; Mo: Molting; M: Mosaic; P: Puckering; SM: Severe mosaic; D: Distortion; CLL: Chlorotic local lesion; B: Blistering; NLL: Necrotic local lesion

Electron Microscopy

Electron micrograph showing rod shaped particles of 280–300 nm in length and breadth being 14–18 nm were observed typical to that of cucumber green mottle mosaic virus *i.e.*, tomato virus reported by Smith, 1972; Hollings *et al.*, 1975; Vani and Verma, 1993; Subbaiah and Gopal, 1997.

References

Ainsworth, G.C. 1935. Mosaic disease of the cucumber. *Ann. Appl. Biol.*, 22: 55–67.

Antignus, Y., Pearlsman, M., Benyoseph, R. and Cohen, S. 1990. Occurrence of a variant of cucumber green mottle mosaic virus in Israel. *Phytoparasitica*, 18(1): 50–56.

Bhargava, K.S. and Joshi, R.D. 1960. Detection of watermelon mosaic virus in U.P. *Curr. Sci.*, 29: 443–444.

Brandes, J. 1957. Einen elektronennikroskopische schnetlmethode zum Nachweis faden and stabschenfomiger viren insbesondere in kartoffeldun-kelkeimen. Nachrbe deut Pflanzenschutzedienst (Brauschweig), 9: 151–152.

Brandes, J. 1964. Identifiziesung von getreckten pflanzenpathogenen viren auf morphologischer Grandlage Mitt. Bioi. Bundesanstalt. Land forst wirtsch, Berllii-Dahlem, 110: 1–130.

Capoor, S.P. and Verma, P.M. 1948. A mosaic disease of *Lagenaria vulgaris* ser. in the Bombay province. *Curr. Sci.*, 17: 274–275.

Hollings, S.M., Komuro, Y. and Tochihara, H. 1975. CGMMV. *CMI/AAB* Description of plant viruses No. 154.

Jagannathan, T. and Ramakrishnan, K. 1971. Studies on virus diseases of cucurbits. *Madras Agric. J.*, 58: 331–337.

Joshi, L. and Thornberry, H.H. 1970. Effect of calcium phosphate on infectivity of TMV. *Phytopath. Z.*, 68: 155–163.

Noordam, D. 1973. *Identification of Plant Viruses: Methods and Experiments*. Oxford and IBH publishing Co., New Delhi, pp. 250.

Rao, A.L.N. and Varma, A. 1984. Transmission studies with cucumber green mottle mosaic virus. *Phytopath. Z.*, 109: 325–335.

Raychaudhari, M. and Varma, A. 1975. Virus diseases of cucurbits in Delhi. In: *Proc. 62nd Indian Sci. Congr.*, 3: 74.

Smith, K.M. 1972. *A Textbook of Plant Virus Diseases*. Academic Press, New York, pp. 688.

Subbaiah, K.V. and Gopal, D.V.R.S. 1997. Characterization and identification of a tomato virus infecting snake gourd (*Trichosanthes anguina* L.) in Andhra Pradesh. *Indian J. Virology*, 13(2): 153–157.

Uppal, B.N. 1934. The absorption and elution of cucumber mosaic virus. *Indian J. Agric. Sci.*, 4: 656–662.

Van Koot, Y. and Van Dotst, H.J.M. 1959. Virus disease of cucumber in Netherlands. *Tijdschr. Plziekt.*, 65: 257–271.

Vani, S. and Varma, A. 1993. Properties of cucumber green mottle mosaic virus isolated from water of river Jamuna. *Indian Phytopath.*, 46(2): 118–122.

Vasudeva, R.S. and Lal, T.B. 1943. A mosaic disease of bottle gourd. *Phytopathology*, 33: 182–191.

Vasudeva, R.S. and Nariani, T.K. 1952. Host range of bottle gourd mosaic virus and its inactivation by plant extracts. *Phytopathology*, 42: 149–152.

Vasudeva, R.S., Raychaudhari, S.P. and Singh, J. 1949. A new strain of cucumis virus-2. *Indian Phytopath.*, 2: 180–185.

Chapter 48

Variation in the Morphological Characters of *Brachiaria mutica* Due to Urban Sewage Water Irrigation

S.T. Girisha* and S. Umesha**

**Department of Botany, University of Mysore, Manasagangothri, Mysore – 570 006*
***Department of Applied Botany, Seed Pathology and Biotechnology, University of Mysore, Manasagangothri, Mysore – 570 006*

ABSTRACT

The effect of urban sewage water irrigation on the morphological characters of the fodder grass *Brachieria mutica* were studied in details. Morphological parameters of *Brachieria mutica* were enhanced due to sewage water irrigation. Morphological features studied were root length, number of clums, length, thickness of clum, internodul length, no of length and width of leaves and leaf area were studied in sewage water irrigated and plants irrigated normally in the departmental garden were studied. All the morphological parameters studied were increased with the; sewage water irrigated plants when compared to that of plants grown in the department garden. This increase depends on the type of irrigation followed. The effect of urban sewage water irrigation with the type of irrigation on the morphological features of the fodder grass *Brachieria mutica* were discussed in the present article.

Keywords: *Brachiaria mutica, Morphological characters, Poaceae, Urban sewage water, Forage grass.*

Introduction

Brachiaria mutica (Forsk) stapf (para grass) a fodder grass belongs to family poaceae, is a perennial runner, which root profusely at nodes with flowering stems. Culms are erect, leafy, hallow, succulent

---

** Author for Correspondence.

and glabrous with hairy nodes, leaf blades are dark green with panicle inflorescence. It grow in almost all environments and are particularly prominent in tropical and semi-tropical areas. It constitute a major part of the diet of grazing animals. A brief literature survey indicates a lot remains to be done regarding the study of this grass because this commercial forage crop irrigated with sewage wastewater, However, there is very little information on the physico-chemical properties of sewage water and its erect on the morphology of forage crops. There are many reports explaining the effect of effluent on the morphological changes in *Oryza sativa* (Mishra and Singh, 1987), *Cenchrus ciliaris* (Pisal *et al.*, 1991), *Forage sorghum* (Chand *et al.*, 1992), *Avena sativa* (Joshi *et al.*, 1993), *Zea mays* (Naresh Kumar and Chaudhary, 1993), *Vigna unguiculata* (Tripathi and Tripathi, 1993) were the associated nutrients present in the effluent in lower concentration increase the biomass. Solgare and Andhyarujina (1989) reported changes in the morphological aspects or vegetation due to sewage irrigation. At high concentrations the effluent effected the overall growth in green gram (Subramani *et al.*, 1995), They also reported that effluent dilution enhanced growth, when effluent concentration was high, the morphometrical parameters showed a decreasing trend in *Vigna unguiculata* (Subramani *et al.*, 1998) and Nikhil (2000) reported that grass species will die for using the effluent for agricultural purpose at a place where *Lens esculenla* crop is cultivated, it should be diluted to 50 per cent concentration to minimize its harmful effect at the time of germination and growth of crop (Anil and Bindu, 2003). A large quantity of sewage water is routinely used to cultivate *B. mutica* and this fodder is being fed to milk yielding cattles. The present investigation has been carried out to evaluate the effect of Mysore city sewage water on the morphological characters of this commercial fodder grass *B. mutica*.

Materials and Methods

Sampling studies were made in Mysore sewage form where they grow this fodder grass by using sewage water. Sewage irrigated *B. mutica* plants were collected randomly from sewage form Mysore and were used in the present study. Plants were collected at different growth stages in different seasons. These plants were analysed for morphological studies while garden grown plants were used as control. Morphological characters such as, height of culm, number of leaves, root length, internodal length, number of culms, culm thickness, leaf length, leaf width (Basappa, 1982) and leaf area (Yoshida *et al.*, 1972), were scored on mature plants. These parameters were scored in 10 plants for each population of ten populations of each experimental site. The experiments were conducted in three consecutive years, each with two consecutive seasons. All the data were statistically analysed for significance means were compared for significance using analysis or variance (ANOVA) and means were compared for significance using Duncan's Multiple Range Test (DMRT: P = 0.05).

Results and Discussion

Morphological growth parameters have been studied at research site in all conditions to evaluate the biomass production of *B. mutica*. The detailed observations of effect of urban sewage water irrigation on the morphological features of *B. mutica* were clearly depicted in Table 48.1. The following observations were recorded.

Root Length

Root length was significantly (P = 0.05) increased from 5.6–8.9 cm in the control plants to 6.3–12.8 cm, when the plants were irrigated with sewage water (Table 48.1).

Number of Culms

In control the number of clums ranges from 6–51, but in sewage water plants it ranged from 25–66 (Table 48.1).

Table 48.1: Effect of Sewage Irrigation on Morphological Feature of *B. mutica*

Sl.No.		Sewage Irrigated Plants			Control Plants		
		Dryland Condition	Irrigated Condition	Water Logged Condition	Dryland Condition	Irrigated Condition	Water Logged Condition
1.	Root Length (cm)						
	Summer seasons	5,90±0.95	7.30±0.82	11.30±0.41	4.30±0.38	6.80±0.31	7.50±06.8
	Rainy seasons	6.30±0.38	8.10±0.78	12.80±0.58	5.60±0.31	7.90±0.48	8.90±0.86
2.	Number of culms						
	Summer seasons	20.60±0.19	35.10±0.38	63.50±0.59	5.25±0.28	44.00±0.58	45.00±0.19
	Rainy seasons	25.00±0.39	42.80±0.19	66.00±0.51	6.25±0.39	50.30±0.61	51.10±0.28
3.	Height of culm (cm)						
	Summer seasons	16.70±0.28	42.0.3180±	88.60±0.68	22.75±0.31	54.00±0.58	55.10±0.31
	Rainy seasons	18.10±0.39	46.500.28±	89.90±0.61	29.00±0.28	56.50±0.68	59.40±0.48
4.	Culm thickness (cm)						
	Summer seasons	1.10±0.93	2.30±0.39	2.80±0.62	1.20±0.19	1.50±0.79	1.90±0.0.49
	Rainy seasons	1.40±0.82	2.60±0.81	3.00±0.63	1.50±0.18	1.80±0.69	2.30±0.58
5.	Inter nodal length (cm)						
	Summer seasons	5.04±0.81	9.07±0.38	11.57±0.58	6.00±0.91	8.03±0.58	8.05±0.39
	Rainy seasons	7.04±0.82	10.43±0.81	11.91±0.39	6.75±0.98	8.00±0.49	9.05±0.48
6.	Number of leaves						
	Summer seasons	2.50±0.81	5.80±0.48	8.90±0.42	2.50±0.12	3.60±0.48	4.00±0.91
	Rainy seasons	3.40±0.59	7.30±0.41	9.00±0.28	4.00±0.32	5.00±0.58	6.60±0.89
7.	Leaf length (cm)						
	Summer seasons	9.09±0.69	20.37±0.39	22.06±0.38	9.25±0.39	15.90±0.59.	17.10±0.98
	Rainy seasons	10.28±0.71	21.75±0.42	24.04±0.39	10.62±0.28	20.50±0.48	21.00±0.81
8.	Leaf width (cm)						
	Summer seasons	1.04±0.79	1.48±0.48	1.64±0.39	1.12±0.21	1.45±0.12	1.88±0.21
	Rainy seasons	1.09±0.82	1.55±0.58	1.99±0.42	1.35±0.24	1.65±0.14	2.00±0.81
9.	Leaf area cm²						
	Summer seasons	5.29±0.19	19.31±0.38	25.30±0.91	6.32±0.31	15.21±0.19	21.21±0.42
	Rainy seasons	8.17±0.28	22.51±0.39	30.65±0.89	7.35±0.29	18.20±0.22	27.72±0.58

Values are the mean ± S.E of four replicates each and two consecutive seasons data were subjected to analysis of variance and compared for significance according to DMRT (P = 0.05).

Culm Height

It is more during rainy season than summer season, ranges from 18 cm to 90 cm in treated plants and in control 29 cm to 59 cm (Table 48.1).

Culm Thickness

It varies from rainy to summer seasons. The clum thickness ranges from 1.40 cm to 3.00 cm in treated and in control and it was reduced 1.50 cm to 2.30 cm during rainy seasons in the sewage water irrigated plants.

Inter Nodal Length

It varies from 7.04 cm to 11.91 cm in treated plants and in control plants it was reduced to 6.75 to 9.05 cm. Increased Inter nodal length was observed during rainy seasons as compared to summer season (Table 481).

Number of Leaves

Luxurient growth of leaves were noticed during rainy season as compare to that of summer season. The number of leaves ranged from 4–9 in sewage treated and it was reduced to 4–6 leaves/culm in the control sets.

Leaf Length

The growth of the plant is better during rainy season, leaf length is more during rainy season as compared to summer seasons. It is less in control conditions, which was ranged from 10.28 cm to 24.04 cm in treated and in control sets it was 10.62 to 21.00 cm (Table 48.1).

Leaf Width

Leaf width in sewage irrigated plants was 1.09 cm to 1.99 cm and was reduced to 1.35 to 2.00 cm in control sets.

Leaf Area

Leaf area was also significantly reduced from 8.17 cm^2–30.65 cm^2 in the sewage treated plants to 7.35–27.72 cm^2 in untreated plants (Table 48.1).

Field experiments were conducted to elucidate the variations in morphological parameters of *B. mutica* grown in municipal wastewater and also in department's garden.

The behaviour of morphological growth characters of root, stem, leaf, flower and caryopsis, in general, are fundamentally constant under normal conditions. Each morphological character when studied in a wide range, proves to be relatively stable. Morphological growth parameters in plants studied at all sites and experimental conditions were computed, and the data obtained have been used to evaluate the production of biomass of *B. mutica*.

Morphological growth parameters like height of culm, nuimber or leaves, leaf area, root length. internodal length, number of clums per plant, culm thickness, leaf length and leaf width were found to be higher in sewage water irrigated plants, in research site during rainy seasons as compared to that of the plants grown in departmental garden. Diluting the polluted water not only decrease effluent phytotoxicity, but also promotes the growth of plants. Our results indicate that the plant grown in sewage irrigated sites during rainy seasons tend to clum increase height and leaf area and the culms bear more number of leaves during rainy seasons than during summer and attain maximum growth, eventually producing more yield. Our observations are similar to the observations reported by Bahadur and Sharma (1989).

Para grass grown in both sewage irrigated and water logged conditions exhibited significantly tall culms. The result of field experiments indicate that there is differential response in their growth parameters to sewage irrigation. When plants were irrigated with sewage water during summer seasons, the concentration of sewage which was high, was found to decrease plant height at research sites (unpublished observations). The shoot and root length, leaf length and leaf area of the grass considerably increased in irrigated and waterlogged conditions during rainy seasons. However, a continuous reduction was observed in these parameters with the increase of sewage concentration and a maximum reduction was observed during summer seasons.

Vegetative growth of plant increased in sewage irrigated and waterlogged conditions during rainy season as compared to the control which may be attributed to growth promoting effects of nitrogen and other mineral elements present in effluents. Morphological growth parameters show a reduced growth at higher concentrations which may be attributed to the excess amount of accumulated nitrogen and other mineral elements present in the sewage during summer seasons. The inhibitory effects at higher concentrations of sewage is due to this excess nitrogen and other mineral elements present, which are harmful to plant growth, there by indirectly affecting the metabolic processes of *B. mutica* during summer seasons. The leaf length and leaf width was more in waterlogged conditions, as these elements get diluted during rainy seasons.

The soil salinity is due to the presence of high amounts of total solids. Salt is one of the principle factors responsible for the deterioration of soil quality. The adverse effect of excessive salt present in effluents may also be responsible for the reduction in plant growth (Srivasthava *et al.*, 1988), Morphological growth parameters like height of the culm, number of culms and leaf area showed a decreasing trend with the increase in sewage concentration, However, all growth parameters like height of culms, number of culms, number of leaves, leaf length and leaf width showed a marked increase during rainy seasons. Thus after proper dilution sewage can safely be used as a fertilizer to crop and fodder plants. The exact mechanism involved in this process needs further investigations and studies are going on to observe the adverse effect of urban sewage water irrigation.

References

Anil, K., Kainà and Bindhu Aggarwal. 2003. Effect of paper mill effluent on the germination, growth and yield of lens gulentamoench. *J. Ind. Poll. Contl.*, 19(2): 301–306.

Bahadur, B. and Sharma, B.K. 1989. Growth responses of *Triticum aestivum* (var. Up 115) to the effluent released from three industrial units at Clutterbuckganj, Bareilly. *Poll. Res.*, 8(2): 63–70.

Basappa, G.P. 1982. Biosystematic studies in the genus *Brachiaria griseb*, Poaceae. *Ph.D. Thesis*, University of Mysore, Mysore.

Chand, M.L.K, Dixit and Arora, S.K. 1992. Yield and quality of forage sorghum as affected by phosphorus fertilization. *J. Indian Soc. Soil Sci.*, 40: 302–306.

Joshi, Y.P., Virendra Singh and Verma, S.S. 1993. Response of Oat (*Avena sativa*) to nitrogen. *Forage Res.*, 19(3–4): 272–277.

Kumar, Nikhil. 2000. Bioreclamation of coal over burden dump a experimental approach. In: *Int. Conf. on Ocean Coal Initiatives*, New Delhi.

Mishra, L.C. and Singh, K.K. 1987. Pollution effect of fertilizer factory effluent on growth and development of corn and rice seedlings. *Conservation and Recycling*, 10(1): 27–35.

Nareshkumar and Chaudhary, M.L. 1993. Effect of phospoorus and zinc on the yield, nitrogen, phosphorus and zinc content of maize (*Zea mays*). *Forage Res.*, 19(2): 176–182.

Pisal, A.A., Patil, S.K., Khot, R.B. and Desale, J.S. 1991. Studies on biofertilizer in the nitrogen economy and yield of anjan grass (*Cenchrus ciliaris*). *Forage Res.*, 17: 55–58.

Salgare, S.A. and Andhyarujina, K.B. 1989. Effect of polluted water of Ulhas river on the rate of growth its vegetation III. *Biojournal*, 1(2): 83– 88.

Srivastsav, N., Neelam and Sahai, R. 1988. Ecophysiological response of certain crop plants of eastern Uttar Pradesh to industrial effluents. *Int. J. Ecol, Environ. Sci.*, 14: 229–248.

Subramani, A., Saravanan, S., Sundaramurthy, P. and Lakshmanachary, A.S. 1998. Impact of fertilizer factory effluent on the morphometrical and biochemical change of cowpea (*Vigna ungiculala*). *Adv. Plant Sci.*, 11(1): 137–141.

Subramani, A., Sundaramoorthy, P. and Lakshmanachary, A.S. 1995. Effect of biological treated distillery effluent on seed germination and seedling growth of greengram (*Vigna radiata* L.). Wilczek Var. CO_2. *Poll. Res.*, 14(1): 37–41.

Tripathi, S.B. and Tripathi, R.K. 1993. Effect of S-levels and sources on yield and nutrient content of cowpea. *Forage Res.*, 19: 141–147.

Yoshida, S., Fordo, J., Cook and Gomez, J. 1972. *Laboratory Manual for Physiological Studies of Rice*, 3rd edition. International Rice Research Institute, Manila, Philippines.

Chapter 49

Relationship of Mosaic Disease of Chilli with its Aphid Vectors

C. Reddy, D.M. Sawant and G.P. Deshmukh

Department of Plant Pathology, Mahatma Phule Krishi Vidyapeeth, Rahuri – 413 722, Maharashtra, India

ABSTRACT

The chilli mosaic virus (Chmv) was transmitted more effectively by *M. persicae, Aphis gossypii* and *A. craccivora.* The vector efficiency of these aphids varied considerably. Adults of *M persicae* in groups of 15 per test plant and given acquisition access for 15 min and inoculation access for 30 min gave maximum infection. Pre-acquisition starvation for 2 hrs was beneficial, while post-acquisition upto 2 hrs yielded very low transmission. The feeding aphids lost infectivity after 1 hr. The studies indicated non-persistent relationship of chilli mosaic virus with *M. persicae.*

***Keywords**: Chilli mosaic virus, Aphid vectors.*

Introduction

Chilli is affected by a number of diseases. Among the viral diseases, the viruses affecting chilli in India are chilli mosaic virus and potato virus Y reported by Jayarajan and Ramakrishnan (1969). Detailed investigations were carried out to determine the virus vector relationship of the chilli mosaic virus and the results obtained are presented in the present article.

Materials and Methods

The virus was collected from diseased chilli plants and maintained on glasshouse grown chilli seedlings by conventional leaf rub method by using carborandum 800 mesh as an abrassive.

The colonies of non viruliferous aphids *viz., A. gossypii, A. craccivora, Myzus persicae* and *Bemisia tabaci* were raised on cotton, groundnut, cabbage and tobacco. Wingless adult females were employed in all the insect transmission experiments. Ten to twelve days young plants of chilli were used as test plants. During fasting period, insects were placed in the test tubes after collecting from their colonies by means of camel hair brush and the open ends of the test tubes were covered with muslin cloth. For acquistion feeding, insects were liberated on detached, diseased chilli leaves kept in petridishes. After a requisite acquisition feeding, the aphids were transferred to the test plants kept in glass chimney. The chimney top was covered with muslin cloth and its basal portion was inserted into the soil. The acquisition of virus by whitefly and their transfer on chilli for transmission feeding was carried out after following method described by Varma (1963). After the requisite transmission feeding period, plants were sprayed with 0.03 per cent monocrotophos to kill the insects and kept in glasshouse under observation for 45 days.

Results and Discussion

Comparative Efficacy of Different Insect to Transmit Chilli Mosaic Virus

Initially transmission tests of the chilli mosaic virus by means of aphids and the whitefly were carried out to ascertain the efficiency of the insect to transmit the chilli mosaic virus. The insects were given pre-acquisition fasting for 2 hrs, acquisition feeding for 10 min and inoculation access for 24 hrs. Fifteen insects were employed per test plant. The results revealed that the virus was transmitted by *M. persicae, A. gossypii, A. craccivora*. Also *M. persicae* transmitted the virus to an extent of 65 per cent and proved to be most efficient vector as compared to other aphid spp. Simons (1969) reported that two strain of PVY transmitted by *M. persicae* produced severe symptoms. Present findings are addition to this study.

Inoculation Threshold

After pre-acquisition fasting for 2 hrs, the aphids were allowed for 15 min acquisition feeding and then were liberated on test seedlings in groups of 15 aphids for infection feeding for different periods varying from 2–60 minutes but the maximum transmission was obtained after an access period of 30 min.

Effect of Pre and Post Acquisition Starvation on Transmission Efficacy of the Virus by *M. persicae*

Aphids were starved for varying periods and then allowed for access feeding for 15 min. These aphids were then released in groups of 15 aphids per test plant and allowed to feed on them for 3 hrs. For each test 20 seedlings were exposed to the aphids. The results revealed that the unstarved aphids transmitted the virus to a very low percentage of plants and maximum transmission was obtained when they starved for 2 hrs before access feeding. Similar results on aphid transmission have been obtained by Powell (1993).

For post acquisition starvation, pre-acquisition fasting for 2 hrs and acquisition feeding for 15 min were given. These aphids were starved for different periods *viz.*, 0,0.5, 1, 2, and 3 hrs before transferring them in groups of 15 aphids per test plant for test feeding for 3 hrs. The results indicated that post-acquisition fasting brought about progressive loss in transmission ability of aphids with the increased deviation of this period.

Number of Aphids Required to Transmit Chilli Mosaic Virus

Varying number *viz.*, 1,2,3,5, 10, 15, 20, 25 and 30 individuals of viruliferous aphids of *M. persicae* were transferred to test plant of chilli and allowed for transmission feeding for 24 hrs. As before, the aphids were allowed pre-acquisition fasting of 2 hrs and acquisition feeding for 10 min. Twenty seedlings were exposed to each group of aphids. It was observed that even a single adult female was able to acquire and transmit the virus. Percentage transmission increased with increase in number of aphids per test plant. Maximum infection was obtained with 15 adult aphids per test plant. Similar results on aphid transmission have been obtained by Perez *et al.* (1992).

However, there was decrease in transmission percentage when 20 to 30 aphids were liberated on each test plant. The reason for this decrease in transmission is not understood.

Acquisition Threshold

Aphids were starved for 2 hrs before access feeding for varying periods *i.e.*, 0.5, 1, 2, 3, 5, 10, 15, 20, 40 and 60 minutes. For each test at least 20 chilli seedlings were exposed to the aphids. *M. persicae* could acquire the virus after access feeding for a minute. The maximum transmission was obtained after an access feeding for 15 min. Access feeding beyond 15 minutes reduced the transmission efficiency of the aphid.

Persistence of the Virus in *M. persicae*

Aphids were starved for 2 hrs and allowed acquisition feeding for 15 minutes. Thereafter individual aphids were transferred to a series of 10 test seedlings at intervals of 10 min each. The results showed that the aphids failed to infect test plants after a period of 60 min.

In the next experiment, the aphids after pre-acquisition starvation of 2 hrs and access feeding of 15 min were transferred to the test seedlings in groups of 15 adults. These were allowed to feed on it on the first series of test plants for 8 hrs and transferred to another series of test plant for 16 hrs each day. The experiment was carried out for 5 days and repeated once. The results obtained showed that the aphids transmitted the virus only to the first test plant of the series on which they had fed for 8 hrs.

Summing up the results, it may be concluded that *M. persicae* is an efficient vector of chilli mosaic virus, transmitting the same more efficiently than the other species i. e. *A. gossypii* and *A. craccivora* tested. The short transmission threshold and retention of infectivity strongly point out the vector virus relationship to be of the non-persistent type.

References

Jayarajan, R. and Ramakrishnan, K. 1969. Potato virus Y on chilli (*Capsicum annum* L.) in Tamil Nadu. *Madras Agric. J.*, 56: 761–766.

Perez, P., Gimeno, D.C., Verdugo, M.S., Soto, M.S., Ponz, F. and Ferreres, A. 1992. Population dynamics of vectors and transmission of potato virus Y in pepper cultivars. *Boletin-de-Sanidad. Vegetal.*, 18: 225–235.

Powell, G. 1993. The effect of pre-acquisition starvation on aphid transmission of potyviruses during observed and electrically recorded stylet penetrations. *Entomologia. Expt. Applicata*, 66: 255–260.

Simons, J.N. 1969. Differential transmission of closely related strains of potato virus Y by the green peach aphid and the potato aphid. *J. Econ. Ent.*, 62: 1088–1096.

Varma, P.M. 1963. Transmission of plant virus diseases by white flies. In: *Proc. Symp. Plant Animal Viruses. Bull. Nat. Inst. Sci., India*. 24: 11–33.

Chapter 50

Regional Agricultural Land Use in Bihar: A Sectoral Analysis

Tara Shankar

Department of Agricultural Economics, Bihar Agriculture College, Sabour – 813 210, Bhagalpur, India

ABSTRACT

During the course of any development programme of national resources, land utilization occupies a special attention. At any particular time, temperate moisture, topography, soil and physical structure determine it. Obviously, land has the characteristic of its fixit in supply and scarcity. Therefore, land use pattern is directly concerned with the problem arising in the process of deciding upon and carrying out into action the optimum use. In a dynamic world, certain modification can occur in the existing pattern of land utilization. It deals with the process of putting various types of land to the optimum use. In brief, a close look of the present regional agricultural land use in three broad sectors and its trends in Bihar will go a long way to recommend the scope of systematic and planned shift in the pattern.

Keywords: *Agro-climatic Zones, Ecological sector, Non-agricultural sector, Agricultural sector.*

Introduction

Land is a pre-requisite resource to the agricultural sector and has been used for various purposes for the well being of society. The pattern of land use is a very dynamic concept as the extent of its use to varied purposes. It also varied longitudinally over space and time to meet the requirement of society. The changes in land use affect the ecosystem of an area in terms of vegetation, agro-climate, land quality itself and the quality of life that can be sustained. Initially, there are technical issues concerning the natural endowment associated with land in any region, and which deal with the agro-climatic factors determining the production possibility frontiers and preservation of land. Within these production frontiers, the social and economic issues guide and govern the actual land use. The socio-economic factors can be viewed from different angles. First, the institutional angle focusing on land as a social asset whose ownership and distribution pattern influence its use and conservation. Second, the commercial exploitation angle focusing on land as a productive resource whose disaggregate enterprise wise allocation determines the actual land use. Third, the sectoral aggregate angle focuses on land as a scarce mid exhaustible resource whose sectoral allocation and utilisation or under utilisation determine the aggregate land use and thereby, the states capability to fulfil the mandate of the present and the future. This article attempts to investigate into the regional agricultural land use in the state from this angle.

The study is based on secondary data. General information and other required data were collected from the office of the Directorate of Statistics and Evaluation, Government of Bihar. District wise data were analysed for four agro-climatic zones of Bihar under three broad sectors, *viz.*

1. Ecological sectors comprising area under forest, permanent pastures grazing land, miscellaneous tree crops and other groves and Barren land.
2. Non-agricultural sector comprising land put to non-agricultural use, and
3. Agricultural sector comprising culturable wastes, net sown area, current fallows and other fallows (Pandey and Tiwary, 1996).

Analytical Techniques

Temporal and spatial analyses were undertaken to have an idea about the structural changes in area under different sectors in the state over the period from 1970–1971 to 1999–2000. For the purpose, the six points of time with average of five consecutive years (quinquennial) data were computed. The study period was divided into six sub-period, namely, period–I (1970–1971 to 1974–1975), period–II (1975–1976 to 1979–1980), period–III (1980–1981 to 1984–1985), period–IV (1985–986 to 1989–1990), period–V (1990–1991 to 1994–995) and period–VI (1995–1996 to 1999–2000).

Physical Features of Bihar

In India, Bihar occupies ninth position in area (9359.56 thousand ha.) and third position with respect to population (82.879 million) with population density of 882 per sq km but has the lowest literacy (47.53 per cent) and second highest poverty (42.6 per cent). Bihar lies between 24°15′ and 27°31′ N latitudes and 83°20′ and 88°19′ E longitudes with an average length of 483 km (E–W) and width of 385 km (N–S).

Agro-climatic Zones or Bihar

Bihar is broadly divided into two natural halves by the river Ganga *i.e.*, north Bihar and south Bihar, where a series of other rivers southward and north ward meet the Ganga (Table 50.1). The north Bihar is divided into two distinct agro-climatic NARP zones (Zone–I and Zone–II) whereas south Bihar as a whole represent Zone–III alone. Based on significant difference in agro-ecological parameters Zone–III (South Bihar) is subdivided into Zone–III A and IIIB. The name of districts and important features of different agro-climatic zones are given in Table 50.1.

Results and Discussion

Regional Variation is Area Under Ecological Sector

Ecological sector comprising categories of forest (F_1), permanent pastures and other grazing lands (P), Miscellaneous trees and other groves (M) and Barren and unculturable land (U). It is observed from Table 50.2 that the proportion of land under ecological sector declined from time period (13.95 per cent) to the third period (8.90 per cent) in zone–I but after the third period the trend was mixed.

Table 50.1: Important Feature of the Different Agro-climatic Zones of Bihar

Agro-climatic Zonos	*Latitude*	*Longitude*	*Total Geographical Area (000' ha)*	*District*
Zone–I North-West Alluvial Plain	25°24'–27°31' N	83°20'–86°30' E	3449.09	E/W Champaran, Saran Siwan, Gopalganj, Siwhar, Muffarpur, Vaishali, Darbhanga, Samastipur, Begusarai, Madhubani, Sitamarhi
Zone–III North-East Alluvial Plain	25°10–26°32' N	86°21–88°19' E	1798.11	Purnia, Katihar, Khagaria, Saharsa, Supaul, Arsria, Madhepura, Kisanganj
Zone–III(a) South Bihar Alluvial Plain (East)	24°22'–25°30' N	85°30–87°30' E	1194.51	Munger, Shekhpura, Jamui, Lakhisarai, Bhagalpur, Banka
Zone–III(b) South Bihar Alluvial Plain (West)	24°15'–25°42' N	83°25'–86°30' E	2917.84	Patna, Gaya, Jehanabad, Nawada, Nalanda, Rohtas, Bhojpur, Aurangabad, Buxer and Bhabhua

Table 50.2: To Change in Area Under Ecological Sector in Different Zones of Bihar (in 000' ha)

Zone	*1970–1971 to 1974–1975*	*1975–1976 to 1979–1980*	*1980–1981 to 1984–1985*	*1985–1986 to 1989–990*	*1990–1991 to 1994–1995*	*1995–1996 to 1999–2000*
Zone–I	481.40 (13.95)	420.66 (12.19)	307.02 (8.90)	325.76 (9.44)	358.26 (10.38)	343.53 (9.96)
Zone–II	203.80 (11.33)	179.99 (10.00)	151.62 (8.43)	160.15 (8.90)	168.69 (9.38)	165.23 (9.18)
Zone–IIIa	341.60 (28.59)	325.65 (27.26)	316.90 (26.52)	313.37 (26.23)	313.00 (26.20)	315.18 (26.38)
Zone–IIIb	555.60 (19.04)	526.66 (18.04)	468.61 (16.06)	463.51 (15.88)	461.66 (15.82)	479.54 (16.43)
Bihar	1582.40 (16.90)	1452.96 (15.52)	1244.15 (13.29)	1262.79 (13.49)	1301.61 (13.90)	1303.49 (13.92)

Figures in parentheses show percentage to respective geographical area.

Source: Directorate of Statistics and Evaluation, Government of Bihar.

The situation is almost similar in Zone–II of the state. The decreasing trend in both the zones was obviously due to shift of substantial land to the non-agricultural sector and after the third period forest, miscellaneous tree crop and other groves, permanent pastures and barren and unculturable land shows fluctuations in both the zones under study. So that the land under ecological sector showed uneven trend during last tree periods in both the zones.

Whereas in zone–III(a) proportion of quinquenoial average of land under this sector continuously declined from 28.59 per cent in first period to 26.20 per cent in fifth period but increased to 26.38 per cent during present situation. Table 50.2 further revealed that the land under ecological sector in Zone–III(b) has followed the similar trend to that of in Zone–III(a). The probable reason might be the decline in area under barren land and permanent pastures in zones–III(a) and III(b), but area under forest and miscellaneous trees showed increasing trend at the present situation. So that the area under ecological sector was observed to be comparatively high in the present situation from the previous period in these two zones.

In the state of Bihar there was decreasing trend in area under ecological sector during first three periods but showed increasing trend during last three periods. The proportion of ecological sector to the geographical area of the state was comparatively low from the first period (*i.e.* 16.90 per cent to 13.92 per cent) under study.

From the aggregate state data, it appears that almost half of the land released from barren and unculturable land might have shifted to the non-agricultural sector during first three periods (1970–1985) and remaining half is being shared by the permanent pastures, miscellaneous trees and forest during last three periods (1985–2000).

Regional Variation in Area Under Non-agricultural Sector

Non-agricultura1 sector comprises area under residential buildings, roads railways or under water *i.e.*, rivers, canals, tanks, and other land put to use other than agriculture we categorised under this head, occupying 17.26 per cent of the total area of the state at the present situation (1995–2000) as compared to 7.51 per cent in the first period (1970–1975). The area under this sector registered a continuous increase during the period under study in all zones and the state as a whole, though the higher increase was observed in zone–I as compared to other zones under investigation (Table 50.3). It may be observed from the table that the increase in area under non-agricultural sector in Zone–I. Zone–II and Zone–III(b) as well as in Bihar were comparatively high in the third period under study whereas in Zone–III(a) the higher increase was observed during present situation.

Table 50.3: To Change in Area Under Non-agriculture Sector in Different Zones of Bihar (in 000' ha)

Zone	*1970–1971 to 1974–1975*	*1975–1976 to 1979–1980*	*1980–1981 to 1984–1985*	*1985–1986 to 1989–990*	*1990–1991 to 1994–1995*	*1995–1996 to 1999–2000*
Zone–I	231.90	348.00	527.03	594.76	604.15	661.01
	(6.70)	(10.00)	(15.28)	(17.24)	(17.51)	(19.16)
Zone–II	131.80	137.00	236.45	293.72	321.21	336.17
	(7.32)	(7.61)	(13.14)	(16.33)	(17.86)	(18.69)
Zone–IIIa	140.20	150.33	159.99	163.02	168.62	194.06
	(11.73)	(12.58)	(11.21)	(13.64)	(14.11)	(16.24)
Zone–IIIb	199.80	254.66	343.93	381.84	389.30	424.89
	(6.84)	(8.72)	(11.78)	(13.08)	(13.34)	(14.56)
Bihar	703.60	889.99	1267.40	1433.34	1483.29	1616.14
	(7.51)	(9.50)	(13.54)	(15.31)	(15.84)	(17.26)

Figures in parentheses show percentage to respective geographical area.

Source: Directorate of Statistics and Evaluation, Government of Bihar.

The area and proportion of land under non-agricultural sector were increasing throughout the period under study. This phenomenon is very serious because it may have ecological implications in the state. The declining area under agriculture sector due to increase in area under non-agricultural use also a threat to agricultural poduction in the state. Further, substantially larger investments will be required to maintain a balance between the ecological and agricultural sectors through reclamation of waste lands which would be transferred to agriculture or ecological sectors.

Regional Variation in Area Under Agricultural Sector

Agricultural sector comprises area under net sown, current fallow, other fallow and culturable waste. Current fallows are those lands which were not cultivated during the current year only. Other fallow include all those lands which were taken up for cultivation but are temporily out of cultivation for a period not less than one year and not more than five years. The reason for keeping such land fallow may be (a) poverty of the cultivators (b) inadequate supply of water (c) malarious climate (d) un-remunerative nature of farming etc. Culturable waste land is that land which is at present not utilised for any purpose, *i.e.*, lying waste and that can be brought wider cultivation for growing additional food grains. Zone wise area under agriculture sector were computed for specified six periods which are presented in Table 50.4.

Table 50.4: To Change in Area Under Agricultural Sector in Different Zone of Bihar (in 000' ha)

Zone	*1970–1971 to 1974–1975*	*1975–1976 to 1979–1980*	*1980–1981 to 1984–1985*	*1985–1986 to 1989–1990*	*1990–1991 to 1994–1995*	*1995–1996 to 1999–2000*
Zone–I	2735.89	2680.43	2614.84	2528.57	2485.68	2444.55
	(79.32)	(77.71)	(75.81)	(73.31)	(72.06)	(72.87)
Zone–II	1462.51	1481.12	1410.04	1344.24	1308.21	1296.71
	(81.33)	(82.37)	(78.41)	(74.75)	(72.75)	(72.11)
Zone–IIIa	712.71	718.45	743.62	718.12	712.29	685.63
	(59.66)	(60.14)	(62.25)	(60.11)	(59.63)	(57.39)
Zone–IIIb	2162.44	2137.50	2105.30	2072.45	2067.03	2013.41
	(79.11)	(73.25)	(72.15)	(71.02)	(70.84)	(69.00)
Bihar	7073.55	7017.50	6873.80	6663.38	6573.21	6440.30
	(75.57)	(74.97)	(73.44)	(71.19)	(70.22)	(68.80)

Figures in parentheses show percentage to respective geographical area.

Source: Directorate of Statistics and Evaluation, Government of Bihar.

It may be observed from the table that the area under agriculture sector was continuously declined from first period (1970–1971) to the present situation (1999–2000) in the state, this was mainly due to increase in area under non-agricultural sector.

When Zone wise analysis in the state, that the Zone–II had comparatively large proportion of area under agricultural sector (81.3 per cent of respective geographical area) in 1970–75 which was declined to 72.11 per cent during 1995–2000. Similar pattern was observed in Zone–I which had second larger proportion of land under agricultural sector in Bihar.

The larger proportion of land under agricultural sector in these two zones was mainly due to comparatively less area under ecological sector than Zones–III(a) and III(b). Zone–III(a) had only

59.66 per cent of area under agricultural sector in First period (1970–1975) and observed declining at the present situation, this was probably due to these land are up land of the marginal fertility characterised by thin soil, rapid run-off, limited soil moisture and possibility of production only by fallowing. Area under agricultural sector showed almost declining trend in all the agro-climatic zones under study, which warrants immediate action to increase agricultural productivity in all the zones for food security in Bihar. We should remember that the Zone–II and Zone–IIIa of the state is considered as the "play ground of rivers," because all the important rivers of central north Bihar unite here to form "Ghughri" or lower Kosi. Thus the area is the collecting bowl for all major rivers of north Bihar. The area suffers most from devastating floods so that considerably percentage of the total area remains unsown. Hence, special efforts need to be made by the government for the protection of canal system.

Summary and Conclusion

As there is no possibility of land shift from the non-agricultural sector to the agricultural sector, the agricultural sector will have serious ecological implications. While, a part of the shift towards the non-agricultural sector can be attributed to urban and industrial expansion, the other part is due to the expansion in irrigation networks, rural roads, market yards, etc., which are basically the supporting infrastructures for agricultural growth. Given the present state of technology, this trend needs to be checked, particularly the land transfer from the ecological sector to the non-agricultural sector. The urbanization, industrial expansion and the supporting infrastructure for agricultural growth have to be made on more conservative norms. The land use change within the ecological sector area under permanent pastures and other grazing land in all the zones of the state depleting firstly, affecting the local village ecology.

In the agricultural sector, the net area sown demonstrated a distressing declining trend in all the zones of the state which would adversely affect the agricultural production in Bihar. This could only be compensated by increasing productivity through a concerted effort to bring all the irrigated areas under intensive cultivation. The culturable waste, current fallow and other fallow land are also declining consistently. This phenomenon will be favourable at least for agricultural sector, but larger investment will be required to maintain a balance between the ecological sector and agricultural sectors through reclamation of such waste land.

Policy Approach

It emerges from the above discussions that there is divergent situations across the zones with respect to land under different sectors. Hence, a generalized approach may not help achieving full utilisation of cultivable and fallow land. In fact, there is a need for disaggregated regional studies into the characteristics and determinants of such land, which may serve as the basis for the formulation of long run region specific strategies for the purpose. However, there are some short run discrepancies which warranted the following policy options:

1. Increasing the area under non-agricultural uses at the cost of net area sown requires special effort for the reversal of trend.
2. The on-going forest expansion and development efforts should be continued vigorously as the forest cover still falls short of the desired proportion.
3. There is a wide scope to bring down the existing cultivable waste through institutional support for increasing the area under crops with appropriate technology.

4. The policies of price support and input subsidies may ensure remunerativeness of crop production even under dry land area.
5. The legalised tenancy may help in finding a landless entrepreneur to make full use of agricultural land.
6. Canal irrigation system should be proper cared by Government and other agencies because canal was the major source of irrigation in the state.

References

Basavaraja, H. and Dabali, S.D. 1999. Land use dynamics in India: An agricultural perspective. *Agril. Situation in India*, p. 199–200.

Choudhary, B.N. 1981. *Land Utilisation in Subaranarekha Basin*. Classical Publishing Company, New Delhi.

Dhindsa, K.S. and Sharma, Anju. 1996. Land use planning: A case study of pluses in Punjab. *Agril. Situation in India*, 12: 52.

Giri, R. 1966. Changes in land use pattern in India. *Indian J. Agril. Econ.*, 21(3): 23–32.

Jha, B.N. 1980. *Problems of Land Utilisation: A Case Study on Koshi-region*. Classical Publications, New Delhi.

Mishra, B.B., Mall J., Choudhary, J. and Singh, R.A 2001. Nutrient mining in different agro-climatic zones of Bihar. *Fertil. News*, 46(11): 21–28.

Pandey, V.K. and Tiwari, S.K. 1996. Regional agricultural land use: A sectoral aggregate view. *Indian J. Agril. Econ.*, 51(1&2).

Rao, V.M. 1985. Land use pattern in AP during 1955–1956 to 1979–1980. *Agril. Situation in India*, 40(7): 617–620.

Singh, Virendra P. and Singh, O.P. 1999. Land utilization pattern and distribution of operational land holding in India: An Analytical view. *Agril. Situation in India*, p. 191–193.

Vidya, C.S and Sikka, B.K 1991. Land utilisation pattern in Himachal Pradesh. *Agril. Situation in India*, 46(7): 539–540.

4. The policies of price support and input subsidies may ensure remunerativeness of crop production even under dry land area.
5. The legalised tenancy may help in finding a landless entrepreneur to make full use of agricultural land.
6. Canal irrigation system should be proper cared by Government and other agencies because canal was the major source of irrigation in the state.

References

Basavaraju, H. and Dabali, S.D. 1999. Land use dynamics in India: An agricultural perspective. *Agril. Situation in India*, p. 199–200.

Choudhary, B.N. 1981. *Land Utilisation in Saharsa and Purnia*. Classical Publishing Company, New Delhi.

Dhindsa, K.S. and Sharma, Anju. 1996. Land use planning: A case study of pulses in Punjab. *Agril. Situation in India*.

Giri, R. 1966. Changes in land use pattern in India. *Indian J. Agril. Econ.*, 21(3): 23–32.

Jha, B.N. 1980. *Problems of Land Utilisation: A Case Study on Kosi Region*. Classical Publications, New Delhi.

Mishra, S.P., Mishra, J., Choudhary, J. and Singh, R.N. 2001. Nutrient mining in different agro-climatic zones of Bihar. *Fertil. News*, 46(11): 13–28.

Pandey, V.K. and Tiwari, S.K. 1996. Regional agricultural land use: A sectoral analysis. *Indian J. Agril. Econ.*, 51(1&2).

Rao, V.M. 1985. Land use pattern in AP during 1955–1956 to 1979–1980. *Agril. Situation in India*, 40(7): 617–622.

Singh, Virendra, P. and Singh, O.P. 1999. Land utilisation pattern and distribution of operational land holding in India: An Analytical view. *Agril. Situation in India*, p. 191–193.

Vidya, C.P. and Sinha, B.K. 1991. Land utilisation pattern in Himachal Pradesh. *Agril. Situation in India*, 46(4): 239–240.

Index

S